IUTAM Symposium on Progress in the Theory and Numerics of Configurational Mechanics

IUTAM BOOKSERIES
Volume 17

Aims and Scope of the Series

The IUTAM Bookseries publishes the proceedings of IUTAM symposia under the auspices of the IUTAM Board.

For other titles published in this series, go to
www.springer.com/series/7695

P. Steinmann

Editor

IUTAM Symposium on Progress in the Theory and Numerics of Configurational Mechanics

Proceedings of the IUTAM Symposium held in Erlangen, Germany, 20–24 October 2008

Springer

P. Steinmann
Chair of Applied Mechanics
University of Erlangen-Nuremberg
Erlangen, Germany

ISBN 978-94-007-3686-3 e-ISBN 978-90-481-3447-2
Springer Dordrecht Heidelberg London New York

Softcover reprint of the hardcover 1st edition 2009

Printed on acid-free paper

Springer is part of Springer Science+Business Media (www.springer.com)

Table of Contents

Preface

Configurational mechanics has attracted quite a bit of attention from various research fields over the recent years/decades. Having been regarded in its infancy of the early years as a somewhat obscure and almost mystic field of research that could only be understood by a happy few of insiders with a pronounced theoretical inclination, configurational mechanics has developed by now into a versatile tool that can be applied to a variety of problems.

Since the seminal works of Eshelby a general notion of configurational mechanics has been developed and has successfully been applied to many problems involving various types of defects in continuous media. The most prominent application is certainly the use of configurational forces in fracture mechanics. However, as configurational mechanics is related to arbitrary material inhomogeneities it has also very successfully been applied to many materials science and engineering problems such as phase transitions and inelastic deformations.

Also the modeling of materials with micro-structure evolution is an important field, in which configurational mechanics can provide a better understanding of processes going on within the material. Besides these mechanically, physically, and chemically motivated applications, ideas from configurational mechanics are now increasingly applied within computational mechanics.

In this regard, in particular the combination of configurational mechanics and the finite element method has a notable impact to computational mechanics. New methods based on configurational mechanics are developing in computational fracture mechanics, structural optimization and adaptivity. These methods include, for example, r- and h-adaptive methods for mesh optimization and refinement.

The IUTAM Symposium on "Progress in the Theory and Numerics of Configurational Mechanics" that took place at the University of Erlangen/Nuremberg, Germany from October 20th to 24th, 2008, shed light on the most recent state of affairs in configurational mechanics. As a result of the inspiring contributions and lively discussions these proceedings emerged. They assemble a number of peer-reviewed articles that deal with the current developments in the already intriguing success story of configurational mechanics.

I am convinced that this volume gives the reader an appropriate overview on some of the fascinating applications of configurational mechanics.

Paul Steinmann
Erlangen, June 2009

Acknowledgements

The IUTAM Symposium on "Progress in the Theory and Numerics of Configurational Mechanics" would not have been possible without the generous financial support by the Cluster of Excellence "Engineering of Advanced Materials" established at the University of Erlangen-Nuremberg, by IUTAM, by Siemens and by Springer. This support is gratefully acknowledged!

On Discontinuities of Material Momentum and Eshelby Stress in Hyperelasticity and Thermoelasticity

Manfred Braun

Abstract The balance of material momentum is applied to a propagating singular surface at which velocity and displacement gradient are discontinuous. The resulting jump condition contains an additional source term if the material is assumed purely hyperelastic. In a thermoelastic material, this imbalance is compensated by an appropriate jump of entropy such that, under adiabatic conditions, the strict balance is retained even on a singular surface.

1 Motivation

The balance of material momentum usually is derived from the balance of physical momentum by a pull-back onto the material manifold [4]. The transformation is performed locally, i.e., on the basis of *differential* balance laws. The balance of physical momentum, in its original form, represents a global balance, valid for an arbitrary material volume $\mathscr{B}$, from which the local version is obtained by use of the divergence theorem. This localization process can be performed only if the field quantities involved are differentiable. Therefore the local balance is less general than the global balance from which it is derived.

The local balance of material momentum can be integrated, of course, over an arbitrary material volume thus being converted to a global balance, as indicated in Fig. 1. It should be kept in mind, however, that the process of globalization does not render the result more general than the local balance from which it has been generated.

The local balance equations of physical and material momenta exhibit a pronounced symmetry as elaborated in great detail by Steinmann [5, 6]. Formally this symmetry can be carried over to the global formulation. However, there is a great

Manfred Braun
Chair of Mechanics and Robotics, University of Duisburg-Essen, 47048 Duisburg, Germany;
e-mail: manfred.braun@uni-due.de

P. Steinmann (ed.), IUTAM Symposium on Progress in the Theory and Numerics of Configurational Mechanics, 1–10.

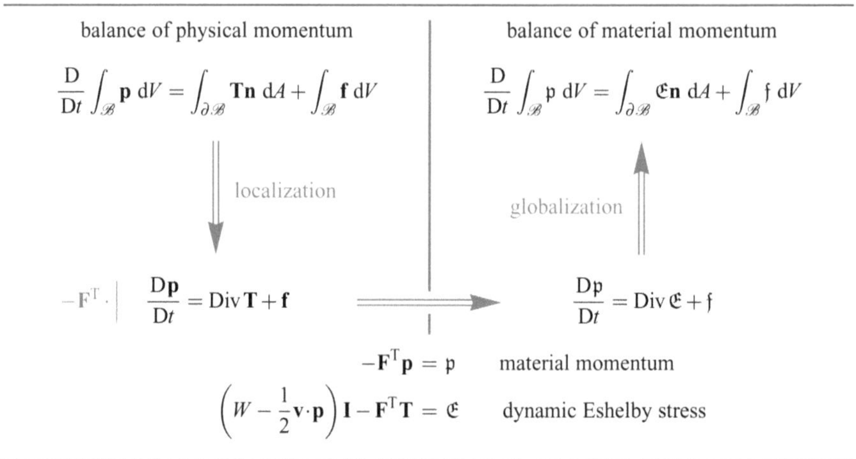

Fig. 1 Global and local balances of physical and material momentum.

difference in the validity of the global balance laws. While the balance of physical momentum holds in general for arbitrary fields without any continuity requirements, the validity of the global balance of material momentum is restricted to the regularity conditions necessary to derive the local version from which it has been generated. In this sense, the balance of material momentum does not represent a *fundamental* law as opposed to the physical balance equations [3].

An important consequence of any global balance law, in addition to the local version in differential form, is the so-called jump condition valid at a surface of discontinuity propagating through the material body. Here it is assumed that the fields have a regular behavior in front of and behind the singular surface and approach definite limit values from either side.

The balance of material momentum should be applicable also to a propagating singular surface. The corresponding jump condition, however, must not be derived from the "global" balance which, in this case, is valid only for continuously differentiable fields thus excluding an application to discontinuous behavior. In order to get the discontinuous version, i.e., a jump condition representing the balance of material momentum at a singular surface, one has to start from the corresponding jump condition of physical momentum and pull it back to the reference configuration.

The derivation and discussion of the jump condition representing the balance of material momentum at a propagating singular surface is the principal goal of this paper. It turns out that the discontinuous version of the balance contains an "imbalance" term which does not necessarily vanish and which is not expected from the continuous analogy. It is shown that the imbalance term depends on the specific interpretation of strain energy. If the material is assumed purely hyperelastic without any recourse to thermodynamic or other effects then the imbalance actually remains nonzero. Thus a propagating shock in a hyperelastic material carries a concentrated source of material momentum with it. If, on the other hand, the material is assumed to be thermoelastic and a nonconductor of heat the strain energy can be identified with internal energy. Then the imbalance vanishes due to the balances of physical momentum and energy.

2 Notations

Since the Lagrangian description is used throughout this paper, all relevant quantities are related to the reference configuration, without indicating it by a special subscript. So the mass density ρ, the volume force $\mathbf{f}$, and the distributed energy source r are meant per unit *reference* volume. Also stress $\mathbf{T}$ and heat flux $\mathbf{q}$ are understood in the sense of Piola, i.e., related to the vectorial surface elements in reference configuration.

According to the Lagrangian description, all fields are considered as functions of the material point $\mathbf{X}$ and time t. Derivatives with respect to time at fixed $\mathbf{X}$ are indicated by $\mathrm{D}/\mathrm{D}t$, and differential operators with respect to the material coordinates $\mathbf{X}$ are denoted by Div, Grad, etc. The motion of the continuum is described by the function $\mathbf{x} = \mathbf{x}(\mathbf{X}, t)$ providing the position $\mathbf{x}$ attained by the material particle $\mathbf{X}$ at time t. Its derivatives deliver the velocity vector $\mathbf{v}$ and the deformation gradient $\mathbf{F}$.

A wave front, i.e., a propagating singular surface, is tracked in the reference configuration. It is oriented by its unit normal vector $\mathbf{N}$. Its normal speed of propagation is denoted by U_{N}. A discontinuous field φ is assumed to be regular on either side of the singular surface and to approach limit values $\varphi^{\pm}$ at the surface, where φ^{+} is attained on that side to which the normal vector is pointing. The jump and the mean value of a quantity φ at the singular surface are defined as

$$[\![\varphi]\!] = \varphi^{+} - \varphi^{-} \quad \text{and} \quad \langle\varphi\rangle = \frac{1}{2}\left(\varphi^{+} + \varphi^{-}\right), \tag{1}$$

respectively. They satisfy the product rules

$$[\![\varphi\psi]\!] = \langle\varphi\rangle[\![\psi]\!] + [\![\varphi]\!]\langle\psi\rangle \quad \text{and} \quad \langle\varphi\psi\rangle = \langle\varphi\rangle\langle\psi\rangle + \frac{1}{4}[\![\varphi]\!][\![\psi]\!] \tag{2}$$

which can be directly verified from the definitions (1). Some further formulas for manipulating jumps and mean values are given in [1].

3 Balances of Physical Momentum and Energy

Let $\rho(\mathbf{X})$ denote the mass density in the reference state and $\mathbf{v}(\mathbf{X}, t)$ the velocity vector at a material point $\mathbf{X}$ at time t. The physical momentum per unit reference volume is $\mathbf{p} = \rho\mathbf{v}$. The global balance of physical momentum is expressed by the equation

$$\frac{\mathrm{D}}{\mathrm{D}t}\int_{\mathcal{B}} \mathbf{p}\,\mathrm{d}V = \int_{\partial\mathcal{B}} \mathbf{T}\mathbf{n}\,\mathrm{d}A + \int_{\mathcal{B}} \mathbf{f}\,\mathrm{d}V \tag{3}$$

valid for any fixed material region $\mathcal{B}$, where $\mathbf{T}$ and $\mathbf{f}$ denote the Piola stress and the volume force density, respectively. The local versions of the balance of physical momentum, valid for regular and for discontinuous fields, are represented by the equations

$$\frac{\mathrm{D}\mathbf{p}}{\mathrm{D}t} = \mathrm{Div}\,\mathbf{T} + \mathbf{f} \quad \text{and} \quad U_{\mathrm{N}}[\![\mathbf{p}]\!] + [\![\mathbf{T}]\!]\mathbf{N} = \mathbf{0}, \tag{4}$$

respectively. If polar effects are omitted the balance of angular momentum is reduced to the symmetry condition

$$\mathbf{T}\mathbf{F}^{\mathrm{T}} = \mathbf{F}\mathbf{T}^{\mathrm{T}} \tag{5}$$

to be satisfied by the Piola stress.

Let $E(\mathbf{X}, t)$ denote the density, per unit reference volume, of the internal energy. The global form of the energy balance reads

$$\frac{\mathrm{D}}{\mathrm{D}t}\int_{\mathcal{B}} \left(E + \frac{1}{2}\mathbf{v}\cdot\mathbf{p}\right)\mathrm{d}V = \int_{\partial\mathcal{B}} \left(\mathbf{T}^{\mathrm{T}}\mathbf{v} - \mathbf{q}\right)\cdot\mathrm{d}\mathbf{a} + \int_{\mathcal{B}} (\mathbf{f}\cdot\mathbf{v} + r)\,\mathrm{d}V, \tag{6}$$

where $\mathbf{q}$ and r denote the nominal heat flux and the internal heat production per unit reference volume. The corresponding local versions of the energy balance can be simplified by combining them with the momentum balance. To this end the local balance of momentum (4) is multiplied by $\mathbf{v}$ or, in the discontinuous case, by $\langle\mathbf{v}\rangle$ and subtracted from the local energy balance. Eventually the local balances of energy valid for regular and for discontinuous fields assume the forms

$$\frac{\mathrm{D}E}{\mathrm{D}t} = \mathbf{T}\cdot\mathrm{Grad}\,\mathbf{v} - \mathrm{Div}\,\mathbf{q} + r \quad \text{and} \quad U_{\mathrm{N}}[\![E]\!] + \langle\mathbf{T}\rangle\cdot\left([\![\mathbf{v}]\!]\otimes\mathbf{N}\right) - [\![\mathbf{q}]\!]\cdot\mathbf{N} = 0, \tag{7}$$

respectively. A single dot is used to indicate the scalar product whether between two vectors, $\mathbf{a}\cdot\mathbf{b} = a_i b^i$, or between two tensors, $\mathbf{A}\cdot\mathbf{B} = A_i{}^{\alpha} B^i{}_{\alpha}$.

The deformation gradient $\mathbf{F}$ and the velocity vector $\mathbf{v}$ satisfy a compatibility condition whose continuous and discontinuous versions are

$$\frac{\mathrm{D}\mathbf{F}}{\mathrm{D}t} = \mathrm{Grad}\,\mathbf{v} \quad \text{and} \quad U_{\mathrm{N}}[\![\mathbf{F}]\!] + [\![\mathbf{v}]\!]\otimes\mathbf{N} = \mathbf{0}, \tag{8}$$

respectively.

4 Material Momentum and Eshelby Stress

The material momentum $\mathfrak{p}$ and the dynamic Eshelby stress $\mathfrak{E}$ are defined as

$$\mathfrak{p} = -\mathbf{F}^{\mathrm{T}}\mathbf{p} \quad \text{and} \quad \mathfrak{E} = -L\mathbf{I} + \mathbf{F}^{\mathrm{T}}\frac{\partial L}{\partial \mathbf{F}}, \tag{9}$$

respectively, with L denoting the Lagrangian. The kinetic energy density is uniquely defined as $\rho v^2/2$. Further we assume that the volume forces, if any, are conservative and arise from an external potential $U(\mathbf{x};\mathbf{X})$, such that

$$\mathbf{f} = -\frac{\partial U}{\partial \mathbf{x}}. \tag{10}$$

The gravity potential, for instance, has the form $U_g = -\rho(\mathbf{X})\mathbf{g}\cdot\mathbf{x}$ to yield the volume force $\mathbf{f} = \rho\mathbf{g}$.

The most important part of potential energy is the internal potential. It may be identified with the strain-energy density which is due to deformation and is a function $W(\mathbf{F};\mathbf{X})$ depending on the deformation gradient and, in non-homogeneous materials, also directly on the material coordinates. Strain energy might also depend on other variables, especially thermodynamic variables, as will be considered below, or on internal variables. Depending on the choice of thermodynamic variables the internal potential allows still different interpretations. Common to all of them is the property that the derivative with respect to the deformation gradient yields the Piola stress, i.e.,

$$\mathbf{T} = \frac{\partial W}{\partial \mathbf{F}}. \tag{11}$$

The Lagrange function is constituted from the kinetic-energy density and the inner and outer potential-energy densities as

$$L(\mathbf{x},\mathbf{v},\mathbf{F};\mathbf{X}) = \frac{1}{2}\rho(\mathbf{X})\mathbf{v}\cdot\mathbf{v} - W(\mathbf{F};\mathbf{X}) - U(\mathbf{x};\mathbf{X}), \tag{12}$$

its partial derivatives with respect to deformation and position being

$$\frac{\partial L}{\partial \mathbf{F}} = -\frac{\partial W}{\partial \mathbf{F}} = -\mathbf{T} \quad \text{and} \quad \frac{\partial L}{\partial \mathbf{x}} = -\frac{\partial U}{\partial \mathbf{x}} = \mathbf{f}. \tag{13}$$

It is assumed that these derivatives exist and have regular behavior.

The balance of material momentum in its continuous version is obtained by using, in addition to the definitions (9), (i) the balance of physical momentum $(4)_1$, (ii) the compatibility condition $(8)_1$, and (iii) the geometric compatibility condition

$$F^i{}_{\alpha,\beta} = F^i{}_{\beta,\alpha} \tag{14}$$

for the second deformation gradient. Eventually one arrives at the balance equation of material momentum in its local form,

$$\frac{\mathrm{D}\mathfrak{p}}{\mathrm{D}t} = \mathrm{Div}\,\mathfrak{E} + \mathfrak{f}, \tag{15}$$

where

$$\mathfrak{f} = \frac{\partial L}{\partial \mathbf{X}} \tag{16}$$

is interpreted as the material inhomogeneity force density.

The discontinuous case can be handled in an analogous way. In addition to the definitions (9) of material momentum and Eshelby stress one needs the corresponding discrete forms of (i) the balance of physical momentum $(4)_2$, (ii) the compatibility condition $(8)_2$, and (iii) the geometrical compatibility condition

$$[\![F^i{}_\alpha]\!] N_\beta = [\![F^i{}_\beta]\!] N_\alpha, \tag{17}$$

which is the discontinuous analogue of the compatibility condition (14). Thus one arrives at the discontinuous version of the balance of material momentum

$$U_{\mathrm{N}} [\![\mathfrak{p}]\!] + [\![\mathfrak{C}]\!] \mathbf{N} = \sigma \mathbf{N}, \quad \sigma = [\![W]\!] - \langle \mathbf{T} \rangle \cdot [\![\mathbf{F}]\!], \tag{18}$$

where the "imbalance" term $\sigma\mathbf{N}$ on the right-hand side still has to be determined.

The continuous version of the balance of material momentum, (15), exhibits the same appearance as the balance of physical momentum, $(4)_1$. The corresponding jump condition (18), however, as long as the imbalance factor σ does not vanish, deviates from the expected form which should resemble $(4)_2$. Whether this is the case or not depends on the exact interpretation of the strain energy W.

5 Hyperelastic Material

Hyperelasticity is considered as a purely mechanical model of an ideally elastic material for which the stress-strain relationship derives from a strain energy density function

$$W = W(\mathbf{F}; \mathbf{X}). \tag{19}$$

The dependence on the deformation gradient $\mathbf{F}$ must be consistent with the principle of material frame indifference. In case of nonhomogeneity the strain-energy density may also directly depend on the material point $\mathbf{X}$, but not on any other quantities like thermodynamic or internal variables.

Discontinuities of the strain-energy density and its derivatives are brought forward by a discontinuity of the deformation gradient. Under suitable regularity conditions, the function (19) may be expanded around the mean value $\langle \mathbf{F} \rangle$ of the deformation gradient at the singular surface. Forming then the discontinuity leads to an expansion with the leading terms

$$[\![W]\!] = A_i{}^\alpha [\![F^i{}_\alpha]\!] + \frac{1}{24} A_i{}^\alpha{}_j{}^\beta{}_k{}^\gamma [\![F^i{}_\alpha]\!] [\![F^j{}_\beta]\!] [\![F^k{}_\gamma]\!] + \cdots, \tag{20}$$

where

$$A_i{}^\alpha = \left. \frac{\partial W}{\partial F^i{}_\alpha} \right|_{\mathbf{F}=\langle \mathbf{F} \rangle} \quad \text{and} \quad A_i{}^\alpha{}_j{}^\beta{}_k{}^\gamma = \left. \frac{\partial^3 W(\mathbf{F}; \mathbf{X})}{\partial F^i{}_\alpha \, \partial F^j{}_\beta \, \partial F^k{}_\gamma} \right|_{\mathbf{F}=\langle \mathbf{F} \rangle} \tag{21}$$

denote the elasticities of first and third order, respectively, evaluated at the mean deformation gradient $\langle \mathbf{F} \rangle$. In a similar way, the mean value of the Piola stress (11) can be represented by an expansion

$$\langle T_i{}^\alpha \rangle = A_i{}^\alpha + \frac{1}{8} A_i{}^\alpha{}_j{}^\beta{}_k{}^\gamma [\![F^j{}_\beta]\!][\![F^k{}_\gamma]\!] + \cdots, \tag{22}$$

again with the elasticities (21) evaluated at the mean deformation gradient $\langle \mathbf{F} \rangle$.

Using these expansions, the imbalance factor σ according to $(18)_2$ is represented by the expansion

$$\sigma = -\frac{1}{12} A_i{}^\alpha{}_j{}^\beta{}_k{}^\gamma [\![F^i{}_\alpha]\!][\![F^j{}_\beta]\!][\![F^k{}_\gamma]\!] - \cdots. \tag{23}$$

So the imbalance factor is at least of third order in the jump $[\![\mathbf{F}]\!]$ of the deformation gradient, which is a measure of the strength of the discontinuity. The general form of the expansion is provided in [1].

6 Thermoelastic Material

As soon as thermodynamics is included the exact meaning of strain energy becomes relevant. The property of being a stress potential has to be retained, but it can be identified either with the internal energy or with the free energy, depending on the choice of the thermodynamical variable.

Since for the internal energy a balance equation has been formulated it might be reasonable to identify strain energy with internal energy. Correspondingly, the entropy S per unit reference volume should be chosen as the thermodynamic variable. Thus the density of strain energy (or internal energy) is considered to be a function

$$W = E(\mathbf{F}, S; \mathbf{X}), \tag{24}$$

whose partial derivatives

$$\frac{\partial E}{\partial \mathbf{F}} = \mathbf{T} \quad \text{and} \quad \frac{\partial E}{\partial S} = \Theta \tag{25}$$

represent the Piola stress and the absolute temperature, respectively. The notation E is kept for the internal energy in order to distinguish it from the strain energy W in the hyperelastic case.

The discontinuous form of the energy balance $(7)_2$ can be rewritten, using the compatibility condition $(8)_2$, as

$$[\![E]\!] - \langle \mathbf{T} \rangle \cdot [\![\mathbf{F}]\!] = \frac{1}{U_{\mathrm{N}}} \mathbf{N} \cdot [\![\mathbf{q}]\!]. \tag{26}$$

Since the strain energy W has been identified with the internal energy E the left hand side is exactly the imbalance factor $(18)_2$. Thus, in the thermoelastic case, there is an imbalance factor

$$\sigma = \frac{1}{U_\mathrm{N}}\mathbf{N}\cdot[\![\mathbf{q}]\!], \tag{27}$$

i.e., the scalar product of the slowness vector $(1/U_\mathrm{N})\mathbf{N}$ of the propagating surface and the jump $[\![\mathbf{q}]\!]$ of the nominal heat flux vector.

If one assumes that heat flux is governed by a constitutive equation like Fourier's law with a finite conductivity, any discontinuity in temperature would be smoothed out immediately. Therefore a sharp surface of discontinuity is possible only in the absence of heat conduction. Under adiabatic conditions the heat flux vanishes and, therefore, also the imbalance factor. Then the jump condition expressing the balance of material momentum assumes the form

$$U_\mathrm{N}[\![\mathfrak{p}]\!]+[\![\mathfrak{C}]\!]\mathbf{N} = 0 \tag{28}$$

which could be expected from its continuous counterpart (15). It should be noted that, in order to obtain the discontinuous version (28) of the balance of material momentum, the energy balance has been used, while the continuous form (15) has been derived without it. The latter holds independently of the thermodynamical interpretation of strain energy and is obtained directly from the balance of physical momentum [4]. The continuous form also holds independently of any assumption about the heat flux while the discontinuous form (28) is restricted to the adiabatic case.

On first sight, it seems strange that the material momentum exhibits an unbalance at a singular surface if the material is assumed hyperelastic, while for a thermoelastic material without heat conduction the unbalance vanishes. Carrying over the analysis of Section 5 to the thermoelastic material one has to expand the internal energy function

$$E = E(\mathbf{F}, S; \mathbf{X}), \tag{29}$$

where both the deformation gradient $\mathbf{F}$ and the entropy S undergo discontinuities at the singular surface. The formulas (20) and (22) have to be adjusted correspondingly to include the additional argument. If the expansion is restricted to the very first terms in $[\![S]\!]$ and $[\![\mathbf{F}]\!]$ the imbalance factor is obtained as

$$\sigma = A'[\![S]\!]-\frac{1}{12}A_{i\ j\ k}^{\ \alpha\ \beta\ \gamma}[\![F^i{}_\alpha]\!][\![F^j{}_\beta]\!][\![F^k{}_\gamma]\!]+\cdots \tag{30}$$

where

$$A' = \left.\frac{\partial E}{\partial S}\right|_{S=\langle S\rangle,\ \mathbf{F}=\langle\mathbf{F}\rangle} \quad \text{and} \quad A_{i\ j\ k}^{\ \alpha\ \beta\ \gamma} = \left.\frac{\partial^3 E}{\partial F^i{}_\alpha\,\partial F^j{}_\beta\,\partial F^k{}_\gamma}\right|_{S=\langle S\rangle,\ \mathbf{F}=\langle\mathbf{F}\rangle} \tag{31}$$

are the absolute temperature and the third-order elasticities, taken at the mean values of entropy and deformation gradient. Since the imbalance factor (30) according to

(28) has to vanish the singular surface must be accompanied by a jump of entropy, which is of third order in the strength of the discontinuity. This is a well-known result in the theory of shock waves [7, p. 176]. If, as in hyperelasticity, the strain energy does not depend on an additional argument that can exhibit a discontinuity at the singular surface, there is no way to compensate these third and higher-order terms in $[\![\mathbf{F}]\!]$, and the imbalance factor does not vanish.

7 Conclusion

The local balance of material momentum is expressed by the equation

$$\frac{\mathrm{D}\mathfrak{p}}{\mathrm{D}t} = \mathrm{Div}\,\mathfrak{C} + \mathfrak{f}, \tag{32}$$

where $\mathfrak{p}$, $\mathfrak{C}$ and $\mathfrak{f}$ denote the material momentum, the dynamic Eshelby stress and the inhomogeneity force, respectively. The equation can be integrated over a finite domain $\mathcal{B}$ whence it assumes the form

$$\frac{\mathrm{D}}{\mathrm{D}t}\int_{\mathcal{B}} \mathfrak{p}\,\mathrm{d}V = \int_{\partial\mathcal{B}} \mathfrak{C}\mathbf{n}\,\mathrm{d}A + \int_{\mathcal{B}} \mathfrak{f}\,\mathrm{d}V. \tag{33}$$

Despite its appearance as a global balance law this equation is not more general than the local balance equation from which it has been generated by integration. In particular the equation (33) must not be used to derive from it a corresponding jump condition using the pill-box argument, since the local balance (32) from which it has been derived does not hold at a surfaces of discontinuity. The jump condition representing the balance of material momentum at a singular surface has to be derived directly, namely from the jump conditions of the physical quantities. It turns out that this jump condition is of the general form

$$U_{\mathrm{N}}[\![\mathfrak{p}]\!]+[\![\mathfrak{C}]\!]\mathbf{N} = \sigma\mathbf{N}, \quad \sigma = [\![W]\!]-\langle\mathbf{T}\rangle\cdot[\![\mathbf{F}]\!] \tag{34}$$

with an imbalance factor σ which does not necessarily vanish. If the global form (33) of the balance of material momentum would hold without restriction this factor had to vanish.

Whether the imbalance factor vanishes or not depends on the specific interpretation of the strain-energy density W. In hyperelasticity strain energy is assumed to depend only on the deformation gradient $\mathbf{F}$ and, in case of nonhomogeneity, also directly on the material point $\mathbf{X}$. In this case, the imbalance factor at a singular surface does not vanish, it is of third order in the strength of the discontinuity. If, on the other hand, strain-energy is identified with the internal energy of a thermoelastic material without heat conduction the imbalance factor vanishes due to the discontinuous form of the energy balance. Unlike the strain energy of a hyperelastic material the internal energy depends on an additional argument, namely the entropy which is also discontinuous at a singular surface. The propagating shock is accompanied by a jump in the entropy, $[\![S]\!]$, which compensates the imbalance term originating from the jump in the deformation gradient, $[\![\mathbf{F}]\!]$, such that the expected form

$$U_{\mathsf{N}}[\![\mathfrak{p}]\!] + [\![\mathfrak{C}]\!]\mathbf{N} = 0 \tag{35}$$

of the discontinuous balance of material momentum is established. Thus both the continuous version (32) of the balance of momentum and its discontinuous counterpart (35) are valid. This, however, does not mean that the global balance (33) holds *in general*, since there might be other kinds of singular behavior that are not covered by the jump condition (35). Therefore the validity of the *global* balance of material momentum remains open.

This sheds some light on the role played by the balance of material momentum within the framework of mechanics or, more general, of physics. Maugin [4] emphasizes that the balance of material momentum is a consequence of the balance of physical momentum while Gurtin [2, p. 2] considers configurational forces "*as basic objects consistent with their own force balance.*" In the present paper, the continuous and discontinuous versions of the balance of material momentum has been *derived* from the balance laws of physical momentum and energy, in accord with Maugin's point of view. This appoach also conforms to the work of Irschik [3]. On the other hand, the discontinuous version of the balance of material momentum severely depends on the interpretation of the internal potential, i.e., the strain-energy density W. While in a hyperelastic material a singular surface is accompanied by an imbalance term the balance is restored by identifying the strain energy with the internal energy of a thermoelastic material. Although the imbalance result of hyperelasticity is formally correct, one tends to abandon the idea of pure hyperelasticity in favor of thermoelasticity, because it is more agreeable to retain the balance. In this sense one is led by the desire for a general balance of material momentum which, however, has to be derived from the *physical* balance laws.

References

1. Braun, M., Balance of material momentum at a shock wave. *Z. Angew. Math. Mech.*, 2009, accepted for publication.
2. Gurtin, M.E., *Configurational Forces as Basic Concepts of Continuum Physics*. Springer, New York/Berlin/Heidelberg, 2000.
3. Irschik, H., On rational treatments of the general laws of balance and jump, with emphasis on configurational formulations. *Acta Mech.* **194**, 2007, 11–32.
4. Maugin, G., *Material Inhomogeneities in Elasticity*. Chapman & Hall, London, 1993.
5. Steinmann, P., On spatial and material settings of hyperelastodynamics. *Acta Mech.* **156**, 2002, 193–218.
6. Steinmann, P., On spatial and material settings of thermo-hyperelastodynamics. *J. Elast.* **66**, 2002, 109–157.
7. Whitham, G.B., *Linear and Nonlinear Waves*. Wiley, New York/London, 1974.

On a Constraint-Based Regularization Technique for Configurational r-Adaptivity and 3D Shape Optimization

Michael Scherer, Ralf Denzer and Paul Steinmann

Abstract This contribution deals with a numerical regularization technique for configurational r-adaptivity and shape optimization based on a fictitious energy constraint. The notion *configurational* refers to the fact that both r-adaptivity and shape optimization rely on an optimization of the potential energy with respect to changes of the (discrete) reference configuration. In the case of r-adaptivity, the minimization of the total potential energy optimizes the mesh and thus improves the accuracy of the finite element solution, whereas the maximization of the total potential energy by varying the initial shape increases the stiffness of the structure. In the context of r-adaptivity, the energy constraint sets the distortion of the mesh to a reasonable limit and improves the solvability of the problem. The application of the energy constraint to a node-based shape optimization is a remedy for well-known problems of node-based shape optimization methods with maintaining a smooth and regular boundary.

1 Introduction

The idea of improving the accuracy of the finite solution by minimizing the potential energy with respect to the node positions of the mesh has already been investigated in the seventies and eighties, see e.g. [3,13–15,23,29]. Whereas the relation between

Michael Scherer
Chair of Applied Mechanics, University of Erlangen-Nuremberg, Egerlandstraße 5, D-91058 Erlangen, Germany; e-mail: michael.scherer@ltm.uni-erlangen.de

Ralf Denzer
Department of Mechanical and Structural Engineering, University of Trento, via Mesiano 77, I-38100 Trento, Italy; e-mail: denzerra@ing.unitn.it

Paul Steinmann
Chair of Applied Mechanics, University of Erlangen-Nuremberg, Egerlandstraße 5, D-91058 Erlangen, Germany; e-mail: paul.steinmann@ltm.uni-erlangen.de

P. Steinmann (ed.), IUTAM Symposium on Progress in the Theory and Numerics of Configurational Mechanics, 11–25.

configurational mechanics and mesh optimization was first pointed out in [10] in the late nineties. Then, with the growing interest in configurational mechanics, the topic has recently been taken up again by several researchers, see e.g. [2,20,24,25,30]. In this contribution, we attempt to attack the problem of element distortion as a consequence of r-adaptivity. Since a positive Jacobian at the Gauss points is a basic requirement of each finite element analysis, an r-adaptive mesh optimization is always implicitly subject to constraints, which was to our knowledge first mentioned in [15]. But this requirement can not avoid the occurrence of arbitrarily distorted elements since the Jacobian is only a local measure for the physical volume. To set the distortion of the entire mesh a more reasonable limit, we introduce an inequality constraint based on a fictitious energy. The fictitious energy measures the distortional deformation of the adapted mesh with respect to the initial mesh. The inequality constraint, the so-called energy constraint, sets an adjustable upper limit to the fictitious energy and thus allows to control the distortion of the mesh.

Beside the application in r-adaptive mesh optimization, the potential energy of a mechanical system can be used as an objective functional for a shape optimization that aims to improve the stiffness of a structure. An almost equivalent approach, a stiffness optimization based on the minimization of the total strain energy within the framework of the nonlinear theory, is presented in [22]. If an energy constraint similar to that that used for the r-adaptive mesh optimization is added to the shape optimization, the coordinates of boundary nodes can directly be chosen as design variables. The constraint eliminates well-known problems of node-based methods with maintaining a smooth and regular boundary. Historically, the node-based approach was a common practice in early works on finite element based shape optimization [16], but the large number of design variables and, more important, the above mentioned problems with the regularity of the boundaries led to the development of other techniques. Well-established is for instance the coupling of a finite element model with a geometry model based on B-splines, Bézier curves/surfaces, or NURBS, see e.g. [8,9]. The main advantage of node-based methods compared to an additional geometry model is the simplicity of model generation, including the straightforward application of boundary conditions. This advantage motivated the return to a node-based approach and the search for a regularization technique that eliminates its main drawback. Recently, a regularization technique for the node-based shape optimization of shell structures using filter techniques has been proposed in [7]. Another concept that is related to the approach of this contribution is introduced in [4]; fictitious forces acting on control nodes of a fictitious linear elastic body are chosen as design variables. Provided that the energy constraint is active at the optimal solution, the shape optimization method presented here can also be interpreted as the deformation of a fictitious elastic body subjected to fictitious forces, which are determined (a posteriori) by the Lagrange multipliers and the gradients of the potential energy and the constraints.

Our contribution is structured as follows. In Section 2, we define the considered class of mechanical problems and introduce the basic equations of the finite element analysis. Sections 3 and 4 illustrate the concepts of the r-adaptive mesh and shape optimization. Section 5 deals with the definition of the fictitious energies for both r-adaptivity and shape optimization. In Section 6, two numerical examples illustrate the effectivity of the proposed regularization technique.

2 Mechanical Problem

We restrict ourselves to linear elastostatic problems that are characterized as follows. A body occupying the open domain $\mathcal{B} \subset \mathbb{R}^2$ is subjected to conservative volume forces b and surface tractions t. The deformation is described via the displacement field $\mathsf{u}(\mathsf{x})$, $\mathsf{x} \in \mathcal{B}$, that is governed by a set of equations consisting of the balance of momentum

$$\mathrm{div}\sigma + \mathsf{b} = 0 \quad \text{in} \quad \mathcal{B} \tag{1}$$

and the boundary conditions

$$\begin{aligned} \mathsf{u} &= 0 && \text{on} \quad \partial\mathcal{B}^u, \\ \sigma \cdot \mathsf{n} &= \mathsf{t} && \text{on} \quad \partial\mathcal{B}^t, \end{aligned} \tag{2}$$

where n denotes the outward normal to the boundary. Note that the Dirichlet boundary conditions are assumed to be homogeneous. The relation between the Cauchy stresses σ and the linear strain tensor ε is given by

$$\sigma = \frac{\partial \psi}{\partial \varepsilon}, \tag{3}$$

where the free energy ψ does not depend on x, i.e. the material is homogeneous. Since the mechanical system is conservative, the equations (1) and (2) are the Euler–Lagrange equations of the principle of minimum potential energy

$$I(\mathsf{u}) \rightarrow \text{Min}, \quad \mathsf{u} = 0 \quad \text{on} \quad \partial\mathcal{B}^u, \tag{4}$$

where the total potential energy reads

$$I(\mathsf{u}) = \int_{\mathcal{B}} \psi \, dv - \int_{\mathcal{B}} \mathsf{u} \cdot \mathsf{b} \, dv - \int_{\partial\mathcal{B}^t} \mathsf{u} \cdot \mathsf{t} \, da. \tag{5}$$

To solve the variational problem and the related boundary value problem numerically, we apply a Ritz method using a finite element approximation. The domain $\mathcal{B}$ is discretized by n_{el} element domains $\mathcal{B}^h = \bigcup_{e=1}^{n_{el}} \mathcal{B}^e$, and the geometry and displacements are approximated elementwise by the shape function $N^{(i)}$ of a reference element $\mathcal{B}^\xi$ parameterized by natural coordinates ξ

$$\mathsf{x}(\xi) = \sum_{i=1}^{n_{en}} \mathsf{x}^{(i)} N^{(i)}(\xi), \quad \mathsf{u}^h(\xi) = \sum_{i=1}^{n_{en}} \mathsf{u}^{(i)} N^{(i)}(\xi), \tag{6}$$

where $\mathsf{x}^{(i)} \in \partial \mathcal{B}^e$ are the position vectors and $\mathsf{u}^{(i)}$ the nodal displacements. Substituting the approximation of the displacements u^h into the energy functional I, we obtain the discrete potential energy

$$I^h(\mathbf{u}) = I(\mathsf{u}^h) = \int_{\mathcal{B}^h} \psi \, dv - \int_{\mathcal{B}^h} \mathsf{u}^h \cdot \mathsf{b} \, dv - \int_{\partial \mathcal{B}^{h,t}} \mathsf{u}^h \cdot \mathsf{t} \, da, \tag{7}$$

which is a quadratic scalar function of the nodal displacements. For notational convenience, we have introduced the column vector $\mathbf{u} = [\mathsf{u}^{(1)}, \ldots, \mathsf{u}^{(n_u)}]^T$ that contains the Cartesian coordinates of all unknown nodal displacements. The necessary (and sufficiency) condition for a minimum of the discrete energy with respect to the nodal displacements is a linear algebraic system of equations that can be stated in the familiar matrix notation

$$\mathbf{r}(\mathbf{u}) = \mathbf{K}\mathbf{u} - \mathbf{f} = \mathbf{0}, \tag{8}$$

where $\mathbf{f}$ denotes the load vector and $\mathbf{K}$ the stiffness matrix.

3 Configurational r-Adaptivity

If the Dirichlet boundary conditions are homogeneous, the discretization error of a finite element solution to the variational problem (4) measured in the energy norm can be expressed in terms of the potential energy as

$$\|\mathsf{u}^h - \mathsf{u}^e\|_E^2 = I(\mathsf{u}^h) - I(\mathsf{u}^e) \geq 0, \tag{9}$$

where u^e is the exact solution to the problem. Hence, a reduction of the discrete potential energy is equivalent to a reduction of the discretization error, and the discrete potential energy can be considered as an error indicator. The concept of configurational r-adaptivity is to minimize this error indicator with respect to the node positions of the mesh. r-Adaptivity is formulated as a problem of nonlinear programming that can be attacked with various well-established tools provided by this mathematical discipline. Further details of this concept are illustrated in the following.

The discrete potential energy is interpreted as a function of the nodal displacements and the node positions

$$I^h = I^h(\mathbf{u}, \mathbf{x}), \tag{10}$$

where the column vector $\mathbf{x} = [\mathsf{x}^{(1)}, \ldots, \mathsf{x}^{(n_x)}]^T$ contains all variable Cartesian node coordinates. The nodal displacements $\mathbf{u}^*$ that solve the discrete mechanical equilibrium condition (8) are considered as an implicit function of the node positions

$$\mathbf{u}^* = \mathbf{u}^*(\mathbf{x}). \tag{11}$$

Substituting this result into eq. (10), we obtain the objective function of configurational r-adaptivity, the potential energy in the state of static equilibrium

$$I^{h*}(\mathbf{x}) = I^h(\mathbf{u}^*(\mathbf{x}), \mathbf{x}), \tag{12}$$

which is solely a function of the node coordinates. But I^{h*} is not minimized with respect to all coordinates. Since every r-adaptivity has to conserve the mechanical problem, the "motion" of boundary nodes is restricted to geometrical constraints. Generally, all boundary nodes have to move along the continuous boundary, and vertex nodes are fixed.

The computation of the gradient of I^{h*} is straightforward since the implicit dependency of the displacements on the node positions drops out [1]

$$\mathbf{R}^{*T} = \frac{\partial I^{h*}}{\partial \mathbf{u}} = \left.\frac{\partial I^h}{\partial \mathbf{u}}\right|_{\mathbf{u}^*,\mathbf{x}} \cdot \frac{\partial \mathbf{u}^*}{\partial \mathbf{x}} + \left.\frac{\partial I^h}{\partial \mathbf{x}}\right|_{\mathbf{u}^*,\mathbf{x}} = \left.\frac{\partial I^h}{\partial \mathbf{x}}\right|_{\mathbf{u}^*,\mathbf{x}}. \tag{13}$$

If the material is homogeneous and the body forces are zero, the components of $\mathbf{R}^*$ that are associated with the nodes can be expressed in terms of the Eshelby stress tensor as

$$\mathsf{R}^{*(i)} = \left.\int_{\mathcal{B}^h} \left[\psi \mathsf{1} - \mathsf{h}^t \cdot \sigma\right] \cdot \nabla_x N^{(i)} \, dv\right|_{\mathbf{u}^*,\mathbf{x}}, \tag{14}$$

and thus are called discrete configurational forces, see e.g. [10,25,28].

Due to several reasons briefly addressed in the following, the minimization of I^{h*} with respect to $\mathbf{x}$ is a challenging numerical problem. Since the Hessian of I^{h*} has in general negative eigenvalues, the objective function is usually nonconvex, and a local minimum is not necessarily a global minimum. Numerical experiments with examples characterized by inhomogeneous stress fields even suggest that a mesh with completely vanishing discrete configurational forces, the necessary condition for a strict (local) minimum, does often not exists. If, as discussed in [27], the optimization is explicitly subject to additional inequality constraints that restrict the deformation of each element, the optimal mesh is usually situated on the boundary of the feasible domain. Only for very coarse meshes, it is usually possible to determine a strict (local) minimum, characterized by inactive constraints and a completely vanishing gradient $\mathbf{R}^*$.

Additional constraints that restrict the distortion of the mesh improve the solvability of the problem. Following this idea, we introduce an energy constraint that sets an adjustable upper limit for a fictitious energy that measures the distortional deformation of the entire adapted mesh with respect to the initial mesh. In contrast to [27], only one constraint is introduced, which, according to our experience, again improves the solvability of the problem. The optimization problem of the r-adaptive mesh optimization including the energy constraint reads

[1] In eq. (13), we exploit that in the state of static equilibrium $\left.\frac{\partial I^h}{\partial \mathbf{u}}\right|_{\mathbf{u}^*,\mathbf{x}} = \mathbf{0}$.

$$\begin{aligned} &\text{minimize} && I^{h*}(\mathbf{x}) \\ &\text{subject to} && \hat{I}^h(\mathbf{x}) - \hat{I}^{\max} \leq 0 \\ & && h_i^{\text{geo}}(\mathbf{x}) = 0,\ i = 1, \ldots, n_{eq}, \end{aligned}$$

where $\hat{I}^h$ denotes the fictitious energy and $\hat{I}^{\max}$ its upper limit, and the geometrical constraints h_i^{geo} ensure the conservation of the mechanical problem.

Because of the similarity of the considered mesh and shape optimization, both problems have the same objective function and include an energy constraint, we first explain the concept of shape optimization before elucidating the design of the fictitious energies.

4 Configurational Shape Optimization

The (negative) potential of the external forces

$$-I^{ext*} = \int_{\mathcal{B}} \mathsf{u}^* \cdot \mathsf{b}\, \mathrm{d}v + \int_{\partial \mathcal{B}^t} \mathsf{u}^* \cdot \mathsf{t}\, \mathrm{d}a \tag{16}$$

is a common objective functional in topology optimization, leading to the so-called minimum compliance problem [6]. If the body forces are zero and the surface tractions are fixed, a reduction of $-I^{\text{ext}*}$ with respect to a set of design variables corresponds to a reduction of the displacements in the direction of the forces and hence a maximization of the stiffness. Moreover, if the Dirichlet boundary conditions are homogeneous, the total potential energy is equal to one half the potential of the external forces

$$I^* = \frac{1}{2} V^* \tag{17}$$

and a maximization of the potential energy is equivalent to a minimization of $-I^{\text{ext}*}$. To improve the stiffness of a structure, we exploit this equivalence and maximize the potential energy with respect to a set of design variables that control the shape of the boundary. An advantage of the potential energy compared to the direct use of the (negative) potential of the external forces as an objective functional is that the computation of the discrete sensitivities is less expensive since, as shown by eq. (13), the implicit dependency of the displacements on the node positions drops out.

The discrete shape optimization problem including the energy constraint reads

$$
\begin{aligned}
&\text{maximize} && I^{h**}(\mathbf{x}_d) \\
&\text{subject to} && \hat{I}^{h*}(\mathbf{x}_d) - \hat{I}^{\max} \leq 0 \\
& && g_j^{**}(\mathbf{x}_d) \leq 0, \; j = 2, \ldots, n_{iq} \\
& && h_i^{**}(\mathbf{x}_d) = 0, \; i = 1, \ldots, n_{eq},
\end{aligned}
\tag{18}
$$

where $\mathbf{x}_d$ is a column vector that contains the design variables. The energy constraint allows to control the admissible shape change of the design by varying the upper limit $\hat{I}^{\max}$ of the fictitious energy $\hat{I}^{h*}$. Additional equality and inequality constraints $h_i^{**} = 0$, $g_j^{**} \leq 0$ describe design restrictions, as for instance a constant volume or given stress limits.

So far, a star added as a superscript indicates the consideration of the implicit relation between the node points and the displacements determined by the mechanical equilibrium condition. The double star signifies that a second equilibrium condition is incorporated that determines the relation between the design variables $\mathbf{x}_d$ and all so-called controlled node coordinates $\mathbf{x}_c$ that are neither fixed nor assigned to the design variables. All interior nodes, for instance, are assigned to the controlled node coordinates $\mathbf{x}_c$ since they have only a very small influence on the potential energy, which is only caused by the discretization error. But fixing all interior nodes would only allow for extremely small geometry changes. To overcome this conflict of objectives, it is a common practice in shape optimization to define a design velocity field that relates the "motion" of the interior nodes to that of the design variables, see e.g. [21, 33]. In this contribution, the controlled node coordinates $\mathbf{x}_c$ are determined as the minimizers of the fictitious energy

$$
\mathbf{x}_c^*(\mathbf{x}_d) = \arg\min_{\mathbf{x}_c} \hat{I}^h(\mathbf{x}_c, \mathbf{x}_d). \tag{19}
$$

Since the fictitious energy is a nonlinear counterpart to a mechanical energy, this approach corresponds to a nonlinear version of the boundary displacement method introduced in [32]. The necessary condition for a minimum of the energy with respect to the controlled node coordinates

$$
\mathbf{R}_c^T = \frac{\partial \hat{I}^h}{\partial \mathbf{x}_c} = \mathbf{0} \tag{20}
$$

can be interpreted as a fictitious equilibrium condition. To indicate that one or both implicit relations defined by the mechanical and fictitious equilibrium condition are incorporated, we add one or two stars as a superscript, i.e.

$$
\begin{aligned}
I^{h**}(\mathbf{x}_d) &= I^h(\mathbf{u}^*(\mathbf{x}_c^*(\mathbf{x}_d), \mathbf{x}_d), \mathbf{x}_c^*(\mathbf{x}_d), \mathbf{x}_d) \\
\hat{I}^{h*}(\mathbf{x}_d) &= \hat{I}^h(\mathbf{x}_c^*(\mathbf{x}_d), \mathbf{x}_d).
\end{aligned}
\tag{21}
$$

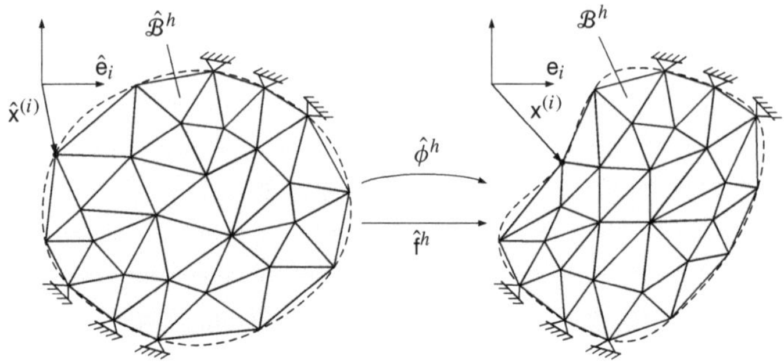

Fig. 1 Kinematics of the fictitious energy approach.

Note that the fictitious energy constraint in the definition of the shape optimization problem (18) sets an upper limit for the energy in the state of fictitious equilibrium. This fact plays an important role for the regularizing effect of the energy constraint, which is explained after the definition of the fictitious energy presented next.

5 Fictitious Energies

In analogy to nonlinear hyperelasticity, the definitions of the fictitious energies are based on an objective strain measure. Using scalar invariants of this objective strain measure, we introduce a local energy that can be interpreted as a fictitious counterpart to the specific free energy. Then, to define the global energies, we perform either a summation of the local energy evaluated at discrete points or an integration of the local energy over the initial design. The first approach yields a purely discrete fictitious energy designed for r-adaptivity and the second approach an energy for shape optimization that has really the character of a discrete total strain energy.

We start with a brief description of the kinematics, illustrated in Figure 1, that in principle corresponds to an ALE kinematics, see e.g. [1,5], with the exception that a third (spatial) configuration is not existent since the mechanical problem is based on the linear theory. In addition to the current discrete domain $\mathcal{B}^h$ used for the analysis of the mechanical problem, we introduce a fixed reference configuration $\hat{\mathcal{B}}$ that corresponds to the initial mesh and design, respectively. Note that Figure 1 refers to the shape optimization since the continuous domains corresponding to $\mathcal{B}^h$ and $\hat{\mathcal{B}}^h$ do not coincide. Using a standard isoparametric finite element formulation for the large strain theory, see e.g. [17, 31], we introduce the C^0-continuous deformation mapping $\phi^h : \hat{\mathcal{B}}^h \to \mathcal{B}^h$. The basic kinematic quantity for the definition of the fictitious energies is the gradient of this deformation mapping with respect to the coordinates of the initial mesh, which can be computed within each element using

$$\hat{\mathsf{f}}^e = \nabla_{\hat{x}} \phi^e = \sum_{i=1}^{n_{en}} \mathsf{x}^{(i)} \otimes \nabla_{\hat{x}} N^{(i)}. \tag{22}$$

To penalize distortional deformations, we define a fictitious local energy

$$\hat{\psi}(\hat{\mathsf{c}}^{\mathrm{iso}}) = \mathsf{1} : \hat{\mathsf{c}}^{\mathrm{iso}} - n_{\mathrm{dim}}, \tag{23}$$

where

$$\hat{\mathsf{c}}^{\mathrm{iso}} = \hat{j}^{\,-2/\mathrm{n_{dim}}}\, \hat{\mathsf{f}}^t \cdot \hat{\mathsf{f}} \quad \text{with} \quad \hat{j} = \det(\hat{\mathsf{f}}) \tag{24}$$

is defined in analogy to the isochoric part of the usual right Cauchy–Green strain tensor. The number of dimensions n_{dim} is subtracted to normalize the energy, i.e. to set $\hat{\psi}(\mathsf{1}) = 0$. Superimposed dilatational deformations are by intention allowed and thus our penalizing fictitious energy $\hat{\psi}$ does not depend on the volumetric part $\mathsf{c}^{\mathrm{vol}} = \hat{j}^{\,2/\mathrm{n_{dim}}}\mathsf{1}$. The fictitious local energy takes its global minimum value if and only if the deformation gradient is a scalar multiple of a rotation tensor, i.e. $\hat{\psi} \geq 0$, where $\hat{\psi} = 0$ if and only if $\mathsf{f} = \alpha \mathsf{q}$ with $\mathsf{q} \in SO(n_d)$ and $\alpha > 0$. Hence, the current configuration is "stress free" if $\hat{\mathsf{f}} = \alpha \mathsf{q}$.

The global energy for the r-adaptive mesh optimization has the general form

$$I^h = \sum_{i=1}^{n_{el}} \sum_{j=1}^{n_\xi} \hat{\psi}(\hat{\mathsf{f}}^e(\xi^{(j)})), \tag{25}$$

where $\xi^{(j)}$ are special evaluation points. To ensure that the energy constraint always guarantees a feasible mesh with positive Jacobians, we use an energy that goes to infinity if only one element within the mesh tends to invert. For the given local energy, this property can be generated by a suitable choice of the evaluation points $\xi^{(j)}$, at least for certain element types. For triangular and tetrahedral elements, one arbitrary point within each element is sufficient since f^e is constant. In the case of bi- and trilinear quadrilateral and hexahedral elements, the local energy has to be evaluated at all vertices of each element. Note that the global energy defined by eq. (25) is closely related to objective functions designed for mesh smoothing, see e.g. [18, 19, 26]. The main difference, apart form scaling and normalization, is that the energy used here measures the deformation with respect to the (discrete) initial configuration, and not with respect to a fictitious mesh consisting of ideal elements, e.g. equilateral triangles or regular tetrahedrons.

Based on the local energy, the discrete global energy of the shape optimization is defined as

$$\hat{I}^h = \int_{\hat{\mathcal{B}}^h} \hat{\psi}(\hat{\mathsf{f}}^h)\, \mathrm{d}\hat{v} = \sum_{i=1}^{n_{el}} \int_{\hat{\mathcal{B}}^e} \hat{\psi}(\hat{\mathsf{f}}^e)\, \mathrm{d}\hat{v}. \tag{26}$$

This definition combined with the boundary displacement method described in Section 4 allows to interpret the proposed shape optimization as the deformation of an artificial body. To derive this interpretation, we consider the first order part of the Karush–Kuhn–Tucker necessary conditions of the optimization problem (18) that reads

$$-\frac{\partial I^{**}}{\partial \mathbf{x}_d} + \lambda_1 \frac{\partial \hat{I}^{h*}}{\partial \mathbf{x}_d} + \sum_{j=2}^{n_{iq}} \lambda_j \frac{\partial g_j^{**}}{\partial \mathbf{x}_d} + \sum_{i=1}^{n_{eq}} \lambda_i \frac{\partial h_i^{**}}{\partial \mathbf{x}_d} = \mathbf{0}, \tag{27}$$

where λ_i and λ_j denote the Lagrange multipliers. If the energy constraint is active and $\lambda_1 > 0$, which is usually the case in practice, the first equation of the necessary conditions can be rearranged and divided by λ_1 to obtain

$$\frac{\partial \hat{I}^{h*}}{\partial \mathbf{x}_d} - \frac{1}{\lambda_1} \frac{\partial I^{**}}{\partial \mathbf{x}_d} + \sum_{j=2}^{n_{iq}} \frac{\lambda_j}{\lambda_1} \frac{\partial g_j^{**}}{\partial \mathbf{x}_d} + \sum_{i=1}^{n_{eq}} \frac{\lambda_i}{\lambda_1} \frac{\partial h_i^{**}}{\partial \mathbf{x}_d} = \mathbf{0}. \tag{28}$$

This result can be interpreted as an equilibrium: the discrete configurational forces and the gradients of the constraints scaled by the quotients of the Lagrange multipliers play the role of discrete external surface forces that are in equilibrium with the discrete forces of an artificial body with the fictitious total strain energy $\hat{I}^h$. Note that these forces act only on the design variables since all fictitious internal forces associated with controlled nodes vanish due to the fictitious equilibrium condition (27).

6 Numerical Experiments

In the following we present two numerical experiments that illustrate the effectivity of the proposed regularization technique. The mechanical problems of both examples are characterized by a homogeneous isotropic linear elastic material with a Young's modulus and a Poisson's ratio of $E = 200\,000\ \mathrm{N/mm^2}$ and $\nu = 0.3$.

To solve the constrained optimization problems of r-adaptivity and shape optimization numerically, we have applied the penalty-barrier algorithm described in [11, 12].

6.1 Cracked Specimen

The mechanical problem of the first numerical example, a cracked plate subjected to tension, is illustrated in Figure 2a. Exploiting the symmetry of the problem, we have modeled only the upper half of the plate. The boundary conditions of the r-adaptive mesh optimization are depicted in Figure 2b. All results presented in the following refer to initial discretizations with regular meshes consisting of linear triangular elements. The energy constraint was active in all performed optimizations.

Figures 3a and 3b show a mesh with 441 nodes before and after the r-adaptive optimization of the node positions. As a result of our optimization strategy, nodes

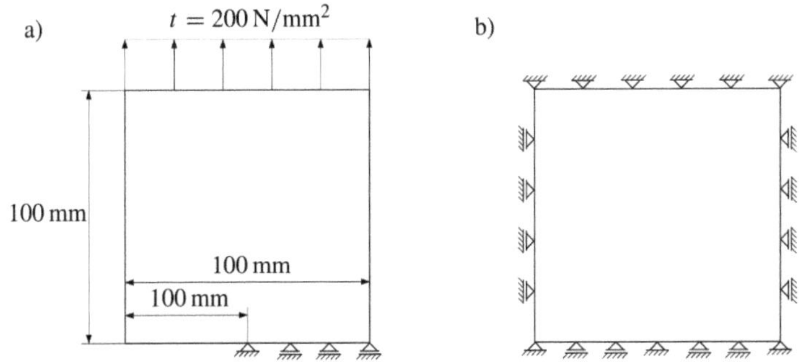

Fig. 2 Cracked specimen: (a) mechanical problem, (b) boundary conditions of the r-adaptive mesh optimization.

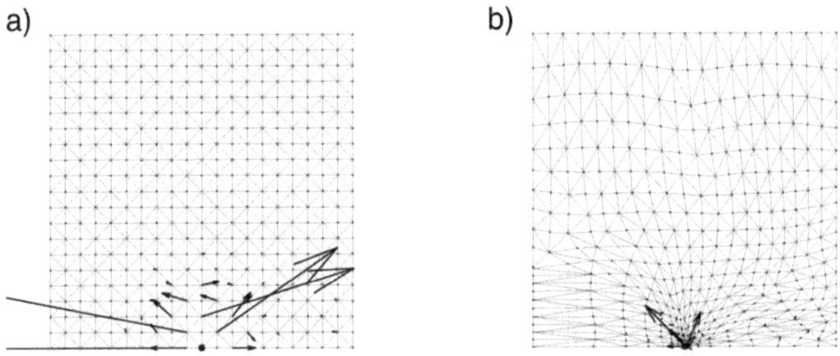

Fig. 3 Cracked specimen: (a) initial discretization, (b) optimized discretization. The arrows in both figures represent discrete configurational forces. Due to the active energy constraint, the configurational forces of the optimized mesh do not completely vanish.

concentrate in the vicinity of the crack tip, where the inhomogeneous stress field of the singularity causes a high discretization error and large (discrete) configurational forces. The energy limit was set to $\hat{I}^{\max} = 0.8\, n_{\text{ev}}$, where n_{ev} is the total number of evaluation points of the fictitious local energy, i.e. n_{ev} is the total number of terms in the sum (25) that defines the fictitious global energy.

In the diagram 4a, the discretization error measured in the energy norm is plotted versus the average global energy $\hat{I}^{\max}/n_{\text{ev}}$. To compute the discretization error based on formula (9), we have approximated the potential energy of the analytical solution by a finite element analysis with a regular mesh consisting of 80000 elements. The qualitative behavior of the three curves representing three discretizations with an increasing number of nodes is very similar. With an increasing energy limit, the discretization error decreases monotonically, but a saturation effect is observed. From a practical point of view, it makes no sense to choose $\hat{I}^{\max}/n_{\text{ev}} > 0.8$ in the given example. For comparison, the diagram also shows the discretization error of an regular mesh with 10000 nodes which is higher than that of the adapted mesh with 841 nodes. The relative decrease of the discretization error with respect to the initial mesh is plotted in Figure 4b. For the finest discretization with 841 nodes, a decrease of 83% is achieved.

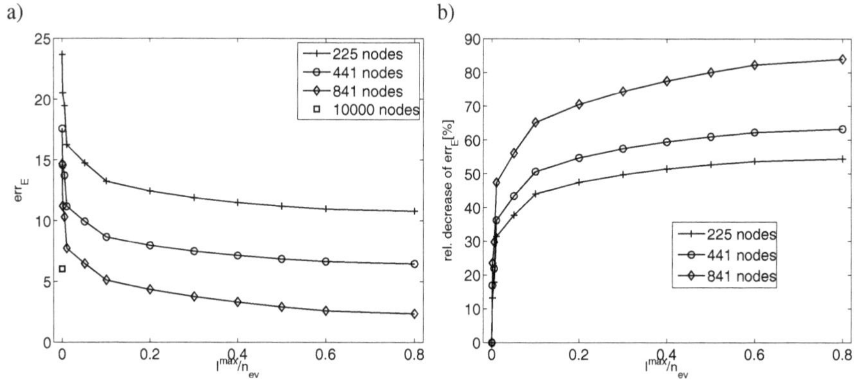

Fig. 4 Cracked specimen: (a) evolution of the energy error for increasing values of the energy limit $\hat{I}^{\max}/n_{\mathrm{ev}}$, (b) evolution of the relative decrease of the energy error (with respect to the initial meshes).

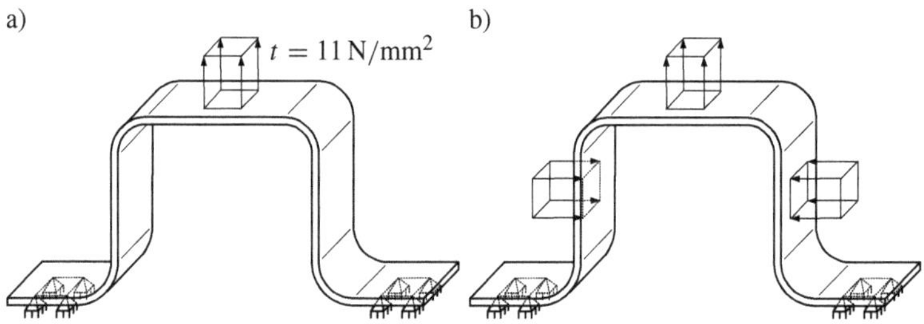

Fig. 5 Sheet metal part: (a) and (b) initial design with two different mechanical boundary conditions.

6.2 Sheet Metal Part

The second numerical experiment is concerned with shape optimization and shows that the energy constraint allows for an optimized design with a smooth shape without the application of an additional geometry model.

Figures 5a and 5b illustrate the considered mechanical problems: a sheet metal part clamped at the two horizontal straps is either subjected to one or three surface loads. The part is discretized by two layers of 648 hexahedral elements with 1115 nodes. The coordinates of all boundary nodes except the nodes on the quadratic areas where the Dirichlet boundary conditions and surface loads of the mechanical problems are applied are chosen as design variables. The shape optimization aims to maximize the stiffness of the part by maximizing the potential energy. Beside the energy constraint, the optimization is subject to the volume constraints $V \leq V_0$, where V_0 denotes the initial volume.

Figures 6a–d show the von Mises stress distribution of the initial design and the optimized design for both mechanical problems. The maximum von Mises stresses

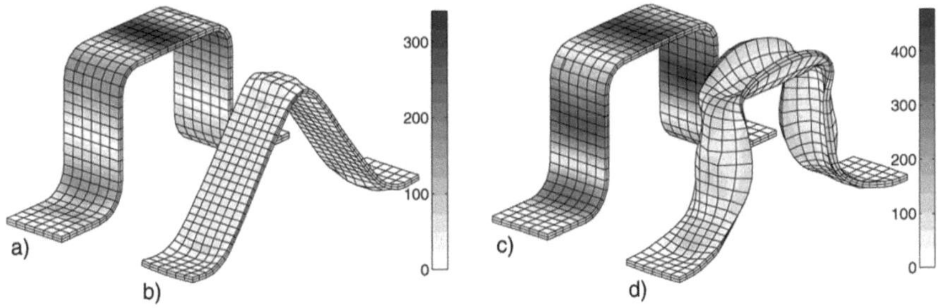

Fig. 6 Sheet metal part: (a) and (b) von Mises stress distribution of the initial and optimized design corresponding to the mechanical problem illustrated in Figure 5a; (c) and (d) von Mises stress distribution of the initial and optimized design corresponding to the mechanical problem illustrated in Figure 5b.

decrease significantly since the optimization improves the equal distribution of the free energy. Figure 6b shows an optimized design one would intuitively expect for the given optimization problem; the improved stiffness is realized by transforming the bending load into a tension load. In Figure 6d the improvement is realized by ribs that increase the moment of inertia of the cross sectional area. The optimization of the stiffness results in a 22 times increase for the design change 6a⇒6b and an 8 times increase for the design change 6c⇒6d. Note that these numerical values refer to the inverse (negative) potential of the external forces as a measure for the stiffness.

The upper limit of the fictitious energy was set to $\hat{I}^{\max} = 0.02\ V_0$. Since the fictitious global energy is defined by a volume integral over the initial design, it is useful to determine the energy limit $\hat{I}^{\max}$ by prescribing the average value $\hat{I}^{\max} / V_0$, where, according to our numerical experience, $\hat{I}^{\max} / V_0 \approx 0.01$ is often a good choice to start with. The energy constraint is active for both mechanical problems, whereas the volume constraint is active only in the second example with 3 surface loads.

7 Conclusions

A new regularization technique for r-adaptivity and shape optimization based on a fictitious energy constraint has been presented. The focus was on the overall concept of the r-adaptive mesh and shape optimization including the applied regularization. Details concerning the the reliability of the algorithms or the numerical effort have not been elucidated yet. Hence, some remarks on the two last named subjects. According to our experience, the reliability of the shape optimization algorithm based on a penalty barrier method is high, whereas that of the r-adaptive mesh optimization needs further improvements. Concerning the numerical effort both algorithms should be improved, especially the r-adaptive mesh optimization since it has to compete against h-adaptive schemes.

Another interesting subject is the question whether the fictitious energy of the shape optimization can be replaced by a linear energy, not in general but for certain applications. The main argument for the nonlinear approach is that shape optimization often goes along with large changes of the initial shape. Therefore, a geometrically nonlinear objective strain measure for the definition of the fictitious energy seems to be natural.

Acknowledgement

The authors would like to thank the "Deutsche Forschungsgemeinschaft" (DFG) for the support of this work under grant STE 544/24.

References

1. Armero, F. and Love, E.: An arbitrary Lagrangian–Eulerian finite element method for finite strain plasticity. *Int. J. Numer. Meth. Engng.* **57**, 2003, 471–508.
2. Askes, H., Kuhl, E. and Steinmann, P., An ALE formulation based on spatial and material settings of continuum mechanics. Part 2: Classification and applications. *Comput. Methods Appl. Mech. Engrg.* **193**, 2004, 4223–4245.
3. Bathe, K.J. and Sussman, T.D., An algorithm for the construction of optimal finite element meshes in linear elasticity. *ASME, AMD, Computer Methods for Nonlinear Solids and Structural Mechanics* **54**, 1983, 15–36.
4. Belegundu, A.D. and Rajan, S.D., A shape optimization approach based on natural design variables and shape functions. *Comput. Methods Appl. Mech. Engrg.* **66**, 1988, 87–106.
5. Belytschko, T., Liu, W.K. and Moran, B., Nonlinear Finite Elements for Continua and Structures. Wiley, 2000.
6. Bendsøe, M.P. and Sigmund, O., Topology Optimization. Springer, 2003.
7. Bletzinger, K.U., Firl, M., Linhard, J. and Wüchner, R., Optimal shapes of mechanically motivated surfaces. *Comput. Methods Appl. Mech. Engrg.*, 2008, published online.
8. Braibant, V. and Fleury, C., Shape optimal design using B-splines. *Comput. Methods Appl. Mech. Engrg.* **44**, 1984, 247–267.
9. Braibant, V. and Fleury, C., An approximation-concepts approach to shape optimal design. *Comput. Methods Appl. Mech. Engrg.* **53**, 1985, 119–148.
10. Braun, M., Configurational forces induced by finite-element discretization. *Proc. Estonian Acad. Sci. Phys. Math.* **46**, 1997, 24–31.
11. Breitfeld, M.G. and Shanno, D.F., A globally convergent penalty-barrier algorithm for nonlinear programming and its computational performance. Technical Report RRR 12-94, RUTCOR, 1994.
12. Breitfeld, M.G. and Shanno, D.F., Computational experience with penalty-barrier methods for nonlinear programming. *Ann. Operations Res.* **62**, 1996, 439–463.
13. Carroll, W.E. and Barker, R.M., A theorem for optimum finite-element idealizations. *Int. J. Solids Structures* **9**, 1973, 883–895.
14. Felippa, C.A., Numerical experiments in finite element grid optimization by direct energy search. *Appl. Math. Modelling* **1**, 1976, 239–244.

15. Felippa, C.A., Optimization of finite element grids by direct energy search. *Appl. Math. Modelling* **1**, 1976, 93–96.
16. Haftka, R.T. and Grandhi, R.V., Structural shape optimization – A survey. *Comput. Methods Appl. Mech. Engrg.* **57**, 1986, 91–106.
17. Hughes, T.J.R., The Finite Element Method. Dover Publications, 1987.
18. Knupp, P.M., Achieving finite element mesh quality via optimization of the Jacobian matrix norm and associated quantities. Part II – A framework for volume mesh optimization and the condition number of the Jacobian matrix. *Int. J. Numer. Meth. Engng.* **48**, 2000, 1165–1185.
19. Knupp, P.M., Algebraic mesh quality measures. *SIAM J. Sci. Computing* **23**, 2001, 193–218.
20. Kuhl, E., Askes, H. and Steinmann, P., An ALE formulation based on spatial and material settings of continuum mechanics. Part 1: Generic hyperelastic formulation. *Comput. Methods Appl. Mech. Engrg.* **193**, 2004, 4207–4222.
21. Lindby, T. and Santos, J.L.T., 2-D and 3-D shape optimization using mesh velocities to integrate analytical sensitivities with associative CAD. *Structural Optimization* **13**, 1997, 213–222.
22. Materna, D. and Barthold, F.J., On variational sensitivity analysis and configurational mechanics. *Comput. Mech.* **41**, 2008, 661–681.
23. McNeice, G.M. and Marcal, P.V., Optimization of finite element grids based on minimum potential energy. *Trans. ASME, J. Engrg. Ind.* **95**(1), 1973, 186–190.
24. Mosler, J. and Ortiz, M., On the numerical implementation of varational arbitrary Lagrangian–Eulerian (VALE) formulations. *Int. J. Numer. Meth. Engng.* **67**, 2006, 1272–1289.
25. Mueller, R., Kolling, S. and Gross, D., On configurational forces in the context of the finite element method. *Int. J. Numer. Meth. Engng.* **53**, 2002, 1557–1574.
26. Munson, T., Mesh shape-quality optimization using the inverse mean-ratio metric. *Math. Program.* **110**, 2007, 561–590.
27. Scherer, M., Denzer, R. and Steinmann, P., Energy-based r-adaptivity: A solution strategy and applications to fracture mechanics. *Int. J. Frac.* **147**, 2007, 117–132.
28. Steinmann, P., Ackermann, D. and Barth, F.J., Application of material forces to hyperelastostatic fracture mechanics. Part II: Computational setting. *Int. J. Solids Structures* **38**, 2001, 5509–5526.
29. Sussman, T. and Bathe, K.J., The gradient of the finite element variational indicator with respect to nodal point co-ordinates: An explicit calculation and application in fracture mechanics and mesh optimization. *Int. J. Numer. Meth. Engng.* **21**, 1985, 763–774.
30. Thoutireddy, P. and Ortiz, M., A variational r-adaption and shape-optimization method for finite-deformation elasticity. *Int. J. Numer. Meth. Engng.* **61**, 2004, 1–21.
31. Wriggers, P., *Nichtlineare Finite-Element-Methoden*. Springer, 2001.
32. Yao, T.M. and Choi, K.K., 3-D shape optimal design and automatic finite element regridding. *Int. J. Numer. Meth. Engng.* **28**, 1989, 369–384.
33. Zhang, S. and Belegundu, A.D., A systematic approach for generating velocity fields in shape optimization. *Structural Optimization* **5**, 1992, 84–94.

Some New Properties of the Eshelby Stress Tensor

Erwan Verron, Malik Aït-Bachir and Philippe Castaing

Abstract The Eshelby stress tensor is known to be an appropriate Continuum Mechanics quantity to capture singularities. Nevertheless, even if its use in the calculation of configurational forces is well-established, its peculiar properties were investigated only recently. Here, some new properties of this tensor are studied. In this way, it is assumed that the evolution of microscopic defects in the material can be predicted at the macroscopic scale by examining the components of the Eshelby stress tensor. More precisely, considering that defects can be modeled by material surfaces oriented in all possible directions and assuming that they are able to evolve in every possible directions, it is shown that the maximum amount of energy which can be released by defects evolution is partially contained in the tensor. In the special case of hyperelasticity, the corresponding optimization problem is established and solved for both isotropic and anisotropic materials.

1 Introduction

The theory of Configurational Mechanics was introduced by Eshelby in [4] when he proposed the concept of energy-momentum tensor and configurational forces in continuum mechanics of solids by studying the driving force of a moving defect. Twenty years later, both Eshelby [5] and Chadwick [2] extended the previous

Erwan Verron
Institut de Recherche en Génie Civil et Mécanique, UMR CNRS 6183, École Centrale de Nantes, BP 92101, 44321 Nantes Cedex 3, France; e-mail: erwan.verron@ec-nantes.fr

Malik Aït-Bachir
Institut de Recherche en Génie Civil et Mécanique, UMR CNRS 6183, École Centrale de Nantes, BP 92101, 44321 Nantes Cedex 3, France; e-mail: malik.ait-bachir@ec-nantes.fr

Philippe Castaing
Centre Technique des Industries Mécaniques, 74 route de la Jonelière, BP 82617, 44326 Nantes Cedex 3, France; e-mail: philippe.castaing@cetim.fr

P. Steinmann (ed.), IUTAM Symposium on Progress in the Theory and Numerics of Configurational Mechanics, 27–35.

theory to finite strain. More recently some authors revisited these studies and established the general framework of Configurational Mechanics also designated as the Eshelbian Mechanics and Mechanics in the Material Space [6, 9, 11, 12].

In the majority of studies involving Configurational Mechanics, only configurational forces are investigated through the calculation of path-independent integrals around inhomogeneities, because these forces are recognized as the driving forces of defect evolution. So, the Eshelby stress tensor only appears in the definition of surface tractions, i.e. after contraction with the outward normal of the contour. Most of these works focus on Fracture Mechanics (see for example [16] and the references herein). Moreover, as proposed in [10], other problems can be analyzed with the help of the Eshelbian framework: dislocations in metal, movement of interfaces in two-phase bodies, etc. Opposite to the case of configurational forces, only few studies are concerned with the peculiar properties of the Eshelby stress tensor. As an example, for the linear theory, the physical significance of the Cartesian components of this tensor were identified only recently by Kienzler and Herrmann [8]. More recently, Verron and co-workers proposed to use some components of the Eshelby stress tensor to predict fatigue damage in elastomers [1, 17, 18].

By generalizing and rationalizing the derivation proposed in [17], the present paper will examine some properties of the Eshelby stress tensor. In this way, the evolution of the microstructure of a given material will be studied by considering the evolution of oriented material surfaces under loading. The approach will exhibit the relevance of the polar decomposition of the tensor and define the significance of this decomposition with regards to microstructural evolution. Finally, this work will be illustrated by considering the extension of a transversely isotropic hyperelastic strip.

2 Evolution of Microstructural Defects by Considering the Eshelby Stress Tensor

2.1 *Formulation of the Problem*

Consider a body defined by its reference configuration ($\mathcal{C}_R$), i.e. a set of particles in the material manifold $\mathcal{M}^3$; it is depicted in Figure 1. One particle P of this set is located at $\mathbf{X}$ in the physical space $\mathcal{E}^3$. Under mechanical loading, the body deforms and occupies a time sequence of physical configurations (in $\mathcal{E}^3$). Let ($\mathcal{C}$) be the body configuration at time t defined by the mapping $\mathbf{x}(\mathbf{X}, t)$ and by its gradient $\mathbf{F}(\mathbf{X}, t)$. As shown in Fig. 1, when loading is removed the body will, in general, occupy a new stress-free configuration ($\mathcal{C}'_R$) defined by the motion gradient $\mathbf{f}(\mathbf{x}, t)$. Both configurations ($\mathcal{C}_R$) and ($\mathcal{C}'_R$) represent *natural* configurations of the body [14]. The physical reasons for a body to possess different natural configurations can be very diverse. Indeed, it is the consequence of structural rearrangements at the microscopic scale, e.g. movement of dislocations, cavitation, cleavage fracture,

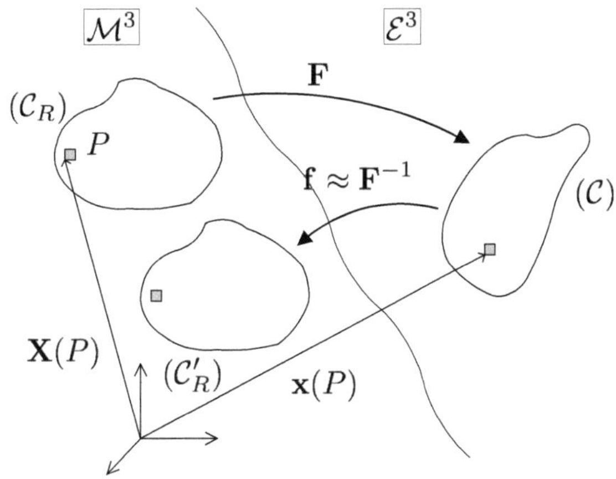

Fig. 1 Deformation of a body: macroscopic scale.

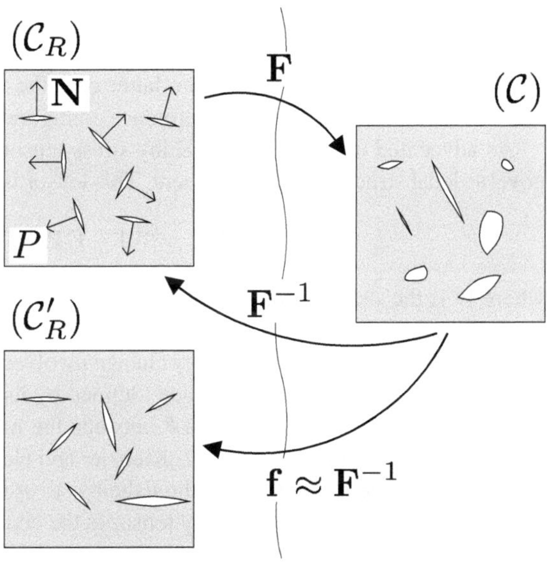

Fig. 2 Idealized microstructural change in the RVE

which manifests in various ways at the mesoscopic scale, e.g. plasticity, damage; or at the macroscopic scale, e.g. crack growth. If the material is perfectly elastic, $\mathbf{f} = \mathbf{F}^{-1}$ and the natural configurations $(\mathcal{C}_R)$ and $(\mathcal{C}'_R)$ are identical. Nevertheless, if irreversible microstructural changes take place during the motion, the gradient $\mathbf{f}$ can be considered close, but not equal, to $\mathbf{F}^{-1}$.

In the latter case, this slight change of configuration is the macroscopic counterpart of a microstructural evolution. In order to illustrate this change, the Representary Volume Element (RVE) which is phenomenologically embedded in the particle P should be considered. The following idealized representation of the RVE is adopted: it contains both bulk material and various defects which are schematized by oriented material surfaces as shown in Figure 2. These surfaces, which can repres-

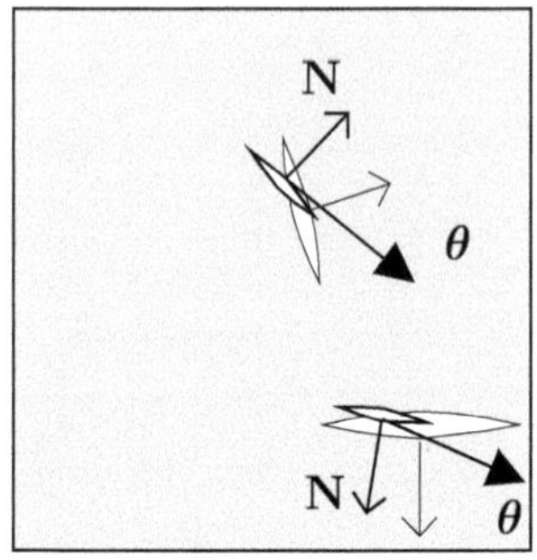

Fig. 3 Evolution of the defects modeled by oriented material surfaces.

ent for example microscopic cracks, are defined by their normal vector $\mathbf{N}$ and it is assumed that their orientations are isotropically distributed. Under motion, the size and shape of the RVE (see the grey square in Figure 2) do not change: the microstructural rearrangement does not change the definition of the RVE, i.e. at the Continuum Mechanics scale, the motion is reversible. Nevertheless, the defects deform reversibly or irreversibly, and in the latter case the microstructural rearrangement leads to the definition of the new reference configuration $(\mathcal{C}'_R)$ in Figures 1 and 2.

As advocated in [3, 13], the Eshelby stress tensor $\boldsymbol{\Sigma}$ is the driving force which governs local structural rearrangement. This tensor is defined by

$$\boldsymbol{\Sigma} = W\mathbf{I} - \mathbf{F}^t\mathbf{P}\,, \tag{1}$$

where W is the strain energy density per unit of undeformed volume which *a priori* depends on both $\mathbf{F}$ and $\mathbf{X}$, $\mathbf{P}$ is the Piola–Kirchhoff stress tensor and $\cdot^t$ denotes the transposition. To quantify the energy change involved during the rearrangement, we consider that a given material surface (defined by its unit normal vector $\mathbf{N}$) is subjected to a material unit translation $\boldsymbol{\theta}$ between the natural configurations $(\mathcal{C}_R)$ and $(\mathcal{C}'_R)$ as shown in Figure 3. In 1997, Kienzler and Herrmann identified the physical significance of the components of the Eshelby stress tensor in the linear context [8]: "[the ij-component] of the Eshelby tensor is the change in the total energy density at a point of an elastic continuum due to a material unit translation in x_j direction of a unit surface with normal in x_i-direction". Extending this definition to finite strain and to arbitrary direction, it can be established that the scalar $\boldsymbol{\theta} \cdot \boldsymbol{\Sigma}\mathbf{N}$ represents the change of energy due to the evolution of the material surface defined by the unit normal vector $\mathbf{N}$ and the material unit translation $\boldsymbol{\theta}$ between the natural configurations $(\mathcal{C}_R)$ and $(\mathcal{C}'_R)$.

To go further in the derivation, a strong assumption is adopted: it is considered that the microstructural change is only due to the evolution of only one set of surfaces (only one orientation of defects, i.e. only one vector $\mathbf{N}$) in only one material translation (only one vector $\boldsymbol{\theta}$) such that the body reduces as much as possible its total energy. Recalling that energy changes are defined positively in $\boldsymbol{\Sigma}$, the problem reduces to the following constraint optimization problem:

$$\max_{\mathbf{N},\boldsymbol{\theta}}\ \theta \cdot (-\boldsymbol{\Sigma})\,\mathbf{N} \tag{2}$$

with two equality constraints

$$\|\mathbf{N}\| = 1, \quad \|\boldsymbol{\theta}\| = 1, \tag{3}$$

and two inequality constraints

$$\mathbf{N} \cdot \boldsymbol{\theta} \geq 0, \quad \boldsymbol{\theta} \cdot (-\boldsymbol{\Sigma})\,\mathbf{N} \geq 0. \tag{4}$$

The two equality constraints Eq. (3) specify that both the normal vector to the material surface and the material translation vector are unit vectors. The inequality constraints Eq. (4) express that the microscopic defects can only open: Eq. $(4)_1$ ensures that $\mathbf{N}$ and $\boldsymbol{\theta}$ are in the same semi-plan, Eq. $(4)_2$ ensures that energy can only decrease due to defects evolution. Finally, the scalar measure of the damage Σ^* can be chosen as the maximum of the function $\theta \cdot (-\boldsymbol{\Sigma})\mathbf{N}$.

Remark. The quantity Σ^* which solves the optimization problem represents the energy involved in the evolution of the oriented material surface (with normal $\mathbf{N}$). The assumptions required to use the Eshelby stress tensor in this context are: (i) the size of defects is very small compared with the size of the RVE, and (ii) the density of defects is also very small, i.e. defects do not interact. It means that this approach is able to localize the damage in the material, but can not be used to calculate the energy release rate of defects because such a quantity obviously includes the size of defects.

2.2 Mathematical Solution

2.2.1 Special Case of an Isotropic Elastic Material

As shown in [3], the Eshelby stress tensor satisfies the following symmetry condition:

$$\boldsymbol{\Sigma}\mathbf{C} = \mathbf{C}\boldsymbol{\Sigma}^t, \tag{5}$$

where $\mathbf{C}$ is the right Cauchy–Green strain tensor. Moreover, recalling that $\boldsymbol{\Sigma}$ can be written in terms of $\mathbf{C}$ and of the second Piola–Kirchhoff stress tensor $\mathbf{S}$,

$$\boldsymbol{\Sigma} = W\mathbf{I} - \mathbf{C}\mathbf{S} \tag{6}$$

and noting that in the special case of isotropic elasticity $\mathbf{C}$ and $\mathbf{S}$ are coaxial and commute, *the Eshelby stress tensor is symmetric*.

So, $\boldsymbol{\Sigma}$ being a symmetric tensor, it possesses three real eigenvalues denoted $(\Sigma_i)_{i=1,3}$ and their corresponding eigenvectors $(\mathbf{V}_i)_{i=1,3}$ which are orthogonal one to each other. In this case, the solution of the optimization problem reduces to:

$$\Sigma^* = \left|\min\left((\Sigma_i)_{i=1,2,3}\,,0\right)\right| \tag{7}$$

and if $\Sigma^* \neq 0$,

$$\mathbf{N} = \boldsymbol{\theta} = \mathbf{V}^*, \tag{8}$$

which is the eigenvector associated with $-\Sigma^*$. In this case, the defects with normal $\mathbf{V}^*$ will grow by extending in a plane orthogonal to $\mathbf{V}^*$.

2.2.2 General Case

More generally, the Eshelby stress tensor is not symmetric, it only satisfies Eq. (5). In order to solve the optimization problem, we consider the right polar decomposition of $\boldsymbol{\Sigma}$:

$$\boldsymbol{\Sigma} = \boldsymbol{\Gamma}\boldsymbol{\Upsilon}, \tag{9}$$

where $\boldsymbol{\Gamma}$ is an orthogonal tensor (a reflection or a rotation tensor depending on the sign of its determinant) and $\boldsymbol{\Upsilon}$ is a positive definite symmetric tensor (when $\boldsymbol{\Sigma}$ is invertible). These tensors are defined by

$$\boldsymbol{\Upsilon} = \sqrt{\boldsymbol{\Sigma}^t\,\boldsymbol{\Sigma}} \quad \text{and} \quad \boldsymbol{\Gamma} = \boldsymbol{\Sigma}\boldsymbol{\Upsilon}^{-1}. \tag{10}$$

In this case the solution vectors are

$$\mathbf{N} = \mathbf{V}^* \quad \text{and} \quad \boldsymbol{\theta} = \boldsymbol{\Gamma}\mathbf{V}^* \tag{11}$$

where $\mathbf{V}^*$ is the eigenvector associated with the largest eigenvalue of $\boldsymbol{\Upsilon}$. If these vectors are not in the same semi-plan, i.e. if they do not satisfy Eq. $(4)_1$, the solution of the whole optimization problem (with constraints) Eqs. (2-4) is equal to 0 similarly to Eq. (7). Otherwise, the maximum of the function $\boldsymbol{\theta} \cdot (-\boldsymbol{\Sigma})\,\mathbf{N}$ is the largest eigenvalue of $\boldsymbol{\Upsilon}$:

$$\Sigma^* = \max\left((\Upsilon_i)_{i=1,3}\right). \tag{12}$$

In this case, the defects with normal $\mathbf{V}^*$ will grow by extending (and being distorted) in a plane of normal $\boldsymbol{\Gamma}\mathbf{V}^*$.

3 Example

First, the case in which the Eshelby stress tensor is symmetric and which corresponds to isotropic elasticity was recently examined in [17]. Authors applied this result to the problem of fatigue loading of rubber materials. Using Eq. (7), they derived a new predictor for rubber fatigue and they demonstrated its ability to reproduce multiaxial loading conditions and to predict macroscopic fatigue crack orientation.

So, the following example will focus on the second result derived in Section 2.2.2, Eqs (11–12). We consider the simple problem of uniaxial extension of a transversely isotropic hyperelastic strip. More precisely, the problem consists in extending a rubber-like thin strip reinforced with long fibers oriented in a given

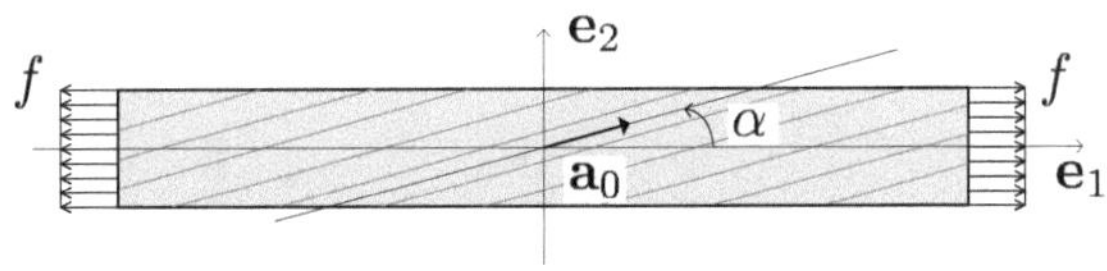

Fig. 4 Extension of a reinforced fiber elastic strip.

direction $\mathbf{a}_0$, as shown in Figure 4. The angle between the fiber direction, i.e. $\mathbf{a}_0$, and the loading direction, i.e. $\mathbf{e}_1$, is denoted α. Following the work of Spencer [15], the strain energy density per unit of undeformed volume for a homogeneous transversely isotropic material W can be written as a function of five invariants: the three classical isotropic invariants

$$I_1 = \mathrm{tr}\mathbf{C}, \quad I_2 = \frac{1}{2}\left[(\mathrm{tr}\mathbf{C})^2 - \mathrm{tr}\mathbf{C}^2\right], \quad I_3 = \det\mathbf{C}, \tag{13}$$

and two additional invariants

$$I_4 = \mathrm{tr}\,(\mathbf{C} : \mathbf{A}), \quad I_5 = \mathrm{tr}\left(\mathbf{C}^2 : \mathbf{A}\right), \tag{14}$$

where $\mathbf{A}$ is the orientation tensor $\mathbf{a}_0 \otimes \mathbf{a}_0$. Here, the simplest transversely isotropic rubber-like constitutive equation is adopted: the material is considered incompressible ($I_3 = 1$) and the strain energy density is assumed to only depends on I_1 and I_4

$$W\,(\mathbf{C}, \mathbf{a}_0) = C_1\,(I_1 - 3) + C_4\,(I_4 - 1)^2, \tag{15}$$

where the two material parameters were set to $C_1 = 1$ and $C_4 = 2$ in order to emphasize the effect of fibers. In the special cases for which the loading direction is parallel ($\alpha = 0°$) or perpendicular ($\alpha = 90°$) to the fiber direction, a given extension can be prescribed and the problem can be solved analytically (see for example [7]). Nevertheless, for other fiber orientations the finite element method should be considered; here we used the software COMSOL Multiphysics. Computations are performed by considering only half of the strip (symmetry with respect to the $x_1 = 0$-axis), the plane stress assumption is adopted (the incompressibility constraint leads to a change in thickness of the strip) and the strip is extended by prescribing the force f in order to obtain an uniform deformation gradient in the whole strip.

Practically, for a given extension ratio λ, simulations with prescribed force f are conducted to determine the value of the force such that $F_{11} = \lambda$. To show quantitative results we plot the evolution of the directions of $\mathbf{V}^*$ (normal vector to the critical material plane) and $\mathbf{\Gamma V}^*$ (material direction of evolution) with respect to the fiber directions in Figure 5. It is to note that for both $\alpha = 0°$ and $\alpha = 90°$, even if the material is not isotropic, the deformation gradient is diagonal and then the Eshelby stress tensor is symmetric. Thus the problem reduces to the special case studied in Section 2.2.1; the normal vector to the material surfaces and the normal to the plane in which they will grow are identical. In other cases, they are different.

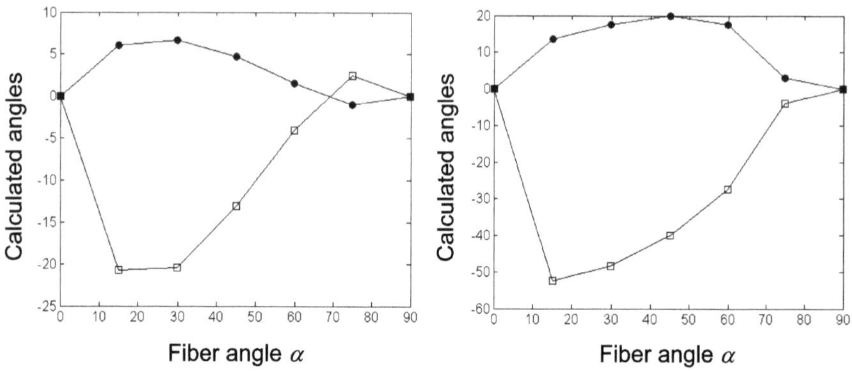

Fig. 5 (•) Orientation of the critical material plane which could release the maximum of energy: $\widehat{(\mathbf{e}_1, \mathbf{V}^*)}$. (□) Orientation of the corresponding material translation $\widehat{(\mathbf{e}_1, \mathbf{\Gamma V}^*)}$. Left-hand side graph: $\lambda = 2$; right-hand side graph: $\lambda = 4$.

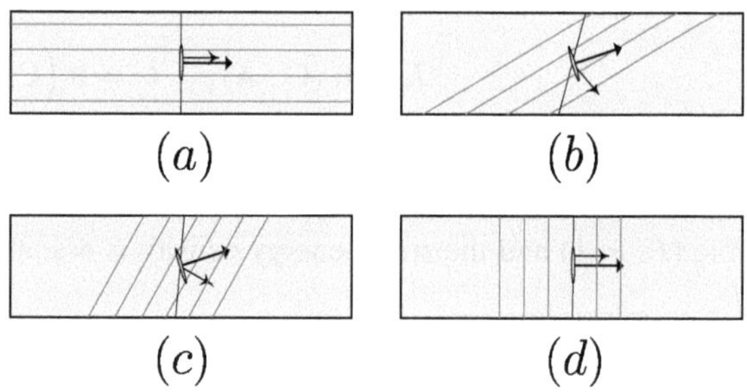

Fig. 6 $\lambda = 4$-extension of a transversaly isotropic hyperelastic strip: (a) $\alpha = 0°$, (b) $\alpha = 30°$, (c) $\alpha = 60°$ and (d) $\alpha = 90°$.

Finally, Figure 6 shows some qualitative results obtained for four fiber orientations. This figure illustrates with simple sketches four particular quantitative results given in Figure 5: for $\lambda = 4$, it exhibits the orientation of the critical material surfaces ($\mathbf{V}^*$) and the direction of the material translation ($\mathbf{\Gamma V}^*$) for four fiber orientation angles.

4 Discussion

The present paper is only a second step towards the understanding of the physical significance of the Eshelby stress tensor components, after the one of Kienzler and Herrmann. The relevance of the polar decomposition of the Eshelby stress tensor was demonstrated by considering that this tensor is the driving force of microstructural evolution. For an idealized RVE, it was shown that the "pure material stress tensor" $\mathbf{\Upsilon}$ can be considered as a measure of the microstructural damage and of the direction of defects that will grow, and also that the "rotational material stress tensor" $\mathbf{\Gamma}$ contains the way these defects will evolve.

References

1. Andriyana, A. and Verron, E., Prediction of fatigue life improvement in natural rubber using configurational stress. *Int. J. Solids Struct.* **44**, 2007, 2079–2092.
2. Chadwick, P., Applications of an energy-momentum tensor in non-linear elastostatics. *J. Elast.* **5**, 1975, 249–258.
3. Epstein, M. and Maugin, G.A., (1990) The energy-momentum tensor and material uniformity in finite elasticity. *Acta Mech.* **83**, 1990, 127–133.
4. Eshelby, J.D., The force on an elastic singularity. *Phil. Trans. R. Soc. Lond. A* **244**, 1951, 87–112.
5. Eshelby, J.D., The elastic energy-momentum tensor. *J. Elast.* **5**, 1975, 321–335.
6. Gurtin, M.E., *Configurational Forces as Basic Concept of Continuum Physics*. Springer, Berlin, 2000.
7. Horgan, C.O. and Saccomandi, G., A new constitutive theory for fiber-reinforced incompressible nonlinearly elastic solids. *J. Mech. Phys. Solids* **53**, 2005, 1985–2015.
8. Kienzler, R. and Herrmann, G., On the properties of the Eshelby tensor. *Acta Mech.* **125**, 1997, 73–91.
9. Kienzler, R. and Herrmann, G., *Mechanics in Material Space*. Springer, Berlin, 2000.
10. Kolling, S., Mueller, R. and Gross, D., A computational concept for the kinetics of defects in anisotropic materials. *Comput. Mat. Sci.* **26**, 2003, 87–94.
11. Maugin, G.A., *Material Inhomogeneities in Elasticity*. Chapman and Hall, London, 1993.
12. Maugin, G.A., Material forces: Concepts and applications. *Appl. Mech. Rev.* **48**, 1995, 213–245.
13. Maugin, G.A., Material mechanics of materials. *Theor. Appl. Mech.*, 27, 2002, 1–12.
14. Rajagopal, K.R. and Srinivasa, A.R., On the role of the Eshelby energy-momentum tensor in materials with multiple natural configurations. *Math. Mech. Solids* **10**, 2005, 3–24.
15. Spencer, A.J.M., *Continuum Theory of the Mechanics of Fibre-reinforced Composite*. Springer-Verlag, Berlin, 1984.
16. Steinmann, P., Application of material forces to hyperelastostatic fracture mechanics. I. Continuum mechanical setting. *Int. J. Solids Struct.* **37**, 2000, 7371–7391.
17. Verron, E. and Andriyana, A., Definition of a new predictor for multiaxial fatigue crack nucleation in rubber. *J. Mech. Phys. Solids* **56**, 2008, 417–443.
18. Verron, E., Le Cam, J.-B. and Gornet, L., A multiaxial criterion for crack nucleation in rubber. *Mech. Res. Commun.* **33**, 2006, 493–498.

On Configurational Aspects of Finite Deformation Inelasticity: A Variational Approach Versus the Transformation of Balance of Momentum

A. Menzel and B. Svendsen

Abstract The main goal of this contribution consists in reviewing two different approaches that both account for the so-called material or rather configurational balance of linear momentum representation. One formulation considered is based on variational strategies, the other framework studied makes use of transforming the standard spatial balance of linear momentum relation. While the similarity of both approaches is obvious for the purely elastic case – in other words, when all quantities of interest are derived from potentials – inelastic response requires, in view of the variational formulation, incremental strategies and, moreover, renders additional contributions to the volume forces that reflect the material's inhomogeneity. The general kinematics framework is kept as simple as possible but embedded into a large strain setting, i.e. a simple continuum is considered, the multiplicative decomposition or rather the introduction of a material isomorphism is adopted, and further hardening effects are neglected.

1 Introduction

Engineering materials are nowadays often designed for particular advanced applications in the various fields of modern technology. Commonly, it is the specific sub- or micro-structure of the material that is shaped and used to influence the overall properties of, for instance, a high-performance composite or metal structure. Either the

A. Menzel
Institute of Mechanics, Faculty of Mechanical Engineering, TU Dortmund, Leonhard-Euler-Str. 5, D-44227 Dortmund, Germany; e-mail: andreas.menzel@udo.edu and
Division of Solid Mechanics, Lund University, P.O. Box 118, SE-221 00 Lund, Sweden; e-mail: andreas.menzel@solid.lth.se

B. Svendsen
Institute of Mechanics, Faculty of Mechanical Engineering, TU Dortmund, Leonhard-Euler-Str. 5, D-44227 Dortmund, Germany; e-mail: bob.svendsen@udo.edu

P. Steinmann (ed.), IUTAM Symposium on Progress in the Theory and Numerics of Configurational Mechanics, 37–46.

production process itself or the deformation of the specimen – to give an example, think of a forming process of a rolled sheet metal – renders the material body considered to possess so-called inhomogeneities, which furthermore also evolve.

The sound modelling and description of such material inhomogeneities as well as heterogeneities dates back to the pioneering work by Eshelby; see the collected papers in [10]. Related forces that drive the evolution of these inhomogeneities have been variously designated in the literature, for example as pseudo forces, material forces, configurational forces, and so forth; for an overview, the reader is referred to the monographs by Hanyga [8], Maugin [11], Šilhavý [22], and Gurtin [4]. The derivation of, for instance, configurational forces can be based on either transformation operations – see, among others, [24] – or variational strategies as discussed in [19,21]. As shown, for example, in [29], the variational formulation of configurational fields and balance relations can be extended to inelastic effects by making use of incremental variational formulations as proposed and elaborated by Hackl [7], Ortiz and Repetto [20], Miehe [17], and Carstensen et al. [1].

A general and well-established continuum approach to model local inelastic deformation in crystalline metals is based on incorporating these inelastic effects by means of an elastic material isomorphism. In the configurational context, this deformation may also locally embody the materials inhomogeneity; see [25,26] and references cited therein. In this regard, the pioneering contribution by Noll [18] already emphasises that, in case of inhomogeneities being present in the body considered, the standard balance of linear momentum relation does not directly capture these material properties. Furthermore, a framework accounting for the related material forces within this nowadays classical inelasticity approach has been proposed in [3, 12], or similarly in [13]. Due to the overall incompatibility of the locally defined material isomorphism, dislocation structures are included in these modelling approaches – either in terms of the corresponding inelastic connection, or in terms of the corresponding dislocation density tensor; see, for example, [15,23]. In general, such dislocation-related quantities can be used to enhance field or evolution equations for the underlying material isomorphism, or additionally be incorporated as arguments into energy potentials; see, for instance, [14] and references cited therein, or [5,6,9,27].

The main purpose of this contribution is to review some essentials of the two different approaches to the formulation of configurational fields and balance relations mentioned above, namely the variational and a transformation-based formulation. In order to simplify the subsequent investigations, we restrict ourselves to simple continua and, moreover, neglect any further hardening mechanisms. It turns out that the two configurational balance of linear momentum relations share common characteristics even for the inelastic case and, concerning future research, a numerical finite-element-based study is of particular interest in order to compare both strategies in more detail. Additional background information on the subsequent outline can also be found in [16,28,30].

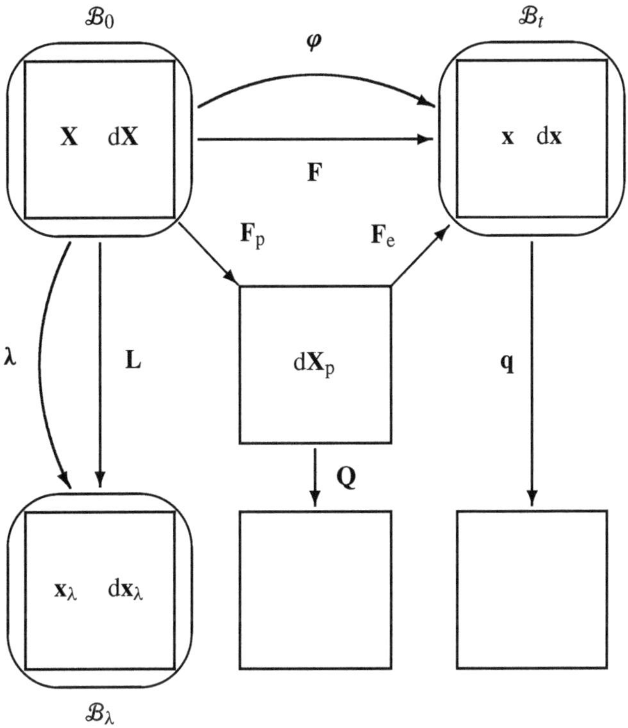

Fig. 1 Essential kinematics: mappings and local transformation quantities.

2 Basic Considerations

To introduce notation and in order to summarise the kinematics background essential to the subsequent elaborations, this section reviews fundamental concepts nowadays well-established in continuum inelasticity. To be specific, we make use of what commonly is denoted as the multiplicative decomposition of the deformation gradient or rather introduce a material isomorphism.

Let the sufficiently smooth motion of a body B be characterised by the mapping $\mathbf{x} = \boldsymbol{\varphi}(\mathbf{X}, t) : \mathcal{B}_0 \times \mathcal{T} \to \mathcal{B}_t$. Here, the referential position of a material particle is represented by $\mathbf{X} \in \mathcal{B}_0$, and its spatial or current position at time t by $\mathbf{x} \in \mathcal{B}_t$. Apart from the deformation gradient $\mathbf{F} = \nabla_{\mathbf{X}} \boldsymbol{\varphi}$, the material behaviour is determined here by the local inelastic deformation $\mathbf{F}_p$. Assuming as well from the configurational point of view that the material behavior is inhomogeneous, the stored energy is given by

$$\psi = \psi_0(\mathbf{F}, \mathbf{F}_p; \mathbf{X}) . \tag{1}$$

In the current work, $\mathbf{F}_p$ is modeled in particular as an elastic material isomorphism. In this case, there exists a reduced form W_0 of the stored energy such that

$$\psi_0(\mathbf{F},\mathbf{F}_\mathrm{p};\mathbf{X}) = W_0(\mathbf{F}\cdot\mathbf{F}_\mathrm{p}^{-1};\mathbf{X}) = W_0(\mathbf{F}_\mathrm{e};\mathbf{X}) \tag{2}$$

holds. This in turn implies a direct coupling of both deformation measures as shown in Figure 1. Representing an energy density, we have

$$W_0 = J_\mathrm{p}\, W_\mathrm{p} = J\, W_t\,, \tag{3}$$

with $J_\mathrm{p} = \det(\mathbf{F}_\mathrm{p})$ and $J = \det(\mathbf{F}) = J_\mathrm{e}\, J_\mathrm{p}$. Besides $\mathbf{F}_\mathrm{p}$ being such an isomorphism, two further assumptions about its constitutive nature are considered. In the first case, $\mathbf{F}_\mathrm{p}$ is assumed to be a material uniformity, i.e.,

$$\psi_0(\mathbf{F},\mathbf{F}_\mathrm{p};\mathbf{X}) = \bar{\psi}_0(\mathbf{F},\mathbf{F}_\mathrm{p}(\mathbf{X}))\,. \tag{4}$$

Here, further material inhomogeneities are excluded. Alternatively, consider the case that $\mathbf{F}_\mathrm{p}$ is materially constant, but the material properties may be heterogeneous, i.e.,

$$\psi_0(\mathbf{F},\mathbf{F}_\mathrm{p};\mathbf{X}) = \widetilde{\psi}_0(\mathbf{F},\mathbf{F}_\mathrm{p}\big|_{\mathrm{const\,in}\,\mathbf{X}};\mathbf{X})\,. \tag{5}$$

For the moment, attention is focused on equation (2); later, we come back to equations (4) and (5). In what follows, similar restrictions on $\mathbf{F}_\mathrm{p}$ will also be made for the dissipation potential.

Besides these basic assumptions on the nature of $\mathbf{F}_\mathrm{p}$, the particular form of the strain energy function is restricted by the principle of material frame-indifference and the material's symmetry properties. To be specific, W_0 must remain invariant under the spatial action of the orthogonal group on the deformation gradient, $W_0(\mathbf{F}_\mathrm{e};\mathbf{X}) = W_0(\mathbf{q}\cdot\mathbf{F}_\mathrm{e};\mathbf{X})$ with $\mathbf{q}^\mathrm{t} = \mathbf{q}^{-1}$, so that the related directional derivative with respect to $\mathbf{q}_\varepsilon = \exp(\varepsilon\,\mathbf{w})\cdot\mathbf{q}$ results in

$$[\,\partial_{\mathbf{F}_\mathrm{e}} W_0\cdot\mathbf{F}_\mathrm{e}^\mathrm{t}\,] : \mathbf{w} = 0 \qquad \forall\,\mathbf{w} = -\,\mathbf{w}^\mathrm{t}\,. \tag{6}$$

When assuming the material body to be isotropic, we similarly obtain the relation that the strain energy function does not sense any material action of the orthogonal group on $\mathbf{F}_\mathrm{e}$, i.e. $W_0(\mathbf{F}_\mathrm{e};\mathbf{X}) = W_0(\mathbf{F}_\mathrm{e}\cdot\mathbf{Q};\mathbf{X})$ with $\mathbf{Q}^\mathrm{t} = \mathbf{Q}^{-1}$. Consequently, the directional derivative as based on $\mathbf{Q}_\varepsilon = \exp(\varepsilon\,\mathbf{W})\cdot\mathbf{Q}$ yields

$$[\,\mathbf{F}_\mathrm{e}^\mathrm{t}\cdot\partial_{\mathbf{F}_\mathrm{e}} W_0\,] : \mathbf{W} = 0 \qquad \forall\,\mathbf{W} = -\mathbf{W}^\mathrm{t}\,. \tag{7}$$

Please note that equation (6) and equation (7) render – when adopting a hyper-elastic stress form – the Kirchhoff and Mandel stresses to be symmetric. Moreover, the kinematics framework reviewed in this section can straightforwardly be extended to so-called higher-gradient continua, which is not discussed in this contribution.

The particular representation of the material isomorphism $\mathbf{F}_\mathrm{p}$ is here assumed to be determined by means of an appropriate evolution equation, the particular form of which is not specified in this work. Hence, $\mathbf{F}_\mathrm{p}$ may in general depend on the material placement $\mathbf{X}$ but – as a consequence of being constrained by the evolution equation itself – cannot be varied independently.

3 Variational Approach

The variational approach adopted in the following captures the material's properties, inhomogeneities, and hereogeneities by means of a so-called incremental potential. As point of departure we start with the balance of entropy, from which field equations in local form can be derived. By introducing the dissipation rate density D_0 as well as the traction vector $\mathbf{t}$ we recall for an isothermal and quasi-static setting that

$$\int_{\mathcal{B}_0} \dot{W}_0 + D_0 \, \mathrm{d}V = \int_{\partial\mathcal{B}_0} \mathbf{t} \cdot \dot{\boldsymbol{\varphi}} \, \mathrm{d}A \,, \tag{8}$$

wherein contributions stemming from additional volume forces as well as singular surfaces are not considered and the notation $\dot{\bullet} = \partial_t \bullet |_{\mathbf{X}}$ characterises the material time derivative. Since the kinetics of dislocation activation and motion are dissipative, the evolution of $\mathbf{F}_\mathrm{p}$ is determined via a dissipation potential

$$P = P_0(\dot{\mathbf{F}}_\mathrm{p}; \mathbf{X}) \tag{9}$$

via the evolution-constitutive relation

$$\partial_{\dot{\mathbf{F}}_\mathrm{p}} R_0 = \mathbf{0} \tag{10}$$

for $\mathbf{F}_\mathrm{p}$ in terms of the rate potential

$$R_0 = \dot{W}_0 + P_0 \,. \tag{11}$$

In order to satisfy the dissipation principle sufficiently, D_0 is assumed convex in $\dot{\mathbf{F}}_\mathrm{p}$, such that $D_0 = \partial_{\dot{\mathbf{F}}_\mathrm{p}} P_0 : \dot{\mathbf{F}}_\mathrm{p} \geq P_0$ (and $D_0 = P_0$ in case of rate-independence). Similarly to the assumptions on the stored energy, the dependency of the dissipation potential on $\dot{\mathbf{F}}_\mathrm{p}$ maybe further reduced to

$$P_0(\dot{\mathbf{F}}_\mathrm{p}; \mathbf{X}) = \bar{P}_0(\dot{\mathbf{F}}_\mathrm{p}(\mathbf{X})) \quad \text{or} \quad P_0(\dot{\mathbf{F}}_\mathrm{p}; \mathbf{X}) = \widetilde{P}_0(\dot{\mathbf{F}}_\mathrm{p}|_{\mathrm{const\,in}\,\mathbf{X}}; \mathbf{X}) \,. \tag{12}$$

In analogy to equation (8) we now compute the variation of the bulk contribution, namely

$$\delta \int_{\mathcal{B}_0} R_0 \, \mathrm{d}V = \int_{\mathcal{B}_0} \partial_{\dot{\mathbf{F}}} R_0 \cdot \delta\dot{\mathbf{F}} \, \mathrm{d}V = \int_{\partial\mathcal{B}_0} \mathbf{t} \cdot \delta\dot{\boldsymbol{\varphi}} \, \mathrm{d}A \,, \tag{13}$$

via (10). Because $\mathbf{F}_\mathrm{p}$ is assumed to vary constitutively with $\mathbf{F}$, it is implicitely included in the variation with respect to $\mathbf{F}$ in equation (13). By making use of the identity

$$\partial_{\dot{\mathbf{F}}} R_0 : \delta\dot{\mathbf{F}} = \nabla_\mathbf{X} \cdot [\, \delta\dot{\boldsymbol{\varphi}} \cdot \partial_{\dot{\mathbf{F}}} R_0 \,] - [\, \nabla_\mathbf{X} \cdot \partial_{\dot{\mathbf{F}}} R_0 \,] \cdot \delta\dot{\boldsymbol{\varphi}} \tag{14}$$

together with the divergence theorem, one obtains

$$\int_{\mathcal{B}_0} [\nabla_\mathbf{X} \cdot \partial_{\dot{\mathbf{F}}} R_0] \cdot \delta\dot{\boldsymbol{\varphi}} \, \mathrm{d}V + \int_{\partial\mathcal{B}_0} [\mathbf{t} - \partial_{\dot{\mathbf{F}}} R_0 \cdot \mathbf{N}] \cdot \delta\dot{\boldsymbol{\varphi}} \, \mathrm{d}A = 0 \tag{15}$$

with $\mathbf{N}$ denoting the outward surface unit normal vector. Accordingly, the corresponding local form reads

$$\nabla_{\mathbf{X}} \cdot \partial_{\dot{\mathbf{F}}} R_0 = \mathbf{0} \ \text{ in } \ \mathcal{B}_0 \qquad \text{and} \qquad \mathbf{t} = \partial_{\dot{\mathbf{F}}} R_0 \cdot \mathbf{N} \ \text{ on } \ \partial \mathcal{B}_0^t \, . \tag{16}$$

Together with the local variational constitutive form (10) for $\mathbf{F}_{\mathrm{p}}$, this last relation represents the principle result of the current approach at the rate level.

The incremental form of the field relation (16) is referred to a finite interval in time, i.e. $\Delta t = t_{n+1} - t_n \geq 0$. In this regard, consider the integration in time of the bulk contribution R_0, which – based on the review summarised above – takes the representation

$$w_0 = \int_{t_n}^{t_{n+1}} R_0 \, \mathrm{d}t \; = \; W_{0\,n+1} - W_{0\,n} + \int_{t_n}^{t_{n+1}} P_0 \, \mathrm{d}t \, . \tag{17}$$

The simplest approach to computationally integrate P_0 in time is provided by an Euler-forward-type scheme, i.e.,

$$\int_{t_n}^{t_{n+1}} P_0 \, \mathrm{d}t \; \approx \; \Delta t \; P_0(\Delta \mathbf{F}_{\mathrm{p}}/\Delta t; \mathbf{X}) = p_0(\mathbf{F}_{\mathrm{p}\,n+1}, \Delta t; \mathbf{X}) \, . \tag{18}$$

In this case, one obtains the incremental form

$$\partial_{\mathbf{F}_{\mathrm{p}\,n+1}} w_0 = \partial_{\mathbf{F}_{\mathrm{p}\,n+1}} W_{0\,n+1} + \partial_{\mathbf{F}_{\mathrm{p}\,n+1}} p_0 = \mathbf{0} \tag{19}$$

of equation (10) for the algorithmic evolution of $\mathbf{F}_{\mathrm{p}}$ as well as the field relation

$$\nabla_{\mathbf{X}} \cdot \partial_{\mathbf{F}_{n+1}} w_0 = \mathbf{0} \ \text{ in } \ \mathcal{B}_0 \quad \text{and} \quad \mathbf{t} = \partial_{\mathbf{F}_{n+1}} w_0 \cdot \mathbf{N} \ \text{ on } \ \partial \mathcal{B}_0^t \tag{20}$$

by analogy with equation (16). Obviously, $\partial_{\mathbf{F}} w_0$ represents the Piola-type stresses.

4 Two Configurational Field Formulations

In this section, we formulate configurational fields and balance relations using two approaches, i.e., the variational approach sketched in the previous section, and an alternative approach as based on the transformation of configurational relations to established balance relations.

4.1 Variational Approach

In the context of the incremental variational formulation of the last section, configurational fields and relations can be obtained by means of (i) superposing a compatible deformation-type mapping onto the reference configuration, and (ii) computing the related derivative of the modified field equations for the mapping coinciding with the identity. A graphical representation of the kinematics at hand is displayed in Figure 1. According to this, we introduce the sufficiently smooth mapping $\mathbf{x}_\lambda = \boldsymbol{\lambda}(\mathbf{X}, t) : \mathcal{B}_0 \times \mathcal{T} \to \mathcal{B}_\lambda$ with $\mathbf{L} = \nabla_\mathbf{X}\boldsymbol{\lambda}$ and $J_\lambda = \det(\mathbf{L})$. The incremental potential energy of the body considered is now referred to this mapping and takes the form

$$\begin{aligned} w_0 &= J_\lambda\, w_\lambda(\mathbf{F}\cdot\mathbf{L}^{-1}, \mathbf{F}_\mathrm{p}(\boldsymbol{\lambda}(\mathbf{X})), \boldsymbol{\lambda}(\mathbf{X})) \qquad \text{such that} \\ \delta\int_{\mathcal{B}_\lambda} w_0\,\mathrm{d}V &= \int_{\mathcal{B}_\lambda} w_\lambda\partial_\mathbf{L} J_\lambda : \delta\mathbf{L} + J_\lambda\partial_{\mathbf{F}\cdot\mathbf{L}^{-1}} w_\lambda : \delta[\,\mathbf{F}\cdot\mathbf{L}^{-1}\,] \\ &\quad + J_\lambda\partial_{\boldsymbol{\lambda}} w_\lambda\cdot\delta\boldsymbol{\lambda}\,\mathrm{d}V = \int_{\partial\mathcal{B}_\lambda}\mathbf{T}\cdot\delta\boldsymbol{\lambda}\,\mathrm{d}A\,. \end{aligned} \tag{21}$$

As mentioned in Section 2, the internal variable $\mathbf{F}_\mathrm{p}$ is included as a quantity constrained by an evolution equation, respectively a variational constitutive form, so that we here do not vary $\mathbf{F}_\mathrm{p}$ independently. Its variation in the reference configuration $\mathcal{B}_0$, however, accounts for the inhomogeneity of the material behaviour.

Based on $\partial_\mathbf{L} J_\lambda = \mathrm{cof}(\mathbf{L})$ and $\delta\mathbf{L}^{-1} = -\,\mathbf{L}^{-1}\cdot\delta\mathbf{L}\cdot\mathbf{L}^{-1}$, one obtains – by analogy with equation (15) and equation (16) – for $\boldsymbol{\lambda}$ coinciding with the identity mapping the following expression

$$\begin{aligned} \int_{\partial\mathcal{B}_0}\mathbf{T}\cdot\delta\boldsymbol{\lambda}\,\mathrm{d}A = \int_{\mathcal{B}_0} &\big[\,w_0\,\mathbf{I} - \mathbf{F}^\mathrm{t}\cdot\partial_\mathbf{F} w_0\,\big] : \nabla_\mathbf{X}\delta\boldsymbol{\lambda} \\ &+ \big[\,\partial_{\mathbf{F}_\mathrm{p}} w_0 : \nabla_\mathbf{X}\mathbf{F}_\mathrm{p} + \partial_\mathbf{X} w_0\,\big]\cdot\delta\boldsymbol{\lambda}\,\mathrm{d}V\,, \end{aligned} \tag{22}$$

respectively the local form

$$\begin{aligned} \nabla_\mathbf{X}\cdot[\,w_0\,\mathbf{I} - \mathbf{F}^\mathrm{t}\cdot\partial_\mathbf{F} w_0\,] - \partial_{\mathbf{F}_\mathrm{p}} w_0 : \nabla_\mathbf{X}\mathbf{F}_\mathrm{p} - \partial_\mathbf{X} w_0 = \mathbf{0} \quad &\text{in } \mathcal{B}_0 \\ \text{and}\quad \mathbf{T} = [\,w_0\,\mathbf{I} - \mathbf{F}^\mathrm{t}\cdot\partial_\mathbf{F} w_0\,]\cdot\mathbf{N} \quad \text{on} \quad &\partial\mathcal{B}_0^T\,. \end{aligned} \tag{23}$$

Apparently, the contribution of which the divergence is computed represents the Eshelby-type stresses and the remaining part of equation $(23)_1$ are the related volume forces. It is also interesting to note that this volume force takes the reduced forms

$$\begin{aligned} \nabla_\mathbf{X}\cdot[\,w_0\,\mathbf{I} - \mathbf{F}^\mathrm{t}\cdot\partial_\mathbf{F} w_0\,] &= \partial_{\mathbf{F}_\mathrm{p}} w_0 : \nabla_\mathbf{X}\mathbf{F}_\mathrm{p} \qquad &\text{for}\ \ R_0 = \dot{\bar{\psi}}_0 + \bar{P}_0\,, \\ \nabla_\mathbf{X}\cdot[\,w_0\,\mathbf{I} - \mathbf{F}^\mathrm{t}\cdot\partial_\mathbf{F} w_0\,] &= \partial_\mathbf{X} w_0 \qquad &\text{for}\ \ R_0 = \dot{\tilde{\psi}}_0 + \tilde{P}_0\,, \end{aligned} \tag{24}$$

in $\mathcal{B}_0$, in case the underlying potentials are additionally restricted, compare equations (4,5,12). In the first case, i.e., when $\mathbf{F}_\mathrm{p}$ represents a material uniformity, its material variation $\nabla_\mathbf{X}\mathbf{F}_\mathrm{p}$ results in an additional configurational body-force. On the other hand, if it is materially constant, and for example the material properties are heterogeneous, no translation invariance of the material behaviour is present, and $\partial_{\mathbf{X}} w_0$ represents an additional configurational body force density.

4.2 Transformation-Based Approach

We turn now to the alternative approach as based on the application of transformation relations to standard balance relations. Consider the quasi-static standard local balance of linear momentum representation for vanishing spatial volume forces, $\nabla_{\mathbf{X}} \cdot \mathbf{P} = \mathbf{0}$ in $\mathscr{B}_0$ and $\mathbf{t} = \mathbf{P} \cdot \mathbf{N}$ on $\partial \mathscr{B}_0^t$, with the Piola stresses being defined in terms of the hyper-elastic form $\mathbf{P} = \partial_{\mathbf{F}} W_0$. The material version of this balance of linear momentum representation is well-established and can be derived from the identification

$$- \mathbf{F}^{\mathrm{t}} \cdot [\, \nabla_{\mathbf{X}} \cdot \mathbf{P} \,] = \nabla_{\mathbf{X}} \cdot \boldsymbol{\Sigma} + \mathbf{B}_0 \quad \text{in } \mathscr{B}_0 \quad \text{and} \quad \mathbf{T} = \boldsymbol{\Sigma} \cdot \mathbf{N} \quad \text{on } \partial \mathscr{B}_0^T \,, \tag{25}$$

wherein $\boldsymbol{\Sigma}$ denotes the Eshelby stresses, $\mathbf{B}_0$ is the related material volume force, and $\nabla_{\mathbf{X}} \cdot \operatorname{cof}(\mathbf{F}) = \mathbf{0}$. The particular energy momentum representation of $\boldsymbol{\Sigma}$ stems from a Piola transformation

$$\boldsymbol{\Sigma} = \mathbf{p} \cdot \operatorname{cof}(\mathbf{F}) = W_0 \, \mathbf{I} - \mathbf{F}^{\mathrm{t}} \cdot \mathbf{P} \quad \text{with} \quad \mathbf{p} = \partial_{\mathbf{F}^{-1}} W_t \tag{26}$$

and the volume forces additionally account for the material's inhomogeneity as well as heterogeneity, namely

$$\mathbf{B}_0 = -\, \mathbf{P}_{\mathrm{p}} : \nabla_{\mathbf{X}} \mathbf{F}_{\mathrm{p}} - \partial_{\mathbf{X}} W_0 \quad \text{with} \quad \mathbf{P}_{\mathrm{p}} = \partial_{\mathbf{F}_{\mathrm{p}}} W_0 \,. \tag{27}$$

Similar to equation (24), the two particular representations of the stored energy as highlighted in equations (4) and (5) render reduced forms for the configurational volume forces, i.e.,

$$\begin{aligned} \mathbf{B}_0 &= -\, \mathbf{P}_{\mathrm{p}} : \nabla_{\mathbf{X}} \mathbf{F}_{\mathrm{p}} && \text{for} \quad W_0 = \bar{\psi}_0 \,, \\ \mathbf{B}_0 &= -\, \partial_{\mathbf{X}} W_0 && \text{for} \quad W_0 = \widetilde{\psi}_0 \,. \end{aligned} \tag{28}$$

Even though this framework renders one and the same balance law to be represented in different forms, its configurational version just derived provides additional insight on the forces driving the material's inhomogeneities. The derivation followed here is direct and so independent of any variational approach. If a variational formulation is available, however, any such framework is particularly attractive from the computational point of view.

5 Comparison and Discussion

The two different approaches to derive the material or rather configurational representation of balance of linear momentum, namely the incremental variational formulation and the transformation-based framework, are related via the formal identification of w_0 with W_0. However, it is important to recall that w_0 does not only represent the incremental part of the strain energy function but also the corresponding dissipation-related contribution. In this regard, the flux term $\partial_{\mathbf{F}} w_0$ formally corresponds to the Piola stresses $\mathbf{P}$ and $\partial_{\mathbf{X}} w_0$ can directly be referred to $\partial_{\mathbf{X}} W_0$. For the particular model at hand, these respective quantities actually are identical, i.e. $\partial_{\mathbf{F}} w_0 = \mathbf{P}$ and $\partial_{\mathbf{X}} w_0 = \partial_{\mathbf{X}} W_0$, as the dissipation potential P_0 has been assumed to depend only on $\dot{\mathbf{F}}_{\mathrm{p}}$ and not on $\dot{\mathbf{F}}$ or $\dot{\boldsymbol{\varphi}}$. The stresses conjugate to $\mathbf{F}_{\mathrm{p}}$, however, are different for both models so that $\partial_{\mathbf{F}_{\mathrm{p}}} w_0 \neq \mathbf{P}_{\mathrm{p}}$.

While this contribution aimed at comparing the two different configurational approaches for the simplest large strain inelastic case, one could also extend these investigations towards more advanced modelling approaches. Apart form elaborating a variational approach with respect to the so-called intermediate configuration – and comparing the results to the transformation of balance of linear momentum, this intermediate configuration – further internal variables and hardening mechanism can be considered. Moreover, the incorporation of $\mathbf{F}_{\mathrm{p}}$ as a field variable, the extension towards non-simple continua as well as neglecting the ansatz of the here assumed kinematic coupling, $\mathbf{F}_{\mathrm{e}} = \mathbf{F} \cdot \mathbf{F}_{\mathrm{p}}^{-1}$, are of special interest for future research. As previously mentioned in the introduction, a detailed finite-element-based comparison of both configurational frameworks would provide further inside into the computational advantages and properties of the two formulations.

References

1. Carstensen, C., Hackl, K. and Mielke, A., Non-convex potentials and microstructures in finite-strain plasticity. *Proc. Roy. Soc. London A* **458**, 2002, 299–317.
2. Epstein, M., The Eshelby tensor and the theory of continuous distributions of dislocations. *Mech. Res. Comm.* **29**, 2002, 501–506.
3. Epstein, M. and Maugin, G.A., The energy-momentum tensor and material uniformity in finite elasticity. *Acta Mech.* **83**, 1990, 127–133.
4. Gurtin, M.E., *Configurational Forces as Basic Concept in Continuum Physics*, Applied Mathematical Sciences, Vol. 137, Springer, Berlin, 2000.
5. Gurtin, M.E., On the plasticity of single crystals: Free energy, microforces, plastic-strain gradients. *J. Mech. Phys. Solids* **48**, 2000, 989–1036.
6. Gurtin, M.E., A gradient theory of single-crystal viscoplasticity that accounts for geometrically necessary dislocations. *J. Mech. Phys. Solids* **50**(1), 2002, 5–32.
7. Hackl, K., Generalized media and variational principles in classical and finite strain elastoplasticity. *J. Mech. Phys. Solids* **45**(5), 1997, 667–688.
8. Hanyga, A., *Mathematical Theory of Non-Linear Elasticity*. Polish Scientific Publishers/Ellis Horwood, 1985.
9. Levkovitch, V. and Svendsen, B., On the large-deformation and continuum-based formulation of models for extended crystal plasticity. *Int. J. Solids Struct.* **43**(24), 2006, 7246–7267.

10. Markenscoff, X. and Gupta, A. (Eds.), *Collected Works of J.D. Eshelby – The Mechanics of Defects and Inhomogeneities*, Solid Mechanics and Its Applications, Vol. 133, Springer, Dordrecht, 2006.
11. Maugin, G.A., *Material Inhomogeneities in Elasticity*, Applied Mathematics and Mathematical Computation, Vol. 3. Chapman & Hall, 1993.
12. Maugin, G.A., Pseudo-plasticity and pseudo-inhomogeneity effects in materials mechanics. *J. Elasticity* **71**, 2003, 81–103.
13. Menzel, A., Denzer, R. and Steinmann, P., On the comparison of two approaches to compute material forces for inelastic materials. Application to single-slip crystal-plasticity. *Comput. Methods Appl. Mech. Engrg.* **193**(48–51), 2004, 5411–5428.
14. Menzel, A. and Steinmann, P., On the continuum formulation of higher gradient plasticity for single and polycrystals. *J. Mech. Phys. Solids* **48**(8), 2000, 1777–1796. Erratum **49**(5), 2001, 1179–1180.
15. Menzel, A. and Steinmann, P., A note on material forces in finite inelasticity. *Arch. Appl. Mech.* **74**, 2005, 800–807.
16. Menzel, A. and Steinmann, P., On configurational forces in multiplicative elastoplasticity. *Int. J. Solids Struct.* **44**(13), 2007, 4442–4471.
17. Miehe, C., Strain-driven homogenization of inelastic microstructures and composites based on an incremental variational formulation. *Int. J. Numer. Methods Engng.* **55**, 2002, 1285–1322.
18. Noll, W., Materially uniform simple bodies with inhomogeneities. *Arch. Rational Mech. Anal.* **27**, 1967, 1–32.
19. Ogden, R.W., *Non-Linear Elastic Deformations*. Dover, 1997.
20. Ortiz, M. and Repetto, E.A., Nonconvex energy minimization and dislocation structures in ductile single crystals. *J. Mech. Phys. Solids* **47**, 1999, 397–462.
21. Podio-Guidugli, P., Configurational balances via variational arguments. *Interfaces and Free Boundaries* **3**, 2001, 223–232.
22. Šilhavý, M., *The Mechanics and Thermomechanics of Continuous Media*. Springer, Berlin, 1997.
23. Steinmann, P., On spatial and material settings of hyperelastostatic crystal defects. *J. Mech. Phys. Solids* **50**(8), 2002, 1743–1766.
24. Steinmann, P., On spatial and material settings of thermo-hyperelastodynamics. *J. Elasticity* **66**, 2002, 109–157.
25. Svendsen, B., A thermodynamic formulation of finite-deformation elastoplasticity with hardening based on the concept of material isomorphism. *Int. J. Plasticity* **14**(6), 1998, 473–488.
26. Svendsen, B., On the modeling of anisotropic elastic and inelastic material behaviour at large deformation. *Int. J. Solids Struct.* **38**(52), 2001, 9579–9599.
27. Svendsen, B., Continuum thermodynamic models for crystal plasticity including the effects of geometrically-necessary dislocations. *J. Mech. Phys. Solids* **50**(6), 2002, 1297–1330.
28. Svendsen, B., On the thermodynamic and variational-based formulation of models for inelastic continua with internal lengthscales. *Comput. Methods Appl. Mech. Engrg.* **193**(48–51), 2004, 5429–5452.
29. Svendsen, B., Continuum thermodynamic and variational models for continua with microstructure and material inhomogeneity. In P. Steinmann and G.A. Maugin (Eds.), *Mechanics of Material Forces*, Advances in Mechanics and Mathematics, Vol. 11, Springer, Berlin, 2005, pp. 173–180.
30. Svendsen, B., Neff, P. and Menzel, A., On some constitutive and configurational aspects of gradient continua with microstructure. *Z. Angew. Math. Mech.*, 2009, doi:10.1002/zamm.200800171.

Configurational Forces Derived from the Total Variation of the Rate of Global Dissipation

Fredrik Larsson, Kenneth Runesson and Johan Tillberg

Abstract Based on the thermodynamic framework for combined configurational and deformational changes, recently discussed in [1], we consider dissipative material response and emphasize the fact that it is possible to identify *explicit* energetic changes due to configurational changes for "frozen" spatial configuration and, in addition, the *configuration-induced* material dissipation. The classical assumption (previously adopted in the literature) is to ignore the latter. In this paper, however, we define configurational forces by considering the *total variation* of the total dissipation with respect to configurational changes. The key task is then to compute the *sensitivity* of the internal variable rates to such configurational changes. We restrict to quasistatic loading under isothermal conditions and elastic-plastic response, and we apply the theory to the simplest possible case of an interface of dissimilar materials in a single bar.

1 Introduction

An important class of configurational changes is defined by the motion/evolution of "singular surfaces" representing discontinuities in the material properties and, consequently, in the state variables when the body is loaded. Typical examples of internal processes are the evolution of defects, phase transformation (microstruc-

Fredrik Larsson
Department of Applied Mechanics, Chalmers University of Technology, SE-41296 Göteborg, Sweden

Kenneth Runesson
Department of Applied Mechanics, Chalmers University of Technology, SE-41296 Göteborg, Sweden; e-mail: keru@chalmers.se

Johan Tillberg
Department of Applied Mechanics, Chalmers University of Technology, SE-41296 Göteborg, Sweden

P. Steinmann (ed.), IUTAM Symposium on Progress in the Theory and Numerics of Configurational Mechanics, 47–59.

tural changes that are mostly accompanied by change of volume and mechanical properties), and internal cracks. Since the literature on configurational mechanics is very rich, we do not intend to give a comprehensive account. We only mention the pioneering work in [2] and the contributions in [3–14]. Some of these works also contain comprehensive overviews with a discussion of various aspects of configurational mechanics ("material" mechanics in Maugin's terminology). A major application is the "driving force" on singular surfaces. A recent comprehensive treatment of "surface energies" is given in [15].

Most studies so far are explicitly confined to elastic material response. While the state of affairs seems to be quite clear in this case, it is not so in the case (local) material dissipation occurs as a result of the configurational change. The formulation of the crack-driving force in such a case and the computational aspects have been discussed in [10, 16–20].

In this contribution we aim at shedding some further light on the energetic consequences of configurational changes for rate-independent dissipative material response. More specifically, we put forward the idea that it is the *total variation* of the appropriately defined dissipation functional for the whole body that represents the "driving force".

The paper is organized as follows: Section 2 introduces definitions and preliminaries related to the unified representation of configurational (material) and deformational (spatial) motions. In Section 3, we establish the pertinent dissipation functional due to imposed configurational changes, and we define generalized configurational forces from the "total variation of the global dissipation", which is the main novel result of the paper. This result is applied in Section 4 to a material singular surface in a simple bar with dissimilar elastic-plastic material properties, and explicit computational results are given in Section 5. Finally, conclusions are given in Section 6.

2 Representation of Configurational and Deformational Motions

2.1 Preliminaries

The spatial (deformational or direct) motion problem (SMP) is expressed in terms of a spatial motion map $\boldsymbol{x} = \boldsymbol{\varphi}(\boldsymbol{X}, t)$, $\mathcal{B}_\mathrm{X} \mapsto \mathcal{B}_\mathrm{x}(t)$. In this paper, we generalize the classical view of motion slightly, cf. Figure 1, since the material configuration is allowed to change with time in the sense that the boundary $\partial\mathcal{B}_\mathrm{X}$ of the body $\mathcal{B}_\mathrm{X}$ may change with time. It is then convenient to take the view that both the "undeformed" (material) and "deformed" (spatial) configurations will undergo time-dependent changes with respect to an *absolute* (fixed, time-invariant) configuration $\mathcal{B}_\xi$. We may then introduce the map $\boldsymbol{X} = \check{\boldsymbol{\Phi}}(\boldsymbol{\xi}, t)$ for the time-dependent "motion" of $\mathcal{B}_\mathrm{X}(t)$ w.r.t. $\mathcal{B}_\xi$, and we introduce the *absolute* spatial motion map $\boldsymbol{x} = \hat{\boldsymbol{\varphi}}(\boldsymbol{\xi}, t)$ for the time-dependent motion (including deformation) of $\mathcal{B}_\mathrm{x}(t)$ w.r.t. $\mathcal{B}_\xi$.

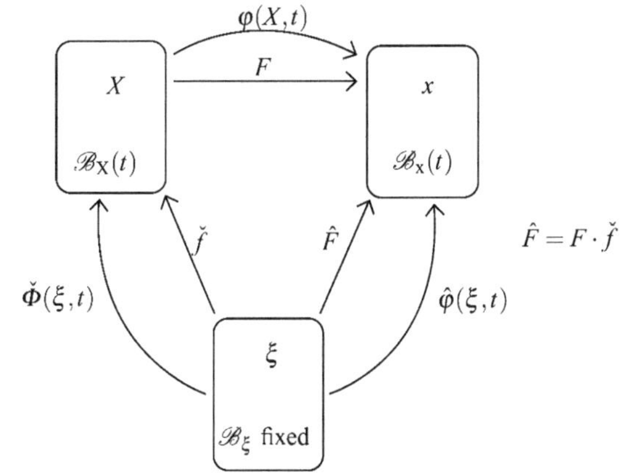

Fig. 1 Spatial (current) configuration, $\mathcal{B}_x(t)$, material configuration, $\mathcal{B}_X(t)$, and absolute (fixed) configuration, $\mathcal{B}_\xi$.

The gradients corresponding to the introduced mappings are $\boldsymbol{F} \overset{\text{def}}{=} \boldsymbol{\varphi} \otimes \boldsymbol{\nabla}_X$, $\hat{\boldsymbol{F}} \overset{\text{def}}{=} \hat{\boldsymbol{\varphi}} \otimes \boldsymbol{\nabla}_\xi$, and $\check{\boldsymbol{f}} \overset{\text{def}}{=} \check{\boldsymbol{\Phi}} \otimes \boldsymbol{\nabla}_\xi$.

2.2 Absolute and Material Time Differentiation

We introduce absolute and material time derivatives of a field f, parametrized in $\mathcal{B}_X$ or $\mathcal{B}_\xi$, as follows:

$$\Delta_t f \overset{\text{def}}{=} \partial_t f|_\xi, \quad \mathrm{D}_t f \overset{\text{def}}{=} \partial_t f|_X \tag{1}$$

where $\partial_t[\bullet] \overset{\text{def}}{=} \partial[\bullet]/\partial t$ denotes the ordinary partial time derivative. In particular, the *material* velocity, $\boldsymbol{v}$, the *absolute* velocity, $\dot{\boldsymbol{x}}$, and the (absolute) *configurational* rate, $\dot{\boldsymbol{X}}$, are defined as

$$\boldsymbol{v} \overset{\text{def}}{=} \mathrm{D}_t \boldsymbol{x} = \partial_t \boldsymbol{\varphi}(\boldsymbol{X}, t), \quad \dot{\boldsymbol{x}} \overset{\text{def}}{=} \Delta_t \boldsymbol{x} = \partial_t \hat{\boldsymbol{\varphi}}(\boldsymbol{\xi}, t), \quad \dot{\boldsymbol{X}} \overset{\text{def}}{=} \Delta_t \boldsymbol{X} = \partial_t \check{\boldsymbol{\Phi}}(\boldsymbol{\xi}, t) \tag{2}$$

Using the chain rule, we obtain the relation

$$\boldsymbol{v} = \dot{\boldsymbol{x}} - \boldsymbol{F} \cdot \dot{\boldsymbol{X}} \tag{3}$$

We shall later make use of the absolute and material time derivatives of a (tensor-valued) *operator* field $y(\boldsymbol{X}(\boldsymbol{\xi}, t), \boldsymbol{x}(\boldsymbol{\xi}, t), \alpha_1(\boldsymbol{\xi}, t), \alpha_2(\boldsymbol{\xi}, t), \ldots, t)$.[1] Let us, for example, consider a volume-specific field, $y(\bullet, t)$ for $\boldsymbol{X} \in \mathcal{B}_X(t)$, for which the following relation holds:

[1] An example is the deformation gradient $\boldsymbol{F}(\boldsymbol{X}, \boldsymbol{x}) = \boldsymbol{x} \otimes \boldsymbol{\nabla}_X$.

$$\mathrm{D}_t y(\bullet, t) = \Delta_t y(\bullet, t) - [y(\bullet, t) \otimes \nabla_{\mathrm{X}}] \cdot \dot{\boldsymbol{X}} \tag{4}$$

Remark. It is also possible to establish a similar relation for a surface-specific field, $\hat{y}(\bullet, t)$ for $\boldsymbol{X} \in \partial \mathcal{B}_{\mathrm{X}}(t)$; however, it is more involved and will be omitted here. □

Suppose that $f_{\mathrm{X}}(\boldsymbol{X}, \boldsymbol{x}, \underline{\alpha}, t)$, for $\boldsymbol{X} \in \mathcal{B}_{\mathrm{X}}(t)$, is a volume-specific quantity and consider a typical conservation quantity F

$$F \stackrel{\text{def}}{=} \int_{\mathcal{B}_{\mathrm{X}}} f_{\mathrm{X}} \, \mathrm{d}V_{\mathrm{X}} \tag{5}$$

The time-derivative of F for time-dependent $\mathcal{B}_{\mathrm{X}}$ is then given as

$$\begin{aligned} \frac{\mathrm{d}}{\mathrm{d}t} F &= \int_{\mathcal{B}_{\mathrm{X}}} \left[\Delta_t f_{\mathrm{X}} + f_{\mathrm{X}} \dot{\boldsymbol{X}} \cdot \nabla_{\mathrm{X}}\right] \mathrm{d}V_{\mathrm{X}} \\ &= \int_{\mathcal{B}_{\mathrm{X}}} \left[\mathrm{D}_t f_{\mathrm{X}} + \left[f_{\mathrm{X}} \dot{\boldsymbol{X}}\right] \cdot \nabla_{\mathrm{X}}\right] \mathrm{d}V_{\mathrm{X}} \\ &= \int_{\mathcal{B}_{\mathrm{X}}} \mathrm{D}_t f_{\mathrm{X}} \, \mathrm{d}V_{\mathrm{X}} + \int_{\partial \mathcal{B}_{\mathrm{X}}} f_{\mathrm{X}} \dot{\boldsymbol{X}} \cdot \boldsymbol{N} \, \mathrm{d}S_{\mathrm{X}} \end{aligned} \tag{6}$$

Further details are given in [1]. It must be noted that (6) holds under the assumption of sufficient smoothness, i.e. material interfaces are introduced at a later stage whereby (6) is valid for each smooth region.

2.3 Global and Localized Dissipation Inequality

The global free energy is given as

$$\Psi \stackrel{\text{def}}{=} \int_{\mathcal{B}_{\mathrm{X}}} \psi_{\mathrm{X}} \, \mathrm{d}V_{\mathrm{X}} \tag{7}$$

where ψ_{X} is the volume-specific free energy. The "configurational flux" of free energy across the boundary due to the configurational motion is given as

$$\Psi^{\mathrm{CONF}} = \int_{\partial \mathcal{B}_{\mathrm{X}}} \psi_{\mathrm{X}}^{\mathrm{CONF}} \dot{\boldsymbol{X}} \cdot \boldsymbol{N} \, \mathrm{d}S_{\mathrm{X}} \tag{8}$$

Moreover, the total mechanical power supply, $\mathcal{W}$, is given as

$$\mathcal{W} = \int_{\mathcal{B}_{\mathrm{X}}} \boldsymbol{b}_{\mathrm{X}} \cdot \boldsymbol{v} \, \mathrm{d}V_{\mathrm{X}} + \int_{\partial \mathcal{B}_{\mathrm{X}}} \boldsymbol{t}_{\mathrm{X}} \cdot \boldsymbol{v} \, \mathrm{d}S_{\mathrm{X}} = \int_{\mathcal{B}_{\mathrm{X}}} \boldsymbol{P} : \mathrm{D}_t \boldsymbol{F} \, \mathrm{d}V_{\mathrm{X}} \tag{9}$$

In order to obtain the last expression in (9), the momentum balance $\boldsymbol{P} \cdot \nabla_X + \boldsymbol{b}_X = \boldsymbol{0}$, was used, where $\boldsymbol{P}$ is the 1st Piola–Kirchoff stress, and $\boldsymbol{b}_X$ is the volume-specific load.

It is *proposed* that the total mechanical dissipation is expressed as

$$\mathcal{D} = \mathcal{W} - \frac{\mathrm{d}}{\mathrm{d}t}\Psi + \Psi^{\mathrm{CONF}} \geq 0 \tag{10}$$

The localized form of (10), in terms of the volume-specific mechanical dissipation rate d_X, becomes the standard expression

$$d_X = \boldsymbol{P} : \mathrm{D}_t \boldsymbol{F} - \mathrm{D}_t \psi_X \geq 0 \tag{11}$$

Next, we introduce the parametrization of the free energy density, $\psi_X(\boldsymbol{F}, \theta, \underline{k}; \boldsymbol{X})$, pertinent to a *dissipative* material response, where $\underline{k}$ represents a set of internal variables. It is noted that a possible explicit dependence of $\boldsymbol{X}$ is included as argument in ψ (representing inhomogeneous variation of material parameters in $\mathcal{B}_X$). Upon evaluating the material time derivative $\mathrm{D}_t \psi_X$, and using standard arguments of the Colemann–Noll type, we obtain the constitutive state equation $\boldsymbol{P} = \partial \psi_X / \partial \boldsymbol{F}$, such that (8) reduces to

$$d_X = \underline{K} \star \mathrm{D}_t \underline{k} \geq 0 \quad \text{with } \underline{K} \stackrel{\text{def}}{=} -\frac{\partial \psi_X}{\partial \underline{k}} \tag{12}$$

where we introduced dissipative stresses $\underline{K}$ that are energy-conjugated to the internal variables $\underline{k}$.[2]

3 Dissipation Functional for Changing Material Configuration

3.1 Preliminaries

We consider the global (mechanical) dissipation inequality for a given *finite* body occupying the current material domain Ω_X with external boundary $\partial\Omega_X$. It is assumed that the body undergoes configurational changes, expressed as $\dot{\boldsymbol{X}} \neq \boldsymbol{0}$, while it is subjected to prescribed loading. For simplicity, we restrict to *isothermal conditions* henceforth. The resulting thermodynamic process is then characterized by time-changes of all the independent thermodynamic fields, $\boldsymbol{F}$ and $\underline{k}$, since they are solutions of the pertinent balance and state equations (equilibrium and constitutive equations) and thus depend on the configurational motion $\boldsymbol{X} = \boldsymbol{\Phi}(\boldsymbol{\xi}, t)$ in an *implicit* manner. As a result, the global dissipation functional $\mathcal{D}$ depends, apart from $\dot{\boldsymbol{X}}$, *potentially* on the fields $\dot{\boldsymbol{x}}$ and $\dot{\underline{k}}$ in the most general case of combined configur-

[2] The "scalar star product" has the appropriate interpretation depending on the tensorial order of variables in the column vector $\underline{k}$.

ational and deformational motion for general loading; hence we denote $\mathcal{D}(\dot{\boldsymbol{X}}, \dot{\boldsymbol{x}}, \underline{\dot{k}})$ the *global dissipation functional*.

3.2 Basic Format of the Dissipation Functional – Configurational and Material Parts

It was shown in [1] that (i) the dissipation functional $\mathcal{D}$ does not depend explicitly on $\dot{\boldsymbol{x}}$, i.e. the parametrization $\mathcal{D}(\dot{\boldsymbol{X}}, \underline{\dot{k}})$ suffices, and (ii) it is natural to split $\mathcal{D}(\dot{\boldsymbol{X}}, \underline{\dot{k}})$ as

$$\mathcal{D}(\dot{\boldsymbol{X}}, \underline{\dot{k}}) = \bar{\mathcal{D}}^{\mathrm{CONF}}(\dot{\boldsymbol{X}}) + \bar{\mathcal{D}}^{\mathrm{MAT}}(\dot{\boldsymbol{X}}, \underline{\dot{k}}) \tag{13}$$

where we introduced

$$\bar{\mathcal{D}}^{\mathrm{CONF}}(\dot{\boldsymbol{X}}) = -\int_{\partial\Omega_{\mathrm{X}}} \left[\psi_{\mathrm{X}} - \psi_{\mathrm{X}}^{\mathrm{CONF}}\right] \boldsymbol{N} \cdot \dot{\boldsymbol{X}} \, \mathrm{d}S_{\mathrm{X}} \tag{14}$$

and

$$\bar{\mathcal{D}}^{\mathrm{MAT}}(\dot{\boldsymbol{X}}, \underline{\dot{k}}) = \int_{\Omega_{\mathrm{X}}} \underline{K} \star \mathrm{D}_t \underline{k} \, \mathrm{d}V_{\mathrm{X}} = \int_{\Omega_{\mathrm{X}}} \underline{K} \star \left[\underline{\dot{k}} - \left[\underline{k} \otimes \boldsymbol{\nabla}_{\mathrm{X}}\right] \cdot \dot{\boldsymbol{X}}\right] \mathrm{d}V_{\mathrm{X}} \tag{15}$$

It appears readily that $\bar{\mathcal{D}}^{\mathrm{MAT}}(\dot{\boldsymbol{X}}, \underline{\dot{k}})$ represents precisely the ordinary material dissipation.

3.3 Total Variation of the Rate of Global Dissipation Due to Configurational Changes

Henceforth, we shall restrict our attention to the situation that a physical process is manifested by fields $\mathrm{D}_t\boldsymbol{x}$ and $\mathrm{D}_t\underline{k}$ that are *brought about solely by configurational changes*. In other words, for a given field $\dot{\boldsymbol{X}}(\boldsymbol{\xi}, t)$, it is possible to solve for all other fields, $\dot{\boldsymbol{x}}\{\dot{\boldsymbol{X}}\}$ and $\underline{\dot{k}}\{\dot{\boldsymbol{X}}\}$, as (implicit) functions of $\dot{\boldsymbol{X}}$ from the momentum equation and the pertinent constitutive relations. It is noted that the *actual physical problem* involves $\dot{\boldsymbol{X}}(\boldsymbol{\xi}, t)$ as part of the total solution, which requires a constitutive relation for $\dot{\boldsymbol{X}}$ in terms of a suitably defined field of "driving forces". However, these driving forces will be of such nature that they can only be determined when the solutions $\dot{\boldsymbol{x}}\{\dot{\boldsymbol{X}}\}$ and $\underline{\dot{k}}\{\dot{\boldsymbol{X}}\}$ are known; hence, the problem of computing the driving forces is indeed nonlinear and must be solved by some sort of iterative procedure in practice.

Next, we shall be concerned with the issue of defining the field of thermodynamically consistent *generalized configurational forces* that are energy-conjugated to a given differential change (variation) of the field $\dot{\boldsymbol{X}}$, henceforth denoted $\mathrm{d}\dot{\boldsymbol{X}}$, in the sense that they represent the total variation of $\mathcal{D}$ with respect to $\dot{\boldsymbol{X}}$. The total differential of $\mathcal{D} \stackrel{\text{def}}{=} \mathcal{D}(\dot{\boldsymbol{X}}, \underline{\dot{k}}\{\dot{\boldsymbol{X}}\})$ can be expressed as

$$\mathrm{d}_{\dot{X}}\mathcal{D} = \bar{\mathcal{D}}^{\mathrm{CONF}}(\mathrm{d}\dot{X}) + \bar{\mathcal{D}}^{\mathrm{MAT}}(\mathrm{d}\dot{X}, \underline{0}) + \bar{\mathcal{D}}^{\mathrm{MAT}}(\mathbf{0}, \dot{\underline{k}}'\{\dot{X};\, \mathrm{d}\dot{X}\}) \qquad (16)$$

where it was used that both $\bar{\mathcal{D}}^{\mathrm{CONF}}$ and $\bar{\mathcal{D}}^{\mathrm{MAT}}$ are linear in their arguments ($\dot{X}$ and $\dot{\underline{k}}$ respectively). In order to carry out the total variation, it is necessary to compute the *sensitivity* fields:

$$\mathrm{d}\dot{x} = \dot{x}'\{\dot{X};\, \mathrm{d}\dot{X}\}, \quad \mathrm{d}\dot{\underline{k}} = \dot{\underline{k}}'\{\dot{X};\, \mathrm{d}\dot{X}\} \qquad (17)$$

which are directional (or Gateaux) derivatives in the classical sense.[3] The sensitivity fields $\dot{x}'$ and $\dot{\underline{k}}'$ must satisfy *global tangent (or sensitivity) relations*, which are derivable from linearization of the equilibrium equation together with the constitutive rate equations for $\dot{\underline{k}}$. It is emphasized that the sensitivity fields, say $\dot{\underline{k}}'\{\dot{X};\, \mathrm{d}\dot{X}\}$, are *linear* in $\mathrm{d}\dot{X}$; however, they represent "spatially global" relations in Ω_{X}. This means, in particular, that the configurational dissipation is not necessarily confined to the (possibly small) part of Ω_{X} where $\mathrm{d}\dot{X} \neq \mathbf{0}$ has been assumed.

4 Model Problem: Bar with Interface Separating Parts with Dissimilar Material Properties

4.1 Problem Formulation – Preliminaries

Consider a bar in a state of unixial stress, with undeformed length L (such that $\Omega_{\mathrm{X}} = [0, L]$) and with unit cross-sectional area (for simplicity), which is subjected to prescribed end displacement at the right end, as shown in Figure 2(a). This is the sole type loading. The material properties (which are assumed elastic-plastic) vary smoothly along the bar with the exception that they may be discontinuous at the "singular" cross-section located at $X = L_1$. Since the state variables (such as $F = x/X$ and $\underline{k}$) are generally discontinuous across the singular cross-section, so is the volume-specific free energy ψ_{X}. The normal on the left part of the bar at $X = L_1$ is denoted $N^-(= N) = 1$; hence, we use the notation $N^+(= -N) = -1$ for the normal in the opposite direction.

4.2 Elastic-Plastic Model – Linearized Format

We adopt a simple elastic-plastic model with isotropic hardening. A linearization of the spatially objective large deformation formulation is obtained by introducing the volume-specific free energy as

[3] A more explicit definition of sensitivities is possible if we define the "unit sensitivity" fields $\dot{x}'^{(i)}$, $i = 1, 2, \ldots, \mathrm{NDIM}$, via the identity $\mathrm{d}\dot{x} = \sum_{i=1}^{\mathrm{NDIM}} \dot{x}'^{(i)}\, \mathrm{d}(\dot{X}_i)$.

$$\psi_{\mathrm{X}}\left(\boldsymbol{F},\boldsymbol{\epsilon}^{\mathrm{p}},k\right)=\psi_{\mathrm{X}}^{\mathrm{e}}\left(\boldsymbol{F},\boldsymbol{\epsilon}^{\mathrm{p}}\right)+\psi_{\mathrm{X}}^{\mathrm{p}}(k)=\frac{1}{2}E\left[\boldsymbol{F}-\boldsymbol{F}^{\mathrm{p}}\left(\boldsymbol{\epsilon}^{\mathrm{p}}\right)\right]^{2}+\frac{1}{2}Hk^{2} \tag{18}$$

where the parts $\psi_{\mathrm{X}}^{\mathrm{e}}\left(\boldsymbol{F},\boldsymbol{\epsilon}^{\mathrm{p}}\right)$ and $\psi_{\mathrm{X}}^{\mathrm{p}}(k)$ represent linear elasticity and isotropic hardening, respectively, with $\boldsymbol{F}^{\mathrm{p}}\left(\boldsymbol{\epsilon}^{\mathrm{p}}\right)\overset{\mathrm{def}}{=}1+\boldsymbol{\epsilon}^{\mathrm{p}}$. The material parameters are E (elasticity modulus) and H (hardening modulus). The thermodynamically consistent stresses are

$$\boldsymbol{P}=\frac{\partial\psi_{\mathrm{X}}}{\partial\boldsymbol{F}}=E\left[\boldsymbol{F}-\boldsymbol{F}^{\mathrm{p}}\left(\boldsymbol{\epsilon}^{\mathrm{p}}\right)\right]=-\frac{\partial\psi_{\mathrm{X}}}{\partial\boldsymbol{\epsilon}^{\mathrm{p}}},\quad K=-\frac{\partial\psi_{\mathrm{X}}}{\partial k}=-Hk \tag{19}$$

We may thus express the (local) volume-specific material rate of dissipation $d_{\mathrm{X}}=\underline{K}\star\mathrm{D}_t\underline{k}$ according to (12) with the matrices $\underline{k}$ and $\underline{K}$ defined as

$$\underline{k}\overset{\mathrm{def}}{=}\begin{bmatrix}\boldsymbol{\epsilon}^{\mathrm{p}}\\ k\end{bmatrix},\quad \underline{K}\overset{\mathrm{def}}{=}\begin{bmatrix}\boldsymbol{P}\\ K\end{bmatrix} \tag{20}$$

The yield criterion with isotropic hardening is expressed as

$$\Phi\left(\boldsymbol{P},K\right)=|P|-[Y+K]=0 \tag{21}$$

where Y is the uniaxial yield stress. We also adopt the (material) evolution rules

$$\mathrm{D}_t\boldsymbol{\epsilon}^{\mathrm{p}}=\frac{\partial\Phi}{\partial\boldsymbol{P}}=\lambda\frac{P}{|P|},\quad \mathrm{D}_t k=\frac{\partial\Phi}{\partial K}=-\lambda \tag{22}$$

where λ is the plastic multiplier.

The continuum tangent relation for P in terms of $\boldsymbol{F}$ and $\underline{k}$ can be written as

$$\mathrm{D}_t\boldsymbol{P}=\frac{\partial\boldsymbol{P}}{\partial\boldsymbol{F}}|_{\mathrm{k,X}}\,\mathrm{D}_t\boldsymbol{F}+\left[\frac{\partial\boldsymbol{P}}{\partial\underline{k}}|_{\mathrm{F,X}}\right]^{\mathrm{T}}\mathrm{D}_t\underline{k}=E\left[\mathrm{D}_t\boldsymbol{F}-\mathrm{D}_t\boldsymbol{\epsilon}^{\mathrm{p}}\right] \tag{23}$$

The consistency condition $\mathrm{D}_t\Phi\leq 0$ at a plastic state gives

$$\lambda=\frac{E}{h}\left\langle\frac{P}{|P|}\mathrm{D}_t\boldsymbol{F}\right\rangle,\quad h\overset{\mathrm{def}}{=}E+H \tag{24}$$

Now, taking the total variation of $\mathrm{D}_t\underline{k}$ with respect to variation of $\dot{\boldsymbol{X}}$, while using (22) and (24), we obtain the sensitivity relation

$$(\mathrm{D}_t\underline{k})'=\underline{\boldsymbol{R}}_{\mathrm{T}}(\mathrm{D}_t\boldsymbol{F})'\quad\text{with}\quad \underline{\boldsymbol{R}}_{\mathrm{T}}\overset{\mathrm{def}}{=}\frac{E}{h}\frac{\langle\frac{P}{|P|}\mathrm{D}_t\boldsymbol{F}\rangle}{|\frac{P}{|P|}\mathrm{D}_t\boldsymbol{F}|}\begin{bmatrix}1\\ -\frac{P}{|P|}\end{bmatrix} \tag{25}$$

which is linear in the sensitivity $(\mathrm{D}_t\boldsymbol{F})'$. The expression in (25) is well-defined for all solutions except at neutral loading defined by $\frac{P}{|P|}\mathrm{D}_t\boldsymbol{F}=0$. In particular, it can be evaluated for the situation $\dot{\boldsymbol{X}}=\boldsymbol{0}$, corresponding to vanishing configurational motion.

Finally, we take the variation of $\mathrm{D}_t \boldsymbol{P}$ upon using (23) and (25) to obtain

$$(\mathrm{D}_t \boldsymbol{P})' = E_{\mathrm{T}}(\mathrm{D}_t \boldsymbol{F})' \quad \text{with } E_{\mathrm{T}} \stackrel{\text{def}}{=} E \left[1 - \frac{1}{h} \frac{\langle \frac{P}{|P|} \mathrm{D}_t \boldsymbol{F} \rangle}{|\frac{P}{|P|} \mathrm{D}_t \boldsymbol{F}|} \right] \tag{26}$$

where E_{T} is the tangent stiffness tensor.

4.3 Explicit (Classical) Configurational Forces at the Material Interface

The explicit expression for $\mathcal{D}^{\mathrm{CONF}}$ was discussed in some detail in [1]. Upon introducing the assumption of vanishing material motion at the exterior boundary, i.e. $\dot{X} = 0$ on $X = 0, L$, we obtain the classical representation

$$\mathcal{G}^{\mathrm{CONF}} \stackrel{\text{def}}{=} [\![\Sigma]\!]|_{X=L_1} = [\![\Psi_{\mathrm{X}}]\!]|_{X=L_1} - [\![F]\!]|_{X=L_1} P \tag{27}$$

Remark. In the special case of linear elasticity, defined by $F^{\mathrm{p}} = 1$, we obtain $\mathcal{G}^{\mathrm{CONF}} = -[\![\Psi_{\mathrm{X}}]\!]|_{X=L_1}$. □

4.4 Implicit (Configurational-Induced) Configurational Forces at the Material Interface – Tangent Problem

The weak format of the (quasistatic) equilibrium equation for the bar under consideration is

$$a(\boldsymbol{x}, \underline{k}; \delta \boldsymbol{x}) \stackrel{\text{def}}{=} \int_{\Omega_{\mathrm{X}}} \frac{\mathrm{d}\delta \boldsymbol{x}}{\mathrm{d}\boldsymbol{X}} \boldsymbol{P}(\boldsymbol{F}, \underline{k}; \boldsymbol{X})\, \mathrm{d}X = 0, \ \forall \delta \boldsymbol{x} \in \mathbb{V}^0 \tag{28}$$

Since $\Omega_{\mathrm{X}} = [0, L]$ (or, more precisely, the location of the singular section) is not time-invariant at configurational changes, the admissible test functions $\delta \boldsymbol{x} \in \mathbb{V}^0$ are time-dependent in the parametrization $\delta \boldsymbol{x}(\boldsymbol{X}, t)$; however, this relation is constrained by the condition that $\delta \boldsymbol{x}(\boldsymbol{X}, t)$ is stationary in the absolute configuration, i.e. $\delta \boldsymbol{x}(\boldsymbol{X}(\boldsymbol{\xi}, t), t) \stackrel{\text{def}}{=} \delta \boldsymbol{x}(\boldsymbol{\xi}, t) = \delta \boldsymbol{x}(\boldsymbol{\xi})$. In particular, this means that

$$\Delta_t \delta \boldsymbol{x} = \boldsymbol{0}, \quad \mathrm{D}_t \delta \boldsymbol{x} = -\frac{\mathrm{d}\delta \boldsymbol{x}}{\mathrm{d}\boldsymbol{X}} \dot{\boldsymbol{X}} \tag{29}$$

$$\Delta_t \left(\frac{\mathrm{d}\delta \boldsymbol{x}}{\mathrm{d}\boldsymbol{X}} \right) = -\left(\frac{\mathrm{d}\delta \boldsymbol{x}}{\mathrm{d}\boldsymbol{X}} \right) \frac{\mathrm{d}\dot{\boldsymbol{X}}}{\mathrm{d}\boldsymbol{X}}, \quad \mathrm{D}_t \left(\frac{\mathrm{d}\delta \boldsymbol{x}}{\mathrm{d}\boldsymbol{X}} \right) = \frac{(\mathrm{D}_t \delta \boldsymbol{x})}{\mathrm{d}\boldsymbol{X}} \tag{30}$$

The purpose is to first solve for the sensitivity field $\dot{\boldsymbol{x}}'\{\bullet,\ \mathrm{d}\dot{\boldsymbol{X}}\} \in \mathbb{V}^0$ from the tangent problem

$$\mathrm{d}_{\dot{X}}\left[\frac{\mathrm{d}}{\mathrm{d}t}a(\bullet;\,\delta\boldsymbol{x})\right]=0,\ \ \forall\delta\boldsymbol{x}\in\mathbb{V}^0 \tag{31}$$

for given variations[4] $\mathrm{d}\dot{\boldsymbol{X}}\in\mathbb{X}^0$, and then to compute $\underline{\dot{k}}'\{\bullet,\,\mathrm{d}\dot{\boldsymbol{X}}\}$. It can be shown (whereby (6) is used) that the tangent relation becomes

$$a_{\mathrm{T}}(\bullet;\,\delta\boldsymbol{x},\dot{\boldsymbol{x}}')=l_{\mathrm{T}}(\bullet;\,\delta\boldsymbol{x},\,\mathrm{d}\dot{\boldsymbol{X}}),\ \ \forall\delta\boldsymbol{x}\in\mathbb{V}^0 \tag{32}$$

where the pertinent tangent forms are defined as

$$a_{\mathrm{T}}(\bullet;\,\delta\boldsymbol{x},\dot{\boldsymbol{x}}')\stackrel{\mathrm{def}}{=}\int_{\Omega_{\mathrm{X}}}\frac{\mathrm{d}\delta\boldsymbol{x}}{\mathrm{d}\boldsymbol{X}}E_{\mathrm{T}}\frac{\mathrm{d}\dot{\boldsymbol{x}}'}{\mathrm{d}\boldsymbol{X}}\mathrm{d}X \tag{33}$$

$$l_{\mathrm{T}}(\bullet;\,\delta\boldsymbol{x},\,\mathrm{d}\dot{\boldsymbol{X}})\stackrel{\mathrm{def}}{=}\int_{\Omega_{\mathrm{X}}}\frac{\mathrm{d}\delta\boldsymbol{x}}{\mathrm{d}\boldsymbol{X}}\left[E_{\mathrm{T}}\boldsymbol{F}\frac{\mathrm{d}\dot{\boldsymbol{X}}}{\mathrm{d}\boldsymbol{X}}+\boldsymbol{r}_{\mathrm{X}}\,\mathrm{d}\dot{\boldsymbol{X}}\right]\mathrm{d}X \tag{34}$$

where $\boldsymbol{r}_{\mathrm{X}}$ is defined as follows:

$$\boldsymbol{r}_{\mathrm{X}}=\Delta E_{\mathrm{T}}\frac{\mathrm{d}\boldsymbol{F}}{\mathrm{d}\boldsymbol{X}}+E\frac{\mathrm{d}\boldsymbol{\epsilon}^{\mathrm{p}}}{\mathrm{d}\boldsymbol{X}}-\frac{\partial\boldsymbol{P}}{\partial\boldsymbol{X}}|_{\mathrm{F},\underline{k}},\quad \Delta E_{\mathrm{T}}\stackrel{\mathrm{def}}{=}E_{\mathrm{T}}-E \tag{35}$$

When the field $\dot{\boldsymbol{x}}'$ has been computed, we obtain $\boldsymbol{v}'=\dot{\boldsymbol{x}}'-\boldsymbol{F}\,\mathrm{d}\dot{\boldsymbol{X}}$ that is discontinuous across the singular surface. Finally, we compute the field $(\mathrm{D}_t\boldsymbol{F})'=\mathrm{d}\boldsymbol{v}'/\mathrm{d}\boldsymbol{X}$ and use (25) to obtain

$$\begin{aligned}\mathcal{G}^{\mathrm{MAT}}&=\int_{\Omega_{\mathrm{X}}^{\mathrm{p}}}\underline{K}^{\mathrm{T}}(\mathrm{D}_t\underline{k})'\,\mathrm{d}X=\int_{\Omega_{\mathrm{X}}^{\mathrm{p}}}\underline{K}^{\mathrm{T}}\underline{\boldsymbol{R}}_{\mathrm{T}}(\mathrm{D}_t\boldsymbol{F})'\,\mathrm{d}X\\&=\int_{\Omega_{\mathrm{X}}^{\mathrm{p}}}\frac{EY}{h}\frac{\langle\frac{P}{|P|}\mathrm{D}_t\boldsymbol{F}\rangle}{|\frac{P}{|P|}\mathrm{D}_t\boldsymbol{F}|}\frac{P}{|P|}(\mathrm{D}_t\boldsymbol{F})'\,\mathrm{d}X\end{aligned} \tag{36}$$

In order to obtain the last expression in (36), we used that $|P|-K=Y$ in the plastic domain $\Omega_{\mathrm{X}}^{\mathrm{p}}$, defined by the current state satisfying the yield criterion, $\Phi=0$.

5 Numerical Example

5.1 Problem Characteristics

The bar in Figure 2(a) is henceforth assumed to have homogeneous material properties in the subdomains (on each side of the singular section). As a consequence $\boldsymbol{F}$ and $\boldsymbol{\epsilon}^{\mathrm{p}}$ will become p.w. uniform and, hence, $\boldsymbol{r}_{\mathrm{X}}=\boldsymbol{0}$. Moreover, linear shape

[4] The required regularity of $\mathbb{X}^0$ depends on the chosen format; here, we assume that $\mathrm{d}\dot{\boldsymbol{X}}$ is continuous and that $\mathrm{d}\dot{\boldsymbol{X}}=\boldsymbol{0}$ on $\partial\Omega_{\mathrm{X,D}}$ for any $\mathrm{d}\dot{\boldsymbol{X}}\in\mathbb{X}^0$.

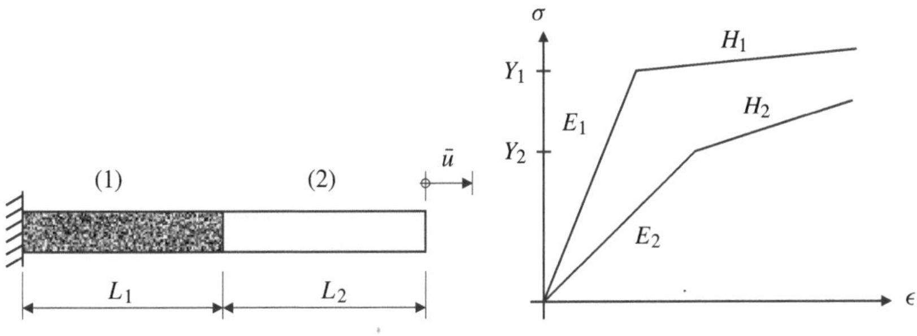

Fig. 2 (a) Bar in uniaxial stress with discontinuity in material properties at interface, (b) linear elastic-hardening stress-strain response curves.

and test functions are chosen for the two elements (indicated in Figure 2(a)) for the material and spatial motions $(\dot{X}(X), \dot{x}(X))$, whereby $(\mathrm{D}_t F)'$ becomes p.w. uniform. The maximal displacement is $\bar{u}_{\max}/L = 0.5 \cdot 10^{-3}$. We choose $L_1/L = L_2/L = 0.5$. The material parameters for the elastic-plastic model are: E, H with E_{ref}, $Y_{\mathrm{ref}} = 0.001 E_{\mathrm{ref}}$, $H_{\mathrm{ref}} = 0.2 E_{\mathrm{ref}}$. They assume different values on each side of the interface, as shown schematically in Figure 2(b).

5.2 *Numerical Results*

A series of computations were carried out to study the development of $\mathcal{G}^{\mathrm{CONF}}$ and $\mathcal{G}^{\mathrm{MAT}}$ with increasing (prescribed) displacement, while it was assumed that there is a mismatch across the interface of (a) elastic and (b) plastic (hardening) properties. In case (a) $H_1 = H_2 = H_{\mathrm{ref}}$, whereas in case (b) $E_1 = E_2 = E_{\mathrm{ref}}$. These results are shown in Figure 3.

For case (a) the following observations are made: Before plastic yielding the only contribution to $\mathcal{G}$ is from the missmatch of the elastic properties. This contribution vanishes, of course, in case (b). In both cases, the contribution from $\mathcal{G}^{\mathrm{MAT}}$ is positive (while $\mathcal{G}^{\mathrm{CONF}}$ is negative); however, this is not a general property since $\mathcal{G}^{\mathrm{MAT}}$ represents the *variation* of the dissipation rate due to configurational changes.

6 Discussion and Conclusions

We have presented a novel investigation in the context of configurational changes for a rate-independent dissipative material; namely, we considered the *total variation* of the rate of global dissipation with respect to the rate of configurational motion. Such a variation may, alternatively, be considered as the evaluation of the *total sensitivity* due to configurational changes, which may be brought about (in their turn) by

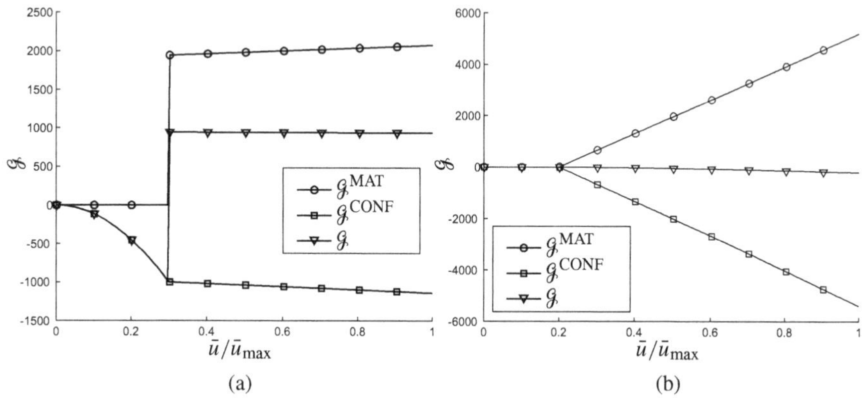

Fig. 3 Development of components $\mathcal{G}^{\mathrm{CONF}}$, $\mathcal{G}^{\mathrm{MAT}}$ and $\mathcal{G} = \mathcal{G}^{\mathrm{CONF}} + \mathcal{G}^{\mathrm{MAT}}$ with increasing loading (prescribed displacement) for (a) mismatch of elasticity: $E_2 = 0.5E_{\mathrm{ref}}$, and (b) mismatch of hardening: $H_2 = 0.5H_{\mathrm{ref}}$.

physical changes. It is then clear that the *material part of the dissipation* is indeed coupled (and sensitive) to configurational changes. However, it is not at all clear how to use this information. The classical approach in defining the "driving force" for configurational changes, advocated in the current literature, is to account only for the explicit part of the total configurational force, denoted $\boldsymbol{\mathcal{G}}^{\mathrm{CONF}}$ in this paper.

The paper discussed essentially two aspects: (1) how to formulate (or construct) the total dissipation in the presence of configurational changes, and (2) how to compute the total sensitivity of the global dissipation via the appropriate formulation of a tangent problem for the standard equilibrium equation. The numerical results obtained for dissimilar material properties across the singular surface in a simple bar showed that a misfit in the elasticity modulus resulted in larger value of $\mathcal{G}^{\mathrm{MAT}}$ than did a misfit in the hardening modulus.

References

1. Runesson, K., Larsson, F. and Steinmann, P., On energetic changes due to configurational motion of standard continua, *Int. J Solids Structures*, 2008, accepted.
2. Eshelby, J., The force on an elastic singularity, *Phil. Trans. Roy. Soc. Lond.*, 1951, 87–112.
3. Abeyaratne R. and Knowles, J., On the driving traction acting on a surface of strain discontinuity in a continuum, *J. Mech. Phys. Solids* **38**, 1990, 345–360.
4. Maugin, G., *Material Inhomogeneities in Elasticity*, Chapman & Hall, London, 1993.
5. Maugin, G., Material forces: Concepts and applications, *Appl. Mech. Rev.* **48**, 1995, 213–245.
6. Maugin, G., Thermomechanics of inhomogeneous-heterogeneous systems: Application to the irreversible progress of two- and three-dimensional defects, *ARI* **50**, 1997, 41–56.
7. Maugin, G., On shock waves and phase-transition fronts in continua, *ARI* **50**, 1998, 141–150.
8. Maugin, G., Thermomechanics of forces driving singular pointsets, *Arch. Mech.* **50**, 1998, 509–519.

9. Maugin, G., On the universality of the thermomechanics of forces driving singular sets, *Arch. Mech.* **69**, 1999, 1–15.
10. Maugin, G. and Trimarco, C., Pseudomomentum and material forces in nonlinear elasticity: variational formulations and application to brittle fracture, *Arch. Mech.* **94**, 1992, 1–28.
11. Maugin, G. and Trimarco, C., The dynamics of configurational forces at phase-transition fronts, *Meccanica* **30**, 1995, 605–619.
12. Gurtin, M., On the nature of configurational forces, *Arch. Rational Mech. Anal.* **131**, 1995, 67–100.
13. Gurtin, M., *Configurational Forces as Basic Concepts of Continuum Physics*, Springer, New York, 2000.
14. Kienzler, R. and Herrmann, G., *Mechanics in Material Space*, Springer, Berlin, 2000.
15. Steinmann, P., On boundary potential energies in deformational and configurational mechanics, *J. Mech. Phys. Solids* **56**, 2008, 772–800.
16. Simha, N., Fischer, F., Kolednik, O. and Chen, C., Inhomogeneity effects on the crack driving force in elastic and elastic-plastic materials, *J. Mech. Phys. Solids* **51**, 2003, 209–240.
17. Simha, N., Fischer, F., Shan, G., Chen, C. and Kolednik, O., J-integral and crack driving force in elastic-plastic materials, *J. Mech. Phys. Solids* **56**, 2008, 2876–2895.
18. Liebe, T., Denzer, R. and Steinmann, P., Application of the material force method to isotropic continuum damage, *Comput. Mech.* **30**, 2003, 171–184.
19. Menzel, A., Denzer, R. and Steinmann, P., Material forces in computational single-slip crystal-plasticity, *Computat. Mater. Sci.* **32**, 1995, 446–454.
20. Nguyen, T., Govindjee, S., Klein, P. and Gao, H., A material force method for inelastic fracture mechanics, *J. Mech. Phys. Solids* **53**, 2005, 91–121.

On Crack Analysis of Functionally Graded Materials with Material Forces

Rolf Mahnken

Abstract Functionally Graded Materials (FGMs) are advanced materials that possess continuously graded properties. Applications of FGMs are on composites, ceramics, alloys and coatings. This work is concerned with the crack analysis of FGMs. To this end we exploit a Clausius–Planck inequality to a migrating control volume. As a consequence of the principle of maximum dissipation the direction of crack propagation is obtained in terms of material forces. In the numerical implementation a staggered algorithm – deformation update for fixed geometry followed by geometry update for fixed deformation – is employed. The corresponding finite element mesh is generated by combining Delaunay triangulation with local mesh refinement. In a numerical example the brittle crack propagation in an FGM is investigated for varying directions of strength gradation within the structures.

1 Introduction

During the past two decades much research of Functionally Graded Materials (FGMs) has been focused on manufacturing, material design and property estimation as well as thermal and structural analysis [24]. Applications of FGMs are on composites, ceramics, alloys and coatings. Compared to conventional materials FGMs have the possibility of tailoring its gradation to maximize its performance.

This work is concerned with the simulation of brittle crack growth in FGMs. On the analysis of stationary cracks for FGMs we refer to the comprehensive review in [21]. The analytical investigations in [3] showed that the asymptotic crack-tip stress field possesses the same square root singularity as in homogeneous materials. Furthermore, Eischen [5] used the traditional eigenfunction expansion technique

Rolf Mahnken
Chair of Engineering Mechanics (LTM), University of Paderborn, Germany;
e-mail: rolf.mahnken@ltm.upb.de

P. Steinmann (ed.), IUTAM Symposium on Progress in the Theory and Numerics of Configurational Mechanics, 61–71.

of Williams to show, that the leading term in 2D crack-tip elastic fields remain square-root singular, which allows to express the stress fields in terms of stress intensity factors (SIFs). Further analytical results for SIFs for some typical fracturing modes in FGMs are presented in [6, 7]. Numerical methods for fracture analysis of FGMs in stationary cracks are presented in [1]. Additional references are given e.g. in [11]. In [15] the concept of material forces is applied to FGMs and the accuracy of the method is discussed for two examples based on comparison with available theoretical and numerical solutions. Concerning the numerical treatment of crack propagation with finite elements in homogeneous structures different possibilities can be distinguished, e.g. insertion of interface elements between existing bulk elements [20], alignment of critical element segments [18], extended finite element method (XFEM) [19] and and meshfree methods [14].

The simulation of crack propagation depends strongly on the three aspects of (1) initiation of crack growth, (2) direction of crack growth and (3) length of crack growth. In this work a Griffith-type crack criterion function is used to account for the first aspect. Furthermore, we employ a dissipation inequality to a time dependent migrating control volume, introduced in [9]. Then, a constitutive equation for crack tip evolution is obtained in terms of material forces from the principle of maximum dissipation, thus rendering the direction of crack growth [18]. The final system of partial differential equations is summarized, and the numerical implementation is briefly outlined. In this way the length of crack growth is a result of a staggered algorithm – deformation update for fixed geometry followed by geometry update for fixed deformation. The corresponding mesh is generated at fixed deformation by combining Delaunay triangulation with local mesh refinement. In a numerical example the brittle crack propagation in an FGM is investigated for varying directions of strength gradation within the structure.

Notations: Square brackets [•] are used throughout the paper to denote 'function of' in order to distinguish from mathematical groupings with parenthesis (•).

2 Kinematics and Balance Equations

2.1 Parameterization of Geometry and Displacement

In Figure 1a we consider a body B with reference configuration $\mathcal{B} \subset \mathbb{E}^{n_{\text{dim}}}$, where $\mathbb{E}^{n_{\text{dim}}}$ denotes the Euclidean space with dimension $n_{\text{dim}} = 2$ or $n_{\text{dim}} = 3$. The boundary has an exterior part $\partial\mathcal{B}_u \cup \partial\mathcal{B}_\sigma$ with $\partial\mathcal{B}_u \cap \partial\mathcal{B}_\sigma = \emptyset$ and the crack is represented by a surface $\Gamma^+ \cup \Gamma^-$. In order to take into account the peculiarity of singular stresses at the crack tip, a cut along a boundary Γ_ϵ^+ is introduced in Figure 1a. Then we concentrate on the reference configuration $\bar{\mathcal{B}} = \mathcal{B} \backslash \mathcal{B}_\epsilon \subset \mathbb{E}^{n_{\text{dim}}}$ as introduced in Figure 1b, where singular stresses are assumed not to occur. A ma-

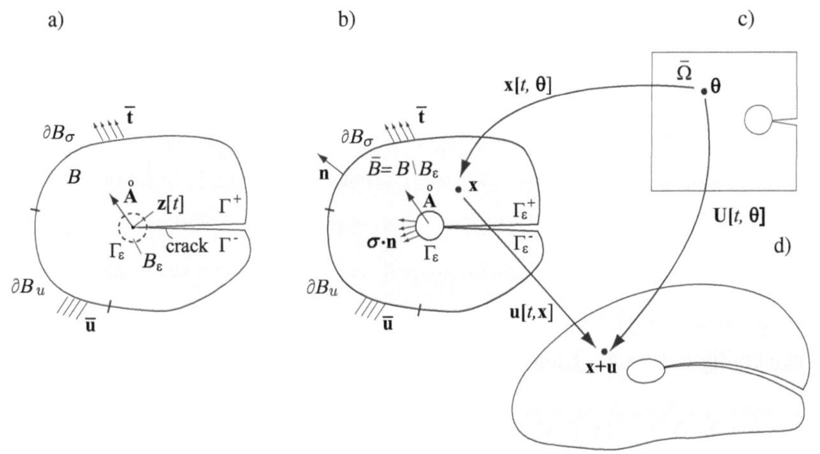

Figure 1 Configurations of body B with a crack: (a) reference configuration $\mathcal{B}$ with sharp crack at time t, (b) reference configuration $\bar{\mathcal{B}}$ with cut at time t, (c) time independent reference structure $\bar{\Omega}$, (d) deformed configuration at time t.

terial point of the structure has a position $\mathbf{x} \in \bar{\mathcal{B}}$. Since during crack propagation the reference configuration $\bar{\mathcal{B}} \subset \mathbb{E}^{n_{\text{dim}}}$ is changing with time $t \in \mathcal{I} = \mathbb{R}^+$, additionally, in Figure 1c we consider the region $\bar{\Omega}$ of a reference structure, which is not dependent on time t. Here a material point has a position $\boldsymbol{\theta} \in \bar{\Omega}$. Additionally, in Figure 1d we have the time dependent current configuration, which is deformed due to loading. By denoting $\mathbf{u} \in \mathbb{E}^{n_{\text{dim}}}$ as the displacement a material point at time t has position $\mathbf{x} + \mathbf{u} \in \mathbb{E}^{n_{\text{dim}}}$.

With reference to all three configurations in Figure 1, analogously to the approach in [18], we introduce the following three mappings

$$1.\; t, \boldsymbol{\theta} \mapsto \mathbf{x}(t, \boldsymbol{\theta}), \qquad 2.\; t, \mathbf{x} \mapsto \mathbf{u}(t, \mathbf{x}), \qquad 3.\; t, \boldsymbol{\theta} \mapsto \mathbf{U}(t, \boldsymbol{\theta}). \tag{1}$$

As outlined in the mapping (1.1) represents a parameterization of the medium in terms of the time t and the time independent position $\boldsymbol{\theta} \in \bar{\Omega}$. In this way it accounts for geometry changes due to crack propagation. The mapping (1.2) represents the time dependent displacement $\mathbf{u}$ parameterized by the time t and the time dependent position $\mathbf{x} \in \bar{\mathcal{B}}$, whereas the mapping (1.3) represents the time dependent displacement $\mathbf{U}$ parameterized by the time t and the time independent position $\boldsymbol{\theta} \in \bar{\Omega}$. Furthermore, we introduce the following time derivatives of Eqs. $(1)_1$ and $(1)_2$

$$1.\; \overset{\circ}{\mathbf{x}} = \frac{\partial \mathbf{x}}{\partial t} = \mathbf{V} \qquad 2.\; \dot{\mathbf{u}} = \frac{\partial \mathbf{u}}{\partial t}. \tag{2}$$

As explained in [16] the symbols ($\circ$) and ($\cdot$) refer to time derivatives w.r.t. to fixed positions $\boldsymbol{\theta}$ and $\mathbf{x}$, respectively. In this way, the vector $\mathbf{V}$ is the *material velocity* and represents the rate of geometry change w.r.t. to the fixed reference structure $\bar{\Omega}$, whereas the vector $\dot{\mathbf{u}}$ represents the velocity of a material point w.r.t. to the non-fixed

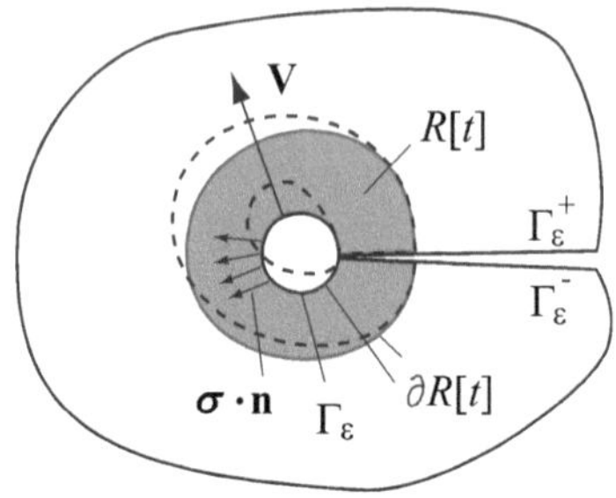

Figure 2 Time dependent migrating control volume $\mathcal{R}[t]$ with material velocity $\mathbf{V}$.

reference configuration $\bar{\mathcal{B}}$. Then in summary we have the relations

$$1.\ \mathbf{v} = \dot{\mathbf{u}} + \mathbf{h} \cdot \mathbf{V}, \quad 2.\ \mathbf{v} \otimes \nabla = \dot{\mathbf{h}} + (\mathbf{h} \cdot \mathbf{V}) \otimes \nabla, \quad 3.\ \frac{d}{dt} dV = (\mathbf{1} : \mathbf{V}) dV, \quad (3)$$

where $\mathbf{h} = \mathbf{u} \otimes \nabla$ is the displacement gradient and dV is a differential volume element [18]. A further fundamental ingredient of a geometrically linear theory is the strain-displacement relation $\boldsymbol{\varepsilon} = 1/2\ (\mathbf{h} + \mathbf{h}^T)$. The spatial and the material velocities are restricted by Dirichlet boundary conditions as $\mathbf{v} = \bar{\mathbf{v}}$ on $\partial \mathcal{B}_u$, $\mathbf{V} = \mathbf{0}$ on $\partial \bar{\mathcal{B}} \backslash \Gamma_\epsilon$. The geometry can change only along the boundary Γ_ϵ due to crack propagation. As indicated in Figures 1a and 1b, the corresponding rate of crack extension at the crack tip boundary is denoted as $\mathbf{V} = \overset{\circ}{\mathbf{A}}$ on Γ_ϵ.

2.2 *Time Dependent Migrating Control Volume*

Balance and state equations of continuum mechanics typically are formulated for control volumes. In order to consider possible geometry changes we follow an approach by Gurtin and Podio-Guidugli [9] using a *migrating control volume*. As indicated in Figure 2 it is a closed subregion $\mathcal{R}[t]$ for which $\partial \mathcal{R}[t]$ evolves smoothly with time t. Following Gurtin and Podio-Guidugli [9] we view the dependence of $\mathcal{R}[t]$ on t as resulting from *the addition and removal of material points*. This viewpoint has important consequences for the formulation of balance and state equations in the next section.

2.3 *Dissipation Inequality*

As noted in [9] the power of the distributed reaction forces $\boldsymbol{\sigma} \cdot \mathbf{n}$ on $\partial \mathcal{R}[t]$ in Figure 2 should account for the power performed in the addition and removal of material at the boundary $\partial \mathcal{R}[t]$ and for the change in material structure as the crack tip evolves. Classically, the standard stress $\boldsymbol{\sigma}$ expends power over the material velocity $\dot{\mathbf{u}}$. But when the control volume migrates, there is no intrinsic material description of the deformed boundary $\mathbf{x}[t] \in \partial\mathcal{R}[t]$, since material is being added and removed. Therefore, following Gurtin and Podio-Guidugli [9] as a velocity for $\mathbf{x}[t] \in \partial\mathcal{R}[t]$ we use the total velocity $\mathbf{v}[\mathbf{x} \in \partial\mathcal{R}[t], t]$ by fixing the structural coordinate $\boldsymbol{\theta}$ and write the boundary power of $\boldsymbol{\sigma}$ in the form $(\boldsymbol{\sigma} \cdot \mathbf{n}) \cdot \mathbf{v}$.

Since material is neither added nor removed from interior points of $\mathcal{R}[t]$, and since there is no change in material structure away from the crack tip, the power of the body force $\mathbf{b}$ has the standard form $\mathbf{b} \cdot \dot{\mathbf{u}}$ [9]. In summary, the power of the mechanical forces is

$$\mathcal{P} = \int_{\partial\mathcal{R}[t]} (\boldsymbol{\sigma} \cdot \mathbf{n}) \cdot \mathbf{v}\, dA + \int_{\mathcal{R}[t]} \mathbf{b} \cdot \dot{\mathbf{u}}\, dV. \tag{4}$$

Let us assume the following functional relationship for the bulk free-energy function with respect to a unit volume:

$$\psi = \psi[\boldsymbol{\varepsilon}[\mathbf{x}[t], t], \mathbf{x}[t]], \tag{5}$$

where ψ is the strain energy function with respect to a unit volume. Note that for a functionally graded material ψ is also explicitly dependent on the position $\mathbf{x}$. The principle of positive dissipation for the migrating control volume is postulated as

$$\mathcal{D} = \mathcal{P} - \frac{d}{dt}\Psi \geq 0. \tag{6}$$

In [16] the above dissipation inequality is rewritten after some algebric manipulations as

$$\mathcal{D} = -\int_{\Gamma_\epsilon} (\boldsymbol{\Sigma} \cdot \mathbf{n}) \cdot \mathbf{V}\, \mathrm{dA} \geq 0. \tag{7}$$

Here, we define the *energy momentum tensor* or *Eshelby tensor*, respectively, introduced by Eshelby [8]

$$\boldsymbol{\Sigma} = \psi \mathbf{1} - \mathbf{h}^T \cdot \boldsymbol{\sigma}. \tag{8}$$

In the limit $\Gamma_\epsilon \to 0$ the dissipation in Eq. (7) reads as

$$1.\ \mathcal{D} = \mathbf{J} \cdot \overset{\circ}{\mathbf{A}} \geq 0, \quad \text{where} \quad 2.\ \mathbf{J} = -\lim_{\Gamma_\epsilon \to 0} \int_{\Gamma_\epsilon} \boldsymbol{\Sigma} \cdot \mathbf{n}\, \mathrm{dA}. \tag{9}$$

Therefore, we can interpret $\mathbf{J}$ in Eq. $(9)_2$ as a material force vector (in the sense of Eshelby), which is work-conjugate to the material velocity $\overset{\circ}{\mathbf{A}}$ at the crack tip.

We remark that in [16] the constitutive relation

$$\boldsymbol{\sigma} = \partial_{\boldsymbol{\varepsilon}} \psi \tag{10}$$

is *derived* from the global dissipation inequality. Furthermore, the balance law of material forces

$$\mathbf{\Sigma} \cdot \nabla + \mathbf{B} = \mathbf{0} \qquad \forall \mathbf{x} \in \bar{\mathcal{B}}, \quad \mathbf{B} = -\mathbf{h}^T \cdot \mathbf{b} - \partial_{\mathbf{x}} \psi. \tag{11}$$

is a consequence of the balance law $\boldsymbol{\sigma} \cdot \nabla + \mathbf{b} = \mathbf{0}$ and the constitutive relation (10). This is conceptually different to approaches in the literature, where the constitutive relation (10) or the balance law of material forces are introduced as a postulate, see e.g. [12, 17, 18] for different approaches.

2.4 Evolution Equation for Crack Propagation

In order to formulate a constitutive equation for the crack growth vector $\overset{\circ}{\mathbf{A}}$ in Eq. $(9)_1$ we follow the approach of [18] by using the principle of maximum dissipation. A Griffith-type crack criterion function is introduced as $\Phi[\mathbf{J}] = ||\mathbf{J}|| - \Gamma_c \leq 0$, where Γ_c is the critical energy release. Next, a Lagrangian is introduced as $\mathcal{L}[\mathbf{J}, \overset{\circ}{\lambda}] = -\mathcal{D} + \overset{\circ}{\lambda} \ \Phi[\mathbf{J}]$. Then, the necessary conditions for a stationary point of the Lagrangian, $\mathcal{L}[\mathbf{J}, \overset{\circ}{\lambda}] \rightarrow$ stationary, renders the evolution vector of crack propagation

$$\overset{\circ}{\mathbf{A}} = \overset{\circ}{\lambda} \ \partial_{\mathbf{J}} \Phi[\mathbf{J}] = \overset{\circ}{\lambda} \frac{\mathbf{J}}{||\mathbf{J}||} \tag{12}$$

along with the crack loading-unloading conditions in Kuhn–Tucker form $\overset{\circ}{\lambda} \geq 0$, $\Phi[\mathbf{J}] \leq 0$, $\overset{\circ}{\lambda} \ \Phi[\mathbf{J}] = 0$.

Since the vector $\mathbf{J}/||\mathbf{J}||$ in Eq. (12) has a unit length, it follows that the multiplier $\overset{\circ}{\lambda}$ in Eq. (12) has the interpretation of the rate of crack growth, i.e. $\overset{\circ}{\lambda} = \overset{\circ}{A} = dA/dt$. As pointed out in [18], the above constitutive equations determine the direction of crack propagation at a point $\mathbf{x} \in \bar{\mathcal{B}}$. However *the specific value of* $\overset{\circ}{\lambda}$ *is undetermined.*

2.5 Summary of Coupled Initial-Boundary Value Problem

From the above derivations we can formulate an initial-boundary value problem for the coupled deformation/crack propagation problem as follows:

$$
\begin{array}{llll}
1. & \boldsymbol{\varepsilon} = \frac{1}{2}\left(\mathbf{h}+\mathbf{h}^T\right) & 2. & \mathbf{h} = \mathbf{u}\otimes\nabla \\
3. & \boldsymbol{\sigma} = \partial_{\boldsymbol{\varepsilon}}\psi & 4. & \boldsymbol{\sigma}\cdot\nabla+\mathbf{b}=\mathbf{0} \\
5. & \boldsymbol{\Sigma} = \psi\mathbf{1}-\mathbf{h}^T\cdot\boldsymbol{\sigma} & 6. & \boldsymbol{\Sigma}\cdot\nabla+\mathbf{B}=\mathbf{0} \\
7. & \overset{\circ}{\mathbf{x}} = \frac{\partial\mathbf{x}}{\partial t}=\mathbf{V} & 8. & \dot{\mathbf{u}} = \frac{\partial\mathbf{u}}{\partial t} \\
9. & \mathbf{v} = \dot{\mathbf{u}}+\mathbf{h}\cdot\mathbf{V} & & \\
10. & \overset{\circ}{\mathbf{A}} = \overset{\circ}{\lambda}\,\frac{\mathbf{J}}{||\mathbf{J}||} & 11. & \Phi[\mathbf{J}] = ||\mathbf{J}||-\Gamma_c\leq 0 \\
12. & \overset{\circ}{\lambda}\geq 0\,,\quad \Phi[\mathbf{J}]\leq 0,\quad \overset{\circ}{\lambda}\;\Phi[\mathbf{J}]=0 & 13. & \mathbf{J} = -\lim_{\Gamma_\epsilon\to 0}\int_{\Gamma_\epsilon}\boldsymbol{\Sigma}\cdot\mathbf{n}\,\mathrm{d}A.
\end{array}
\tag{13}
$$

Here we have the strain-displacement relation $(13)_1$, where the displacement gradient is defined in $(13)_2$. Equation $(13)_3$ is the elastic constitutive relation for the Cauchy stress tensor satisfying the balance equation Eq. $(13)_4$. Equation $(13)_5$ is the definition for the Eshelby stress tensor satisfying the balance equation Eq. $(13)_6$. Equation $(13)_9$ expresses the spatial velocity $\mathbf{v}$ in terms of the material velocity $\mathbf{V}$ Eq. $(13)_7$ and the displacement velocity $\dot{\mathbf{u}}$ Eq. $(13)_8$. The vector of crack propagation is defined in Eq. $(13)_{10}$ in terms of the material force vector defined in Eq. $(13)_{13}$, and a Griffith-type crack criterion function is formulated in Eq. $(13)_{11}$, along with the crack loading-unloading conditions $(13)_{12}$. Additionally, appropriate Neumann and Dirichlet boundary conditions have to be formulated for the Cauchy stress tensor $\boldsymbol{\sigma}$ and the spatial and material velocity $\mathbf{v}$ and $\mathbf{V}$. Furthermore initial conditions $\mathbf{v}(t=0)=\mathbf{v}_0$, $\mathbf{V}(t=0)=\mathbf{V}_0$ are prescribed.

2.6 Solution of Coupled Initial-Boundary Value Problem

The initial-boundary value problem (13) accounting for both, evolution of deformation and evolution of crack propagation, in practice can hardly be solved in a monolithic way. Therefore, following Miehe and Gürses [18] a staggered algorithm – deformation update for fixed geometry followed by geometry update for fixed deformation – is employed within each time increment, $\Delta^{n+1}t = {}^{n+1}t - {}^{n}t$, $n = 1,\ldots,N_t$. On the numerical treatment of the deformation problem and determination of material forces as post-processing we refer to [2, 23]. The domain integral method is used for efficient evaluation of the J-integral vectors, see [4, 13, 15]. Let α denote the relaxation counter for the staggered algorithm, then, the crack propagation criterion $\Phi[{}^{n+1}_{\alpha}\mathbf{J}] = ||{}^{n+1}_{\alpha}\mathbf{J}|| - \Gamma_c \leq 0$ is checked at the crack tip. For $\Phi > 0$ a crack relaxation step is performed, based on the evolution vector of crack propagation (12)

$$
{}^{n+1}_{\alpha}\Delta\mathbf{A} = \Delta\lambda\,\frac{{}^{n+1}_{\alpha}\mathbf{J}}{||{}^{n+1}_{\alpha}\mathbf{J}||}. \tag{14}
$$

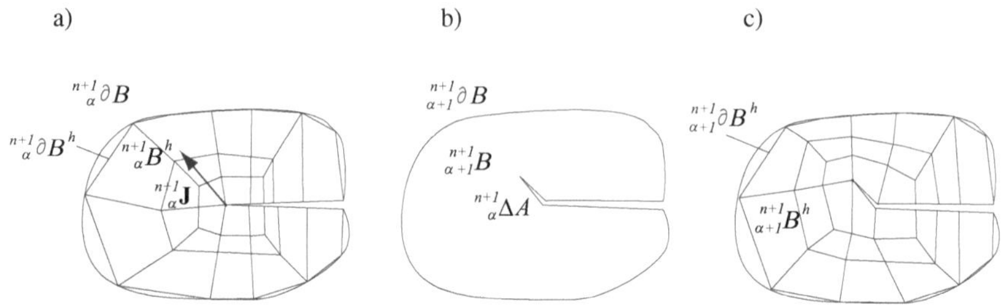

Figure 3 Geometry update for fixed deformation: (a) material force $^{n+1}_{\alpha}\mathbf{J}$ at crack tip, (b) crack propagation $^{n+1}_{\alpha}\Delta\ A$ and update of boundary $^{n+1}_{\alpha+1}\partial\mathcal{B}$, (c) discretization $^{n+1}_{\alpha+1}\mathcal{B}^h = \cup_{e=1}^{^{n+1}_{\alpha+1}N_{el}} \mathcal{B}^e$ after remeshing.

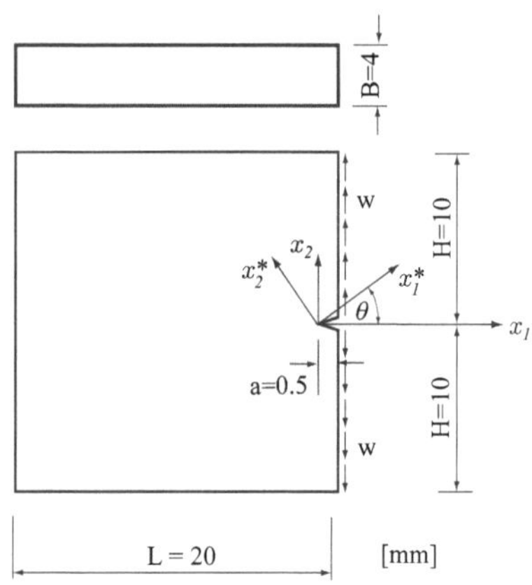

Figure 4 Compact tension specimen: geometry and loading, rotated coordinate x_1^* representing the gradation with angle θ.

Then, as illustrated in Figure 3 the crack propagates by an increment $^{n+1}_{\alpha}\Delta A$. This in turn renders an update of the geometry $^{n+1}_{\alpha+1}\mathcal{B}$. The corresponding mesh is generated by combining Delaunay triangulation with local mesh refinement, [10, 16, 22].

3 Compact Tension Specimen with Graded Material

We consider a compact tension specimen. with geometry according to Figure 4. The applied loading corresponds to a vertical displacements $\pm w$ at the boundary $x_1 = 0.5$. The material gradation within the structure is expressed by the coordinate x_1^*, which is rotated by an angle θ relative to the coordinate x_1 as shown in Figure 4. Then the corresponding Young's modulus is an exponential function of the form $E[x_1^*] = E_1 e^{\beta x_1^*}$. Further details on the material parameters are given in [16].

Two different angles for θ have been investigated: $\theta = 0°$ and $\theta = 90°$. For the case $\theta = 0°$ the material is symmetric with respect to the horizontal x_1-axes, such that a crack starting from the notch propagating horizontally through the ct-

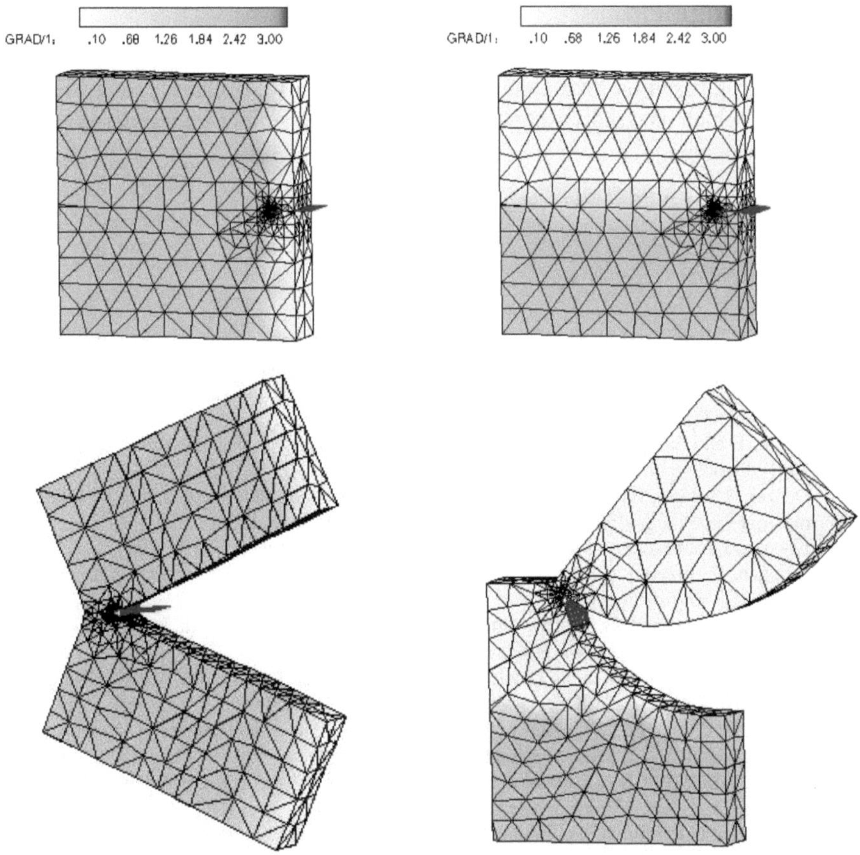

Figure 5 Compact tension specimen: crack propagation for $\theta = 0°$ (left) and $\theta = 90°$ (right).

specimen is expected theoretically. Tetrahedral elements with linear shape functions have been used.

The results of both simulations are shown in Figure 5 for the cases $\theta = 0°$ and $\theta = 90°$. From Figure 5, left, it is observed, that the algorithm captures the theoretical horizontal crack pattern very well. Furthermore the illustration in Figure 5 renders a strong upwards crack for gradation with angle $\theta = 90°$. More results concerning the robustness of the algorithm with respect to spatial and time discretization are presented in [16].

Acknowledgment

This paper is based on investigations of the collaborative research center SFB/TR TRR 30, which is kindly supported by the Deutsche Forschungsgemeinschaft (DFG), Germany.

References

1. Anlas, G., Santare, M.H. and Lambros, J., Numerical calculation of stress intensity factors in functionally graded materials. *Int. J. Fracture* **104**, 2000, 131–143.
2. Braun, M., Configurational forces induced by finite-element discretization. *Proc. Estonian Acad. Sci. Phys. Math.* **46**, 1997, 24–31.
3. Delale, F. and Erdogan, F., Fracture mechanics of functionally graded materials. *J. Appl. Mech.* **50**, 1983, 609–614.
4. Denzer, R., Barth, F.J. and Steinmann, P., Studies in elastic fracture mechanics based on the material force method. *Int. J. Num. Meth. Eng.* **58**, 2003, 1817–1835.
5. Eischen, J.W., Fracture of non-homogeneous materials. *Int. J. Fracture* **34**, 1987, 3–22.
6. Erdogan, F., Fracture mechanics of functionally graded materials. *Composit. Eng.* **5**, 1995, 753–770.
7. Erdogan, F. and Wu, B.H., The surface crack problem for a plate with functionally graded materials. *ASME J. Appl. Mech.* **64**, 1997, 449–456.
8. Eshelby J.D., The force on an elastic singularity. *Philos. Trans. Roy. Soc., Math. Phys. Sci. A* **244**, 1951, 87–112.
9. Gurtin, M.E. and Podio-Guidugli, P., Configurational forces and the basic laws for crack propagation. *J. Mech. Phys. Solids* **44**(6), 1996, 905–927.
10. Joe, B., GEOMPACK, a software for the generation of meshes using geometric algorithms. *Adv. Eng. Software,* **13**(5/6), 1991, 325–331.
11. Kim, J.H. and Paulino, G.H., Finite element evaluation of mixed mode stress intensity factors in functionally graded materials. *Int. J. Num. Meth. Eng.* **53**, 2002, 1903–1935.
12. Le, K.C. and Stumpf, H., 1990, Variational principles of nonlinear fracture mechanics. *Acta Mech.* **83**, 1990, 25–37.
13. Li, F.Z., Shih, C.F. and Needleman, A., A comparison of methods for calculating energy release rates. *Engr. Frac. Mech.* **21**, 1985, 405–421.
14. Lu, Y.Y., Belytschko, T. and Tabbara, M., Element-free Galerkin method for wave-propagation and dynamic fracture. *Comp. Meth. Appl. Mech. Eng.* **126**(1–2), 1995, 131–153.
15. Mahnken, R., Material forces for crack analysis of functionally graded materials in adaptively refined meshes. *Int. J. Fracture* **147**, 2008, 269–283.
16. Mahnken, R., Geometry update driven by material forces for simulation of brittle crack growth in functionally graded materials. *Int. J. Num. Meths. Eng.*, 2007, accepted.
17. Makowski, J., Stumpf, H. and Hackl, K., The fundamental role of nonlocal and local balance laws of material forces in finite elastoplasticity and damage mechanics. *Int. J. Solids Structures*, **43**, 2006, 3940–3959.
18. Miehe, C. and Gürses, E., A robust algorithm for configurational-force-driven brittle crack propagation with r-adaptive mesh alignement. *Int. J. Num. Meths. Eng.* **72**, 2007, 127–155.
19. Moes, N., Dolbow, J. and Belytschko, T., A finite element method for crack growth without remeshing. *Int. J. Num. Meths. Eng.* **46**(1), 1999, 133–150.
20. Pandolfi, A. and Ortiz, M., An efficient adaptive procedure for three-dimensional fragmentation simulations. *Eng. Comput.* **18**, 2002, 148–159.
21. Paulino, G.H., Jin, Z.H. and Dodds, R.H., Failure of functionally graded materials. In: Karihaloo, B., Knauss, W.G. (Eds.), *Comprehensive Structural Integrity*, Volume 2, Elsevier Science, 2003, pp. 607–644.

22. Staten, M.S., Selective refinement of two and three-dimensional finite element meshes. Thesis for the Degree of MSc., Department of Civil Engineering, Brigham Young University, 1996.
23. Steinmann, P., Ackermann, D. and Barth, F.J., Application of material forces to hyperelastic fracture mechanics, Part II: Computational setting. *Int. J. Solids Structures* **38**, 2001, 5509–5526.
24. Surush, S. and Mortensen, A., Fundamentals of functionally graded materials. Institute of Materials, London, 1998.

Momentum and Material Momentum in Superconductors

C. Trimarco

Abstract The electric current of a superconducting solid can be associated with the flow of an ideal compressible fluid. A momentum density can be naturally associated with this flow. In deformable superconductors, the presence of finite deformations affects the electromagnetic fields and possibly the flow of the fluid. One of the result is that to the supercurrent in the undeformed configuration of the solid is associated the *material* (or *configurational) momentum* rather than the momentum density.

1 Introduction

Superconductivity is still a challenge for theoreticians and experimentalists after almost 100 years from its discovery. The main features of a superconductor can be summarised in two. First, an electric current can last theoretically for thousands years without any voltage applied. Second, the magnetic flux is fully ejected from a material in a superconducting state (Meissner effect). This effect leads to strong magnetic interactions: levitation is one of the most impressive examples. Unfortunately, these phenomena occur at very low temperature and one of the challenges is to reproduce them at room temperature, in order to exploit their great potentiality in technological applications. *Warm* superconductors (at 35°Kelvin degrees) have been discovered recently (Bednortz and Müller, Nobel prize in 1986). Nowadays, the upper limit for temperature is estimated around 195°K. However, the microscopic mechanism for superconductivity proposed by Bardeen, Cooper and Schrieffer (Nobel prize in 1972) apparently does not describe satisfactorily this second generation of warm superconductors. Thus, these aspects still represent a theoretical challenge for future research.

C. Trimarco
Department of Applied Mathematics, University of Pisa, Via F. Buonarroti, 1, I-56127 Pisa, Italy; e-mail: trimarco@dma.unipi.it

P. Steinmann (ed.), IUTAM Symposium on Progress in the Theory and Numerics of Configurational Mechanics, 73–82.

The successful explanation of the classical superconductivity provided by Bardeen, Cooper and Schrieffer [3] essentially relies on Fröhlich ideas, according to which elastic wave interactions (phonons) favour the flow of an electric current in some circumstances [6–8, 13, 15]. The mechanism is usually described by the following oversimplified argument. A local polarisation of the lattice cell, due to the passage of an electron, favours the passage of a second electron, which is accelerated by the electric field generated by this polarisation. In all other circumstances, elastic waves are responsible for the resistivity of the conductor. The aforementioned pairs of electrons, the Cooper pairs, are responsible for the supercurrents. These electron pairs, whose total spin is zero, behave statistically as bosons, which are known to condensate at low temperatures. Thus, quantum mechanics is the proper framework of the microscopic theories of superconductivity [5–10, 13, 15, 16].

In this framework, a complex-valued function of space and time is associated with a charged particle. This function, in fact the wave function of quantum mechanics, is governed by the Schrödinger equation. This equation is derived from a classical Hamiltonian formulation after a preliminary introduction of the canonical momenta of the particle and of the electromagnetic fields [5, 11, 15, 16]. Surprisingly, the Schrödinger equation addresses known equations of continuum mechanics for ideal compressible fluids, whose density and velocity fields are defined through the amplitude and phase of the wave function. Due to the presence of the electromagnetic fields, the velocity field of the introduced fluid turns out to depend on the vector potential in such a way that the flow is an irrotational flow in the absence of the magnetic induction. These results do not hold generally for large number of particles. However, in some circumstances and at very low temperatures, groups of charged particles like the aforementioned Cooper pairs behave as indistinguishable particles that are governed by the Bose–Einstein Statistical laws [6] and a single wave function can be associated with all of them, in accordance with the principles and rules of quantum mechanics [5–10, 13–16].

The Ginzburg–Landau (GL) theory of superconductors provides an equation that is essentially a nonlinear Schrödinger equation for this macroscopic wave function [10, 13–16]. The latter plays the role then of an order parameter in the phase transition from the normal to the superconductive state. The GL equation can be extended to the dynamical case and eventually leads to the aforementioned Eulerian compressible fluid, which is now associated with the supercurrent [8, 13, 26]. In this context, the electromagnetic vector potential, which is usually regarded as an auxiliary field, acts as a primary field [1, 14]. This role of primary field persists in the macroscopic framework. In fact, the London's law for electric currents in superconductors, according to which the supercurrent is proportional to the vector potential, naturally emerges if the supercurrent is associated with the momentum of the fluid [13, 15–18].

In deformable bodies, the vector potential combines with the deformation gradient in such a way that the form invariance of the Maxwell equations is preserved in any configuration of the solid body [19–21, 23–27]. With reference to the flow of the aforementioned Eulerian fluid, the supercurrent in the undeformed configuration of the body turns out to be associated with the *material* or *configurational momentum* of the flow, rather than with its momentum in this configuration. This remark suggests a proper reformulation of London's equation in deformable superconductor.

2 Electric Currents

The persistency of the flow of an electric current in a superconductor in the absence of electric fields conflicts with the Ohm's law, but is consistent with the following law suggested by Becker et al. [4] and known as the acceleration equation for point wise charged particles:

$$\frac{\partial}{\partial t}\mathbf{j}_s = \frac{Nq^2}{m}\mathbf{E}. \tag{1}$$

$\mathbf{j}_s$ denotes the supercurrent density, $\mathbf{E}$ the electric field, N the number of particles involved, q and m the electric charge and the mass, respectively, of each particle. The reader is addressed to [26] for notation and further details. Consider the following equation:

$$\frac{\partial}{\partial t}(\Lambda \operatorname{curl}\mathbf{j}_s + \mathbf{B}) = 0. \tag{2}$$

where Λ is a phenomenological quantity and $\mathbf{B}$ the magnetic induction, which satisfy the Maxwell equation

$$\operatorname{div}\mathbf{B} = 0, \tag{3}$$

This equation introduces the vector potential through the equation

$$\operatorname{curl}\mathbf{A} = \mathbf{B}. \tag{4}$$

Equations (1) and (2) are similar the one another, if we assume that $Nq^2/m \equiv \Lambda^{-1}$. This can be readily checked through the curl of Equation (1) and by exploiting the Maxwell equation

$$\operatorname{curl}\mathbf{E} + \frac{\partial}{\partial t}\mathbf{B} = 0. \tag{5}$$

However, there is a remarkable difference between (1) and (2), which is pointed out hereafter. Integration in time of Equation (2) entails the equation

$$\Lambda \operatorname{curl}\mathbf{j}_s + \mathbf{B} = \mathbf{c}. \tag{6}$$

where $\mathbf{c}$ is a vector-valued constant in time. London's conjecture can be summarized in assuming $\mathbf{c} = \mathbf{0}$, whence the two London laws follow straightforwardly

$$\Lambda\,\mathbf{j}_s = -\mathbf{A} + \nabla g, \tag{7}$$

$$\Lambda\frac{\partial}{\partial t}\,\mathbf{J}_s = -\mathbf{E} + \nabla f, \tag{8}$$

g and f represent arbitrary functions of the position $\mathbf{x}$ and time. $\mathbf{x} \in V \equiv V\{\text{simply connected open set}\} \subset E_3 \equiv \{\text{Euclidean space}\}$, $t \in \mathbb{R}$

A further London's conjecture is to identify f with the electromagnetic scalar potential φ. According to this conjecture, the right hand side of Equation (8) identically vanishes in the presence of electrostatic fields. As a result, a supercurrent cannot be accelerated by electrostatic fields, differently from what is predicted by the Becker's acceleration Equation (1) [18]. It will be shown in the next section that the identification $f \equiv \varphi$ is hinted by the wave solution of the Maxwell equations.

3 Maxwellian Fields and London's Conjectures

The Maxwell equations in S.I. units are found in classical textbooks [4, 11]. Here, we only recall the relationships between the electric field and the electromagnetic potentials φ and $\mathbf{A}$, which stem from Equations (3–5)

$$\mathbf{E} = -\nabla\varphi - \frac{\partial}{\partial t}\mathbf{A}. \tag{9}$$

The Maxwell equations can be uniquely written in terms of these potentials if additional conditions are introduced, the gauge conditions. Here, we refer to the classical Lorenz–Lorentz gauge condition [4, 11, 12]

$$\operatorname{div}\mathbf{A} - \frac{1}{c^2}\frac{\partial}{\partial t}\varphi = 0, \tag{10}$$

which addresses the uncoupled wave equations in a vacuum

$$\left(\Delta - \frac{1}{c^2}\frac{\partial}{\partial t}\right)\varphi = -\frac{1}{\varepsilon_0}\rho_e, \tag{11}$$

$$\left(\Delta - \frac{1}{c^2}\frac{\partial}{\partial t}\right)\mathbf{A} = -\mu_0\,\mathbf{j}. \tag{12}$$

The symbol Δ denotes the Laplace operator, ρ_e and $\mathbf{j}$ are the electric charge and the electric current, respectively, ε_0 and μ_0 the electric permittivity and the magnetic permeability of a vacuum. $c^{-2} = \varepsilon_0\mu_0$.

Analogous equations hold true for the Maxwellian fields [4, 11]. For a point wise moving electric charge, whose velocity is v, Equation (12) reads

$$\left(\Delta - \frac{1}{c^2}\frac{\partial}{\partial t}\right)\mathbf{A} = -\mu_0\rho_e\,\mathbf{v}, \tag{13}$$

where ρ_e denotes here the charge density (in distributional sense) of the charged particle of interest. Equations (11) and (12) entail the solution

$$\mathbf{A} = c^{-2}\varphi\mathbf{v}. \tag{14}$$

In this context, the following London further assumption for supercharges is enlightening:

$$\Lambda \rho_s = -c^{-2} \varphi, \tag{15}$$

ρ_s being the supercharge density. This assumption provides a link between the vector potential and the momentum of the particle, through Equation (14). Permittivity and permeability of the superconductor are assumed here to coincide with those of a vacuum. In accordance with these assumptions, the following decoupled equations for φ, $\mathbf{A}$, $\mathbf{E}$, ρ_s, $\mathbf{j}_s$, and for the magnetic field $\mathbf{H} \equiv (\mu_0^{-1})\mathbf{B}$ stem from Equations (11) and (13):

$$\left(\Delta - \frac{1}{c^2} \frac{\partial}{\partial t} - \frac{1}{\lambda^2} \right) (\varphi, \mathbf{A}, \mathbf{E}, \mathbf{H}, \rho_s, \mathbf{j}_s) = 0. \tag{16}$$

$\lambda \equiv (\mu_0 \Lambda)^{1/2}$ represents London's penetration depth.

It is worth recalling that in the mixed regime of coexistence of normal charges ρ_n and currents $\mathbf{j}_n = \mathbf{j} - \mathbf{j}_s$ with supercharges and supercurrents, from the Maxwell equations a conservation law can be derived only for the total charge $\rho = \rho_n + \rho_s$ [17]. A conservation law for supercharges is in principle an additional assumption, which can be stated as

$$\frac{\partial}{\partial t} \rho_s + \operatorname{div} \mathbf{j}_s = 0. \tag{17}$$

Eventually, note that in the stationary case, Equation (16) for $\mathbf{H}$ reduces to

$$\left(\Delta - \frac{1}{\lambda^2} \right) \mathbf{H} = 0. \tag{18}$$

This equation accounts for the celebrated Meissner–Ochsenfeld effect [6, 13, 15, 18].

4 Time Dependent Ginzburg–Landau Equation: Variational Approach

According to Ginzburg and Landau's view, a second order thermodynamical phase transition drives a material from the normal to the superconductive state. In this respect, these authors introduce a complex-valued order parameter $\psi(\mathbf{x})$, which governs the phenomenon around the zero transition point, and a free energy density that depends on ψ and on its spatial gradient [2, 6–10, 15, 19, 25–27]. Here, this description is extended to the dynamical case, in which dissipation is ignored. In this dynamical framework, a Lagrangian density replaces the free energy density and this Lagrangian is assumed to depend on $\psi(\mathbf{x}, t)$ and on its space and time derivatives as follows:

$$L_{em} = \frac{\gamma^2}{2} (\nabla_A \psi) \circ (\overline{\nabla_A \psi})$$

$$+\frac{\gamma}{2i}\left[\psi\left(\overline{\frac{\partial\psi}{\partial t}}\right)_{\varphi}-\left(\frac{\partial\psi}{\partial t}\right)_{\varphi}\bar{\psi}\right]+\hat{F}[(\psi,\bar{\psi},(\text{curl }\mathbf{A})], \tag{19}$$

where the over bar denotes the complex conjugate of the quantity of interest. Specifically,

$$(\nabla_A\psi)\equiv\left(\nabla-\iota\,\frac{\delta}{\gamma}\mathbf{A}\right)\bar{\psi}, \tag{20}$$

$$(\overline{\nabla_A\psi})\equiv\left(\nabla+\iota\,\frac{\delta}{\gamma}\mathbf{A}\right)\bar{\psi}, \tag{21}$$

$$\left(\frac{\partial\psi}{\partial t}\right)_{\varphi}\equiv\left(\frac{\partial}{\partial t}+\iota\,\frac{\delta}{\gamma}\varphi\right)\psi, \tag{22}$$

and

$$\left(\overline{\frac{\partial\psi}{\partial t}}\right)_{\varphi}\equiv\left(\frac{\partial}{\partial t}-\iota\,\frac{\delta}{\gamma}\varphi\right)\bar{\psi}. \tag{23}$$

ι denotes the imaginary unit. δ and γ are phenomenological constants. If the dependence of $\hat{F}$ on (curl $\mathbf{A}$) is quadratic, such as assumed in the GL theory, the Euler–Lagrange (EL) equations associated with the Lagrangian (19) are

$$\frac{1}{\mu_0}\,\text{curl}(\text{curl }\mathbf{A})=\frac{\partial\hat{F}}{\partial\mathbf{A}}\equiv\frac{\delta\gamma}{2\iota}(\bar{\psi}\nabla\psi-\psi\nabla\bar{\psi})-(\psi\bar{\psi})\mathbf{A}\delta^2, \tag{24}$$

$$\frac{\gamma^2}{2}(\nabla_A)\circ[(\nabla_A\psi)]-\frac{\partial\hat{F}}{\partial\bar{\psi}}=\frac{\gamma}{\iota}\left(\frac{\partial\psi}{\partial t}\right)_{\varphi}. \tag{25}$$

Note that Equation (24) corresponds to one of the Maxewll equations. Henceforth, its right hand side represents an electric current, in fact the supercurrent

$$j_s=\frac{\delta\gamma}{2\iota}\,(\bar{\psi}\nabla\psi-\psi\nabla\bar{\psi})-(\psi\bar{\psi})\mathbf{A}\delta^2 \tag{26}$$

upon which we will comment in the next sections. Equation (25) represents a time dependent version of the GL equation if $\hat{F}$ is replaced by its polynomial expansion

$$\hat{F}\approx-c(\psi\bar{\psi})^2+\frac{d}{2}(\psi\bar{\psi})^4+\frac{1}{2\mu_0}(\text{curl }\mathbf{A})^2 \tag{27}$$

and if the coefficients c and d are understood as depending on the temperature and possibly on the infinitesimal strain [2]. The extended GL Equation (25) can be also viewed as a nonlinear Schrödinger equation [8, 15, 26].

5 Euler Equations for Compressible Fluids: Momentum of the Supercurrent

Look for a solution of the following form for Equation (25):

$$\psi = \rho^{1/2}(\mathbf{x}, t) \exp[(\iota/\gamma)\theta(\mathbf{x}, t)] \tag{28}$$

so that $\rho = (\psi\bar{\psi})$ represents the mass density of the supercurrent. Substitution of ψ into Equation (25) leads to the following equations, which are the real and imaginary part of this equation, respectively:

$$\frac{\partial \rho}{\partial t} + \operatorname{div}(\rho \boldsymbol{v}) = 0, \tag{29}$$

$$\frac{\partial \boldsymbol{v}}{\partial t} + (\nabla \boldsymbol{v})\boldsymbol{v} = -\left(\frac{\gamma}{2}\right)^2 (\nabla \Sigma) + \delta(\mathbf{E} + \boldsymbol{v} x \mathbf{B}), \tag{30}$$

where

$$\boldsymbol{v} \equiv (\nabla\theta - \mathbf{A}\delta) \tag{31}$$

and

$$\Sigma \equiv \frac{1}{2}\left(\rho\nabla\left(\frac{1}{\rho}\right)\right)^2 + \operatorname{div}\left(\rho\nabla\left(\frac{1}{\rho}\right)\right) + (c - d\rho). \tag{32}$$

Details for the derivations of these equations are found in [26]. One can easily recognise in Equations (29) and (30) the Euler equations of an ideal compressible fluid, whose density is ρ and the velocity $\boldsymbol{v}$. Note that the quantity Σ plays the role of the enthalpy.

It is worth remarking that Equations (29) and (30) are coupled with the Maxwell equations. Specifically, Equation (31) leads to

$$\operatorname{curl} \boldsymbol{v} = \mathbf{B}. \tag{33}$$

The latter shows in evidence that the flow is an irrotational flow only in the absence of electromagnetic fields, in which case θ plays the role of a velocity potential.

Substitution of expression (28) into Equation (26) leads to the following explicit formula for the supercurrent:

$$\mathbf{j}_s = \delta\rho((\nabla\theta - \delta\mathbf{A}). \tag{34}$$

Equation (34), which is reminding of London's law such as expressed by Equation (7), establishes a link between the supercurrent with the momentum density of the fluid

$$\mathbf{p} \equiv \rho\boldsymbol{v} = \rho(\nabla\theta - \delta\mathbf{A}). \tag{35}$$

All this reasoning is expounded for a rigid superconductor at rest.

6 Material (or Configurational) Momentum of the Supercurrent

Assume now that the superconducting material is a deformable, possibly moving solid. In this case, it is suitable to write all equations and fields of interest (the mechanical and the electromagnetic ones) in the undeformed reference configuration of the material, which is denoted hereafter by $\mathrm{V_R}$. The related current configuration is denoted by V. If $\mathbf{X} \in \mathrm{V_R}$ and $\mathbf{x} \in V, \mathbf{x} = \chi(\mathbf{X}, t)$ is the motion of a material point. As usual, $\mathbf{F}$ denotes the deformation gradient, $\mathbf{F}^T$ its transpose, $J = (\det \mathbf{F}) > 0, \mathbf{v} \equiv \dot{\mathbf{x}}$ the velocity of the material point. $\mathbb{B} = \mathbf{F}\mathbf{F}^T$ is the left Cauchy–Green deformation tensor [22, 28]. The electromagnetic potentials in $\mathrm{V_R}$ are

$$\mathbf{A}_R(\mathbf{X}, t) = \mathbf{F}^T\ \mathbf{A}(\mathbf{x}(\mathbf{X}, t), t) \tag{36}$$

and

$$\Phi(\mathbf{X}, t) = \varphi(\mathbf{x}(\mathbf{X}, t), t) - \mathbf{v} \circ \mathbf{A}. \tag{37}$$

These potentials satisfy the following equations that correspond to Equations (4) and (9) properly written in the reference configuration:

$$\mathbf{B}_R = \mathrm{Curl}\ \mathbf{A}_R, \tag{38}$$

$$\mathbf{E}_R = -\dot{\mathbf{A}}_R - \nabla_R \Phi. \tag{39}$$

Curl and (∇_R) are the differential curl and gradient operator in $\mathrm{V_R}$. The superposed dot denotes the material (or Lagrangian) time derivative.

Introduce the order parameter $\Psi(\mathbf{X}, t)$ in the undeformed configuration and assume that

$$\Psi = \hat{\psi}\ J^{1/2}, \tag{40}$$

where $\hat{\psi} = \hat{\psi}(\mathbf{x}(\mathbf{X}, t), t)$. In the following, the symbol ψ will denote both $\hat{\psi}$ and ψ by abusing notation. In analogy with Equation (28), we can express Ψ in polar form as

$$\Psi(\mathbf{X}, t) = \rho_0^{1/2}(\mathbf{X}, t) \exp\ [(\iota/\gamma)\Theta(\mathbf{X}, t)]. \tag{41}$$

According to these assumptions, the quantity $\Psi\bar{\Psi} = \rho_0 = J\rho$ possibly represents the density of the supercharges in the reference configuration.

Taking into account Equation (36) and recalling that $\nabla_R = \mathbf{F}^T\nabla$, Equation (26) can be written in terms of Ψ as

$$\mathbf{j}_s = \frac{\delta\gamma}{2\iota} J^{-1/2}\, \mathbf{F}^{-T} \left[\bar{\Psi}\nabla_R \left(\Psi J^{-1/2}\right) - \Psi\nabla_R \left(\bar{\Psi} J^{-1/2}\right) - \mathbf{A}_R\ \Psi\bar{\Psi}\delta^2\right], \tag{42}$$

which simplifies as follows:

$$\mathbf{j}_s = \frac{\delta\gamma}{2\iota} J^{-1}\, \mathbf{F}^{-T} \left[\bar{\Psi}\nabla_R\Psi - \Psi\nabla_R\,\bar{\Psi} - \mathbf{A}_R\ \Psi\bar{\Psi}\delta^2\right]. \tag{43}$$

Substitution of Equation (41) into Equation (43) provides the equation

$$J\mathbf{F}^T\mathbf{j}_s = \delta\rho_0(\nabla_R\Theta - \mathbf{A}_R\delta). \tag{44}$$

As $\Theta(\mathbf{X}, t) = \theta[\mathbf{x}(\mathbf{X}, t), t)$ and $\nabla_R\Theta = \mathbf{F}^T(\nabla\theta)$, Equation (44) also reads

$$J\mathbf{F}^T\mathbf{j}_s = \delta\rho_0\,\mathbf{F}^T(\nabla\theta - \delta\mathbf{A}) \equiv \delta\rho_0\,\mathbf{F}^T\nu, \tag{45}$$

where $\rho(\equiv J^{-1}\rho_0)$ and $\boldsymbol{\nu}$ satisfy Equations (29–31), as $\psi(\mathbf{x}, t)$ satisfies (24) and (25). These results encourage to identifying the supercurrent in the undeformed configuration as

$$\mathbf{J}_s \equiv \frac{\delta\gamma}{2\iota}\left[\bar{\Psi}\nabla_R\Psi - \Psi\nabla_R\bar{\Psi} - \mathbf{A}_R\Psi\bar{\Psi}\delta^2\right] = \delta\rho_0(\nabla_R\Theta - \delta\mathbf{A}_R). \tag{46}$$

Accordingly, as $\mathbf{J}_s \equiv J\mathbf{F}^T\mathbf{j}_s$, the following *material momentum* of configurational mechanics is associated with $\mathbf{J}_s$ in a natural way:

$$\mathbf{p}_R = J\mathbf{F}^T\mathbf{p} \equiv J\mathbf{F}^T\rho\boldsymbol{\nu} = \rho_0(\nabla_R\theta - \mathbf{A}_R\delta). \tag{47}$$

Equation (47) shows in evidence that the quantity $(\nabla_R\Theta - \mathbf{A}_R\delta)$, differently from the quantity $(\nabla\theta - \delta\mathbf{A}) \equiv \boldsymbol{\nu}$, cannot be regarded as a velocity field of an Eulerian fluid. In addition, if we assume that in a deformable and moving superconductor London's law (7) holds true in the undeformed configuration, then the supercurrent in the actual configuration is

$$\Lambda\mathbf{j}_s = J^{-1}\mathbb{B}\left(-\mathbf{A} + \nabla\hat{G}\right). \tag{48}$$

This formula is based on the general transformation rule for the electric current in the reference configuration of the body

$$\mathbf{J} \equiv J\mathbf{F}^T(j + \rho_e\mathbf{v}),$$

which satisfies the related Maxwell equation in this configuration [21–23, 25–29].

7 Concluding Remarks

London's relation between the supercurrent and the electromagnetic vector potential survives in a deformable superconductor. Specifically, for a given deformation the supercurrent is linearly related with the vector potential through the left Cauchy–Green deformation tensor.

It is also worth remarking that, in an extended time-dependent GL theory for superconductors, the vector potential plays the role of a velocity field of an Euler compressible fluid. This occurrence allows attributing a momentum and a material momentum density to the supercurrent. The latter result suggests that possible interactions of the supercurrent with electromagnetic waves or light should occur through their material momentum or *pseudomomentum* ($\mathbf{P} \times \mathbf{B}$), rather that through the momentum ($\varepsilon_0 \mathbf{E} \times \mathbf{B}$) [20, 23, 24].

References

1. Aharonov, Y. and Bohm, D., *Phys. Rev.*, **115**(3), 1959, 485.
2. Angilella, G.C.N., Balestrino, G., Cermelli, P., Podio-Guidugli, P. and Varlamov, A.A., *Eur. Phys. J. B* **26**, 2002, 67.
3. Bardeen, J., Cooper, L.N. and Schrieffer, J.R., *Phys. Rev.* **108**, 1957, 1175.
4. Becker, R., Heller, G. and Sauter, F., *Z. Phys.* **85**, 1933, 772.
5. Becker, R. and Sauter, F., *Electromagnetic Interactions Vol. 2*, Blackie & Sons, 1964.
6. Blatt, J.M., *Theory of Superconductivity*, Academic Press, 1964.
7. Bogoliubov, N.N., *The Theory of Superconductivity*, Gordon & Breach, 1962.
8. Fröhlich, H., *Proc. Phys. Soc.* **87**, 1966, 330.
9. Giaquinta, G. and Mancini, N.A., *Riv. Nuovo Cimento* **1**(9), 1978, 1.
10. Ginzburg, V.L. and Landau, L.D., *Zh. Eksper. Teor. Fiz.* **20**, 1950, 1064.
11. Jackson, J.D., *Classical Electrodynamics*, J. Wiley & Sons, 1962.
12. Jackson, J.D., *Amer. J. Phys.* **70**(9), 2002, 917.
13. Jones, W. and March, N.H., *Theoretical Solid State Physics, Vol. 2*, Dover Publishers, 1985.
14. Josephson, B.D., *Rev. Modern Phys.* **36**, 1964, 216.
15. Kittel, C., *Introduction to Solid State Physics*, John Wiley & Sons, 1986.
16. Landau, L.V. and Lifschitz, E., *Méchanique Quantique*, Mir, 1974.
17. von Laue, M., *Theory of Superconductivity*, Academic Press, 1952.
18. London, F. and London, H., *Proc. Roy. Soc. A* **149**, 1935, 71.
19. Maugin, G.A., *C.R. Acad. Sci. Paris, Sér. II* **314**, 1992, 889.
20. Maugin, G.A., *Material Inhomogeneities in Elasticity*, Chapman & Hall, 1993.
21. Nelson, D.F., *Electric, Optic and Acoustic Interactions in Dielectrics*, John Wiley, 1979.
22. Ogden, R.W., *Non-Linear Elastic Deformations*, J. Wiley, 1984.
23. Trimarco, C., *Tech. Mech.* **22**(3), 2001, 175.
24. Trimarco, C., *Proc. Estonian Acad. Phys. Math.* **56**(2), 2007, 116.
25. Trimarco, C., *Int. J. Fracture* **147**, 2007, 13.
26. Trimarco, C., *Il Nuovo Cimento C* **32**, 2009, 1.
27. Trimarco, C. and Maugin, G.A., in *Configurational Mechanics of Materials*, R. Kienzler and G.A. Maugin (Eds.), Springer-Verlag, 2001, pp. 129–172.
28. Trusdell, C.A. and Noll, W., *Handbuch der Physik, Bd. III/3*, Springer-Verlag, 1965.

Dislocations, Microforce and Micromomentum in Second Order Finite Elasto-Plasticity

Sanda Cleja-Ţigoiu

Abstract The paper deals with thermodynamic restrictions for physical force, microforce and micromomentum, compatible with the imbalanced free energy condition, written for isothermal processes, within the constitutive framework of elasto-plastic materials with continuously distributed dislocations, which are mathematically modeled by the existence of the plastic connection with non-zero torsion and non-zero curvature.

1 Introduction

The paper deals with macroscopic constitutive description for elasto-plastic materials with continuously distributed dislocations, involving physical quantities such as *disclination* and *couple forces*. There is an extensive literature devoted to the theories of continuous distributions of dislocations, proposed in different models [2,6,8,10]. The exposure of coordinate-free differential geometry concepts, like connection, torsion, curvature, appropriate for the material uniformity is presented by Noll in [12].

In our description three configurations will be considered:

- k a fixed, initial (reference) configuration of the body $\mathscr{B}$,
- $\chi(\cdot, t)$ the actual configuration at time t, related to the motion function χ of the continuous body,
- $\mathcal{K}_t$ an anholonomic configuration, the so called *configuration with torsion*, defined through the second order plastic deformation $(\mathbf{F}^p, \overset{(p)}{\boldsymbol{\Gamma}}_k)$, *the plastic distortion* and *plastic connection* with non-zero torsion. Here $\mathbf{F}^p$ is an invertible second order tensor, while $\overset{(p)}{\boldsymbol{\Gamma}}_k$ is a third order field.

Sanda Cleja-Ţigoiu
University of Bucharest, Faculty of Mathematics and Computer Science, 14 Academiei, 010014 Bucharest, Romania; e-mail: tigoiu@fmi.unibuc.ro

P. Steinmann (ed.), IUTAM Symposium on Progress in the Theory and Numerics of Configurational Mechanics, 83–93.

In the proposed framework the presence of continuously distributed dislocations are described either by

(a) a plastic connection with non-zero torsion, but with zero curvature (a Bilby's connection in [1]), i.e. the non-zero Burgers vector exists (see for instance in Teodosiu [16]), since the first order compatibility condition is not satisfied, or by
(b) a plastic connection with non-zero torsion, as well as with non-zero curvature (a Kondo's connection, see [9]), which means also the presence of the *disclination*.

The kinematics of the elasto-plastic processes involve non-local effects and rise from the decomposition rule of the second order pair $(\mathbf{F}, \boldsymbol{\Gamma})$, defined for any motion by $\mathbf{F} = \nabla\chi$ and $\boldsymbol{\Gamma} = \mathbf{F}^{-1}\nabla\mathbf{F}$. $\boldsymbol{\Gamma}$ has zero torsion and could be considered to be a measure of the gradient of the deformation gradient, $\nabla\mathbf{F}$.

The decomposition rule

$$(\mathbf{F}, \boldsymbol{\Gamma}) := (\mathbf{F}^e, \overset{(e)}{\boldsymbol{\Gamma}}_{\mathcal{K}}) \circ (\mathbf{F}^p, \overset{(p)}{\boldsymbol{\Gamma}}_k), \quad \Longleftrightarrow$$

$$\mathbf{F} = \mathbf{F}^e\mathbf{F}^p, \quad \boldsymbol{\Gamma} = (\mathbf{F}^p)^{-1} \overset{(e)}{\boldsymbol{\Gamma}}_{\mathcal{K}} [\mathbf{F}^p, \mathbf{F}^p] + \overset{(p)}{\boldsymbol{\Gamma}}_k, \tag{1}$$

means multiplicative decomposition of the deformation gradient into elastic and plastic distortions, and the relationship between connections (see also Cross [4]). Here the following notation for a third order field generated by a connection, say $\boldsymbol{\Gamma}$, and by second order tensors, for instance $\mathbf{F}_1, \mathbf{F}_2$,

$$(\boldsymbol{\Gamma}[\mathbf{F}_1, \mathbf{F}_2]\mathbf{u})\mathbf{v} = (\boldsymbol{\Gamma}(\mathbf{F}_1\mathbf{u}))\,\mathbf{F}_2\mathbf{v}, \quad \forall \mathbf{u}, \mathbf{v} \in \mathcal{V}, \tag{2}$$

has been introduced for vectors $\mathbf{u}$, $\mathbf{v}$.

In our notation $\boldsymbol{\Gamma}$ represents the matrix of the coefficients in a coordinate system for the connection. We make the difference between $\nabla_{\boldsymbol{\Gamma}}\mathbf{F}$, the covariant derivative of $\mathbf{F}$, say $\mathbf{F} = F^i_j \mathbf{i}_i \otimes \mathbf{e}^j$ $(\{\mathbf{i}_j\}$ is a Cartesian basis), calculated with respect to the *connection* $\boldsymbol{\Gamma}$ *with metric property*, instead of the Chistoffel symbol used in the Riemannian geometry,

$$\nabla_{\boldsymbol{\Gamma}}\mathbf{F} = \left(\frac{\partial F^i_j}{\partial x^k} + F^i_m \Gamma^m_{jk}\right) \mathbf{i}_i \otimes \mathbf{e}^j \otimes \mathbf{e}^k, \tag{3}$$

and the derivative of $\mathbf{F}$ in a coordinate system x^i, in the direction $\mathbf{u}$

$$(\nabla\mathbf{F})\mathbf{u} = \frac{\partial F^i_j}{\partial x^k} u^k \mathbf{i}_i \otimes \mathbf{e}^j. \tag{4}$$

Here the rule of derivation with respect to the configuration with torsion is given, for a field $\mathbf{F}$, by

$$\nabla_{\mathcal{K}}\mathbf{F} = \nabla\mathbf{F}(\mathbf{F}^p)^{-1}. \tag{5}$$

2 Plastic Connection

First a new representation theorem for the plastic connection with metric property is proved, certain incompatibility tensors being involved.

Let us remark that two geometrical structures have been introduced on the anholonomic configuration $\mathcal{K}$

$$
\begin{aligned}
&((\mathbf{F}^p)^{-1}, \overset{(p)}{\boldsymbol{\Gamma}}_{\mathcal{K}}), \quad && \text{with a metric tensor} \quad \mathbf{c}^p := (\mathbf{F}^p)^{-T}(\mathbf{F}^p)^{-1} \\
&(\mathbf{F}^e, \overset{(e)}{\boldsymbol{\Gamma}}_{\mathcal{K}}), && \text{with a metric tensor} \quad \mathbf{C}^e := (\mathbf{F}^e)^T(\mathbf{F}^e).
\end{aligned} \tag{6}
$$

The connections $\overset{(p)}{\boldsymbol{\Gamma}}_{\mathcal{K}}$ and $\overset{(p)}{\boldsymbol{\Gamma}}_{k}$ are related through

$$
\overset{(p)}{\boldsymbol{\Gamma}}_{\mathcal{K}} := -\mathbf{F}^p\, \overset{(p)}{\boldsymbol{\Gamma}}_{k}\, [(\mathbf{F}^p)^{-1}, (\mathbf{F}^p)^{-1}]. \tag{7}
$$

Ax. The connection $\overset{(p)}{\boldsymbol{\Gamma}}_{\mathcal{K}}$ *has metric property* (see Schouten [13]), with respect to the metric tensor $\mathbf{c}^p$, i.e. $(\nabla_{\mathcal{K}}\, \mathbf{c}^p)\mathbf{u} = (\overset{(p)}{\boldsymbol{\Gamma}}_{\mathcal{K}}\, \mathbf{u})^T \mathbf{c}^p + \mathbf{c}^p(\overset{(p)}{\boldsymbol{\Gamma}}_{\mathcal{K}}\, \mathbf{u}),\ \forall \mathbf{u}$.

Theorem 1. *Under the hypotheses: plastic connection has metric property and the composition rule for second order elastic and plastic deformations,*

(1) The plastic and elastic connections are represented through

$$
\begin{aligned}
\overset{(p)}{\boldsymbol{\Gamma}}_{\mathcal{K}}\, \tilde{\mathbf{u}} &= \mathbf{F}^p(\nabla_{\mathcal{K}}(\mathbf{F}^p)^{-1})\tilde{\mathbf{u}} + (\mathbf{c}^p)^{-1}(\boldsymbol{\Lambda}\tilde{\mathbf{u}} \times \mathbf{I}) \\
\overset{(e)}{\boldsymbol{\Gamma}}_{\mathcal{K}}\, \tilde{\mathbf{u}} &= (\mathbf{F}^e)^{-1}(\nabla_{\mathcal{K}}(\mathbf{F}^e))\tilde{\mathbf{u}} + (\mathbf{c}^p)^{-1}(\boldsymbol{\Lambda}\tilde{\mathbf{u}} \times \mathbf{I}).
\end{aligned} \tag{8}
$$

with $\boldsymbol{\Lambda}$ *a second order tensor.*

(2) The non-metricity measure $\mathbf{Q}^e_{\mathcal{K}}$*, of the elastic connection is given by*

$$
\mathbf{Q}^e_{\mathcal{K}}\tilde{\mathbf{u}} = -\mathbf{C}^e(\mathbf{c}^p)^{-1}(\boldsymbol{\Lambda}\tilde{\mathbf{u}} \times \mathbf{I}) - [(\boldsymbol{\Lambda}\tilde{\mathbf{u}} \times \mathbf{I})]^T(\mathbf{c}^p)^{-1}\mathbf{C}^e, \tag{9}
$$

starting from the definition

$$
\mathbf{Q}^e_{\mathcal{K}}\mathbf{u} \equiv (\overset{(e)}{\boldsymbol{\Gamma}}_{\mathcal{K}}\, \mathbf{u})^T \mathbf{C}^e + \mathbf{C}^e(\overset{(e)}{\boldsymbol{\Gamma}}_{\mathcal{K}}\, \mathbf{u}) - (\nabla_{\mathcal{K}}\, \mathbf{C}^e)\mathbf{u}.
$$

(3) The second order torsion tensor $\mathcal{N}^p_{\mathcal{K}}$ *(*$\equiv \mathcal{N}^e_{\mathcal{K}}$ *) is expressed by*

$$
\mathcal{N}^p_{\mathcal{K}} = \mathbf{F}^p curl_{\mathcal{K}}(\mathbf{F}^p)^{-1} + (\mathbf{c}^p)^{-1}\big(tr\, \boldsymbol{\Lambda}\mathbf{I} - (\boldsymbol{\Lambda})^T\big), \tag{10}
$$

with Cartan torsion $\mathbf{S}_{\mathcal{K}}$ *and* $\mathcal{N}^p_{\mathcal{K}}$ *related by* $(\mathbf{S}^p_{\mathcal{K}}\tilde{\mathbf{u}})\tilde{\mathbf{v}} = \mathcal{N}^p_{\mathcal{K}}(\tilde{\mathbf{u}} \times \tilde{\mathbf{v}})$.

We use the following notations

$$\overset{(p)}{\mathcal{A}}_{\mathcal{K}}\,\tilde{\mathbf{u}} := \mathbf{F}^p(\nabla_{\mathcal{K}}(\mathbf{F}^p)^{-1})\tilde{\mathbf{u}}, \quad \overset{(e)}{\mathcal{A}}_{\mathcal{K}}\,\tilde{\mathbf{u}} := (\mathbf{F}^e)^{-1}(\nabla_{\mathcal{K}}(\mathbf{F}^e))\tilde{\mathbf{u}}, \tag{11}$$

for the so called Bilby's connections.

Let a connection, say for the elastic connection $\overset{(e)}{\boldsymbol{\Gamma}}_{\mathcal{K}}$, be given. The *torsion* is defined as the skew-symmetric part of the connection

$$(\mathbf{S}^e_{\mathcal{K}}\mathbf{v})\mathbf{u} \equiv (\overset{(e)}{\boldsymbol{\Gamma}}_{\mathcal{K}}\,\mathbf{v})\mathbf{u} - (\overset{(e)}{\boldsymbol{\Gamma}}_{\mathcal{K}}\,\mathbf{u})\mathbf{v}. \tag{12}$$

Remarks 1. In the case of small plastic distortion, i.e. $\mathbf{F}^p = \mathbf{I} + \beta^p$, with $|\beta^p| \ll 1$, formula (10) is reduced to the expression proposed by Nye for the torsion-curvature tensor. $\boldsymbol{\Lambda}$ measures the influence of the non-zero curvature.

2. $\overset{(p)}{\boldsymbol{\Gamma}}_{\mathcal{K}}$ and $\overset{(p)}{\boldsymbol{\Gamma}}_{k}$, both of them have metric property if and only if $\mathbf{F}^p \in Orth$.

3. If $\overset{(p)}{\boldsymbol{\Gamma}}_{k}$ has zero Riemann curvature, then it has metric property. Thus all other appropriate configurations have metric properties.

As a consequence of the adopted definitions in (1), the kinematic relationships between the rates of the elastic and plastic distortions and the velocity gradient $\mathbf{L}$

$$\mathbf{L} = \mathbf{L}^e + \mathbf{F}^e\mathbf{L}^p(\mathbf{F}^e)^{-1}, \quad \mathbf{L} = \nabla\mathbf{v}$$

$$\mathbf{L}^e = \dot{\mathbf{F}}^e(\mathbf{F}^e)^{-1}, \quad \mathbf{L}^p = \dot{\mathbf{F}}^p(\mathbf{F}^p)^{-1}. \tag{13}$$

3 Balance Equations

Two types of forces are involved in the model, *physical forces* represented by on one hand by $\mathbf{T}$, non-symmetric Cauchy stress and $\boldsymbol{\mu}$, the stress momentum and on the other hand by *microforces* $\Upsilon^p_{\mathcal{K}}$ and *micromomenta* $\boldsymbol{\mu}^p_{\mathcal{K}}$.

We introduce the relationship between the stress momentum $\boldsymbol{\mu}$, written with respect to the actual configuration and those represented in the configuration with torsion, $\boldsymbol{\mu}_{\mathcal{K}}$,

$$\boldsymbol{\mu}_{\mathcal{K}} = (\det\mathbf{F}^e)(\mathbf{F}^e)^T\boldsymbol{\mu}[(\mathbf{F}^e)^{-T}, (\mathbf{F}^e)^{-T}]. \tag{14}$$

Based on the adopted kinematics and *power conjugated forces*, we introduce the following assumption:

Ax. The expression of the *internal power is postulated* to be given in terms of physical and micro forces and their appropriate conjugated rates

$$(\mathcal{P}_{int})_{\mathcal{K}} = \frac{1}{\rho}(\mathbf{T}+\mathbf{T}^*)\cdot\mathbf{L}^e + \frac{1}{\rho_{\mathcal{K}}}\boldsymbol{\Upsilon}^p_{\mathcal{K}}\cdot\mathbf{L}^p + \frac{1}{\rho_{\mathcal{K}}}\boldsymbol{\mu}^p_{\mathcal{K}}\cdot\nabla_{\mathcal{K}}\mathbf{L}^p +$$

$$+\frac{1}{\rho_{\mathcal{K}}}\boldsymbol{\mu}_{\mathcal{K}}\cdot\mathcal{L}_{\mathbf{L}^p}(\overset{(e)}{\mathcal{A}_{\mathcal{K}}}) + \frac{1}{\rho_{\mathcal{K}}}\boldsymbol{\mu}^\lambda_{\mathcal{K}}\cdot\left(\frac{d}{dt}(\boldsymbol{\Lambda})\times\mathbf{I}\right). \tag{15}$$

Here

$$\mathcal{L}_{\mathbf{L}^p}(\overset{(e)}{\mathcal{A}_{\mathcal{K}}}) = \frac{d}{dt}\overset{(e)}{\mathcal{A}_{\mathcal{K}}} - \mathbf{L}^p\overset{(e)}{\mathcal{A}_{\mathcal{K}}} + \overset{(e)}{\mathcal{A}_{\mathcal{K}}}[\mathbf{I},\mathbf{L}^p] + \overset{(e)}{\mathcal{A}_{\mathcal{K}}}\mathbf{L}^p =$$

$$= (\mathbf{F}^e)^{-1}\nabla_\chi\mathbf{L}[\mathbf{F}^e,\mathbf{F}^e] - \nabla_{\mathcal{K}}\mathbf{L}^p, \tag{16}$$

and represents the elastic contribution at the second order velocity gradient, while the rate of elastic distortion $\mathbf{L}^e$ is replaced from the kinematic relationship (13).

Based on the principle of the virtual power appropriate to finite elasto-plastic models, formulated by Cleja-Tigoiu in [3], the balance equations for macro and micro forces have been derived within the second order theory, when the plastic connection is flat, i.e. the Riemann curvature tensor is vanishing. The plastic connection expressed through the formula (8) becomes flat when $\boldsymbol{\Lambda} = 0$.

(1) The macro (local) balance equations

$$\rho\mathbf{a} = \operatorname{div}\mathbf{T} + \rho\mathbf{b}_f, \quad \text{in} \quad \chi(\mathcal{P},t) \equiv \mathcal{P}_t,$$

$$\mathbf{T}^* = \operatorname{div}\boldsymbol{\mu} + \rho\mathbf{B}_m, \tag{17}$$

with the boundary condition $\mathbf{Tn} = \mathbf{t}$, $\boldsymbol{\mu}\mathbf{n} = \mathbf{M}$ on $\partial\mathcal{P}_t$, where $\mathbf{n}$ is the unit normal vector. ρ and $\rho_{\mathcal{K}}$ are the mass densities in the actual configuration and in the configuration with torsion, respectively.

(2) The microbalance equations

$$\boldsymbol{\Upsilon}^p_{\mathcal{K}} = \operatorname{div}(\boldsymbol{\mu}^p_{\mathcal{K}}) + \rho_{\mathcal{K}}\mathbf{B}^p_m, \quad \text{in} \quad \mathcal{K}(\mathcal{P},t), \tag{18}$$

with $\boldsymbol{\mu}^p_{\mathcal{K}}\mathbf{n} = \mathbf{M}^p$ on $\partial\mathcal{K}(\mathcal{P},t)$, have to be satisfied by the forces, in any deformation process.

If the curvature tensor associated with the plastic connection is non-zero, then variation of the internal power is produced also by the non-zero rate of $\boldsymbol{\Lambda}$, as it has been postulated via the formula (15). Thus the microbalance equations have to be completed by

$$\operatorname{div}(\boldsymbol{\mu}^\lambda_{\mathcal{K}}) + \rho_{\mathcal{K}}\mathbf{B}^\lambda_m = 0, \quad \text{in} \quad \mathcal{K}(\mathcal{P},t), \tag{19}$$

$\boldsymbol{\mu}^\lambda_{\mathcal{K}}\mathbf{n} = \mathbf{M}^\lambda$ on $\partial\mathcal{K}(\mathcal{P},t)$.

Following Gurtin's idea in [7, 8], the second law of thermomechanics is formalized for any deformation process, through the *free energy imbalance*

$$(\mathcal{P}_{int})_{\mathcal{K}} - \dot{\psi}_{\mathcal{K}} \geq 0, \tag{20}$$

with respect to the configuration with torsion, where $\psi_{\mathcal{K}}$ is the free energy density.

4 Free Energy Imbalance. Thermodynamic Restrictions

The thermodynamic restrictions within the constitutive framework can be derived, starting from different forms postulated for the free energy density function. We account for the dislocation and disclination by considering *the free energy* function in $\mathcal{K}$ in the form

$$\psi = \psi_{\mathcal{K}}(\mathbf{C}^e, \tilde{\mathbf{S}}^e_{\mathcal{K}}, (\mathbf{F}^p)^{-1}, sym \overset{(p)}{\mathcal{A}}_{\mathcal{K}}, \mathbf{\Lambda} \times \mathbf{I}), \tag{21}$$

here $((\mathbf{\Lambda} \times \mathbf{I})\mathbf{u})\mathbf{v} := \mathbf{\Lambda u} \times \mathbf{v}$ and

$$(\tilde{\mathbf{S}}^e_{\mathcal{K}}\tilde{\mathbf{u}})\tilde{\mathbf{v}} = (\overset{(e)}{\mathcal{A}}_{\mathcal{K}}\ \tilde{\mathbf{u}})\tilde{\mathbf{v}} - (\overset{(e)}{\mathcal{A}}_{\mathcal{K}}\ \tilde{\mathbf{v}})\tilde{\mathbf{u}} = (\mathbf{F}^e)^{-1} curl_{\mathcal{K}} \mathbf{F}^e(\tilde{\mathbf{u}} \times \tilde{\mathbf{v}}),$$

$$\overset{(e)}{\mathcal{A}}_{\mathcal{K}} = \frac{1}{2}\tilde{\mathbf{S}}^e_{\mathcal{K}} + sym_R \overset{(e)}{\mathcal{A}}_{\mathcal{K}}, \tag{22}$$

with the last third order field defined by

$$(sym_R \overset{(e)}{\mathcal{A}}_{\mathcal{K}}\ \tilde{\mathbf{u}})\tilde{\mathbf{v}} := \frac{1}{2}\big((\overset{(e)}{\mathcal{A}}_{\mathcal{K}}\ \tilde{\mathbf{u}})\tilde{\mathbf{v}} + (\overset{(e)}{\mathcal{A}}_{\mathcal{K}}\ \tilde{\mathbf{v}})\tilde{\mathbf{u}}\big). \tag{23}$$

In our constitutive framework we consider the relationships

$$\mathbf{C}^e = (\mathbf{F}^p)^{-T}\mathbf{C}(\mathbf{F}^p)^{-1}, \quad \mathbf{C} = \mathbf{F}^T\mathbf{F},$$

$$\overset{(p)}{\mathcal{A}}_{\mathcal{K}} = -\mathbf{F}^p\ \overset{(p)}{\mathcal{A}}_k\ [(\mathbf{F}^p)^{-1}, (\mathbf{F}^p)^{-1}], \quad \overset{(e)}{\mathcal{A}}_{\mathcal{K}} = \mathbf{F}^p(\mathbf{\Gamma}_k - \overset{(p)}{\mathcal{A}}_k)[(\mathbf{F}^p)^{-1}, (\mathbf{F}^p)^{-1}],$$

$$\tilde{\mathbf{S}}^e_{\mathcal{K}} = \tilde{\mathbf{S}}^p_{\mathcal{K}} = -\mathbf{F}^p\tilde{\mathbf{S}}^p_k[(\mathbf{F}^p)^{-1}, (\mathbf{F}^p)^{-1}], \tag{24}$$

as well as

$$sym_R \overset{(p)}{\mathcal{A}}_{\mathcal{K}} = -\mathbf{F}^p(sym\ \overset{(p)}{\mathcal{A}}_k)[(\mathbf{F}^p)^{-1}, (\mathbf{F}^p)^{-1}], \tag{25}$$

that allow us to pass to the constitutive representation for the free energy density, written with respect to the reference configuration

$$\psi = \psi_{\mathcal{K}}(\mathbf{C}^e, \tilde{\mathbf{S}}^e_{\mathcal{K}}, (\mathbf{F}^p)^{-1}, sym \overset{(p)}{\mathcal{A}}_{\mathcal{K}}, \mathbf{\Lambda} \times \mathbf{I}),$$

$$= \bar{\psi}_k(\mathbf{C}, \tilde{\mathbf{S}}^p_k, \mathbf{F}^p, sym_R\ \overset{(p)}{\mathcal{A}}_k, \mathbf{\Lambda} \times \mathbf{I}), \tag{26}$$

Proposition 1. *The rate of stress free energy density is calculated from (21) as follows*

$$\dot{\psi}_{\mathcal{K}} = \partial_{\mathbf{C}^e}\psi_{\mathcal{K}} \cdot \dot{\mathbf{C}}^e + \partial_{\tilde{\mathbf{S}}^e_{\mathcal{K}}}\psi_{\mathcal{K}} \cdot \frac{d}{dt}\tilde{\mathbf{S}}^e_{\mathcal{K}} - (\mathbf{F}^p)^{-T}\partial_{(\mathbf{F}^p)^{-1}}\psi_{\mathcal{K}} \cdot \mathbf{L}^p +$$

$$+\,\partial_{sym_R \overset{(p)}{\mathcal{A}}_{\mathcal{K}}}\psi_{\mathcal{K}} \cdot \left(\frac{d}{dt} sym_R \overset{(p)}{\mathcal{A}}_{\mathcal{K}}\right) + \partial_{\mathbf{\Lambda}\times\mathbf{I}}\psi_{\mathcal{K}} \cdot (\dot{\mathbf{\Lambda}} \times \mathbf{I}), \tag{27}$$

in terms of rate of kinematic quantities, which have to be replaced by their appropriate expressions.

First of all we recall the formula

$$\frac{d}{dt}\overset{(p)}{\mathcal{A}}_{\mathcal{K}} = -\nabla_{\mathcal{K}}\mathbf{L}^p + \mathbf{L}^p\overset{(p)}{\mathcal{A}}_{\mathcal{K}} - \overset{(p)}{\mathcal{A}}_{\mathcal{K}}[\mathbf{I},\mathbf{L}^p] - \overset{(p)}{\mathcal{A}}_{\mathcal{K}}\mathbf{L}^p, \tag{28}$$

and thus the rate of $\tilde{\mathbf{S}}^e_{\mathcal{K}}$ can be derived

$$\frac{d}{dt}(\tilde{\mathbf{S}}^e_{\mathcal{K}}) = -2skw_R\{\nabla_{\mathcal{K}}\mathbf{L}^p\} + 2skw_R\{\mathbf{L}^p\overset{(p)}{\mathcal{A}}_{\mathcal{K}} - \overset{(p)}{\mathcal{A}}_{\mathcal{K}}[\mathbf{I},\mathbf{L}^p] - \overset{(p)}{\mathcal{A}}_{\mathcal{K}}\mathbf{L}^p\}. \tag{29}$$

The time derivative of the elastic strain, defined in (24) and which appears in (27), is replaced by

$$\dot{\mathbf{C}}^e = 2\,(\mathbf{F}^e)^T\{\mathbf{L}\}^s\mathbf{F}^e - 2\,\{\mathbf{C}^e\mathbf{L}^p\}^s, \tag{30}$$

in terms of the velocity gradient and the rate of plastic distortion.

Remark. When we eliminate the rate of the elastic distortion from (27) via (30), then only $\mathbf{L}$ and $\mathbf{L}^p$ and their appropriate differentials, $\nabla_\chi\mathbf{L}$ and $\nabla_{\mathcal{K}}\mathbf{L}^p$, enter the internal power and the rate of the free energy density.

Proposition 2. *The free energy imbalance yields for any virtual process, if the inequality written below*

$$(\mathcal{P}_{int})_{\mathcal{K}} - \dot{\psi}_{\mathcal{K}} = 2\partial_{\tilde{\mathbf{S}}^e_{\mathcal{K}}}\psi_{\mathcal{K}} \cdot skw_R\{\overset{(p)}{\mathcal{A}}_{\mathcal{K}}[\mathbf{I},\widetilde{\mathbf{L}}^p] + \overset{(p)}{\mathcal{A}}_{\mathcal{K}}\widetilde{\mathbf{L}}^p - \widetilde{\mathbf{L}}^p\overset{(p)}{\mathcal{A}}_{\mathcal{K}}\} +$$

$$+\left\{\frac{1}{\rho}(\mathbf{F}^e)^{-1}\{\mathbf{T}+\mathbf{T}^*\}^s(\mathbf{F}^e)^{-T} - 2\partial_{\mathbf{C}^e}\psi_{\mathcal{K}}\right\} \cdot [(\mathbf{F}^e)^T\{\widetilde{\mathbf{L}}\}^s\mathbf{F}^e - \{\mathbf{C}^e \cdot \widetilde{\mathbf{L}}^p\}^s] +$$

$$+\frac{1}{\rho}\{\mathbf{T}+\mathbf{T}^*\}^a \cdot \{\widetilde{\mathbf{L}} - \mathbf{F}^e\widetilde{\mathbf{L}}^p(\mathbf{F}^e)^{-1}\}^a + \left\{\frac{1}{\rho_{\mathcal{K}}}\mathbf{\Upsilon}^p_{\mathcal{K}} + (\mathbf{F}^p)^{-T}\partial_{(\mathbf{F}^p)^{-1}}\psi_{\mathcal{K}}\right\} \cdot \widetilde{\mathbf{L}}^p +$$

$$+\frac{1}{\rho_{\mathcal{K}}}\boldsymbol{\mu}_{\mathcal{K}} \cdot (\mathbf{F}^e)^{-1}(\nabla_\chi\widetilde{\mathbf{L}})[\mathbf{F}^e,\mathbf{F}^e] - \nabla_{\mathcal{K}}\widetilde{\mathbf{L}}^p] + \frac{1}{\rho_{\mathcal{K}}}\boldsymbol{\mu}^p_{\mathcal{K}} \cdot \nabla_{\mathcal{K}}\widetilde{\mathbf{L}}^p +$$

$$+\left(\frac{1}{\rho_{\mathcal{K}}}\boldsymbol{\mu}^\lambda_{\mathcal{K}} - \partial_{\mathbf{\Lambda}\times\mathbf{I}}\psi_{\mathcal{K}}\right) \cdot (\delta\mathbf{\Lambda}\times\mathbf{I}) + 2\partial_{\tilde{\mathbf{S}}^e_{\mathcal{K}}}\psi_{\mathcal{K}} \cdot skw_R\{\nabla_{\mathcal{K}}\tilde{\mathbf{L}}^p\} +$$

$$+\,\partial_{sym_R \overset{(p)}{\mathcal{A}}_{\mathcal{K}}} \psi_{\mathcal{K}} \cdot sym_R\{\nabla_{\mathcal{K}}\widetilde{\mathbf{L}}^p + \overset{(p)}{\mathcal{A}}_{\mathcal{K}}[\mathbf{I}, \widetilde{\mathbf{L}}^p] + \overset{(p)}{\mathcal{A}}_{\mathcal{K}}\widetilde{\mathbf{L}}^p - \widetilde{\mathbf{L}}^p \overset{(p)}{\mathcal{A}}_{\mathcal{K}}\} \geq 0 \tag{31}$$

holds for any $\widetilde{\mathbf{L}} \equiv \nabla_\chi \widetilde{\mathbf{v}}$, $\nabla_\chi \widetilde{\mathbf{L}}$, *and for given* $\widetilde{\mathbf{L}}^p$, $\nabla_{\mathcal{K}}\widetilde{\mathbf{L}}^p$, $\delta\boldsymbol{\Lambda}$ *restricted to certain admissibility conditions.*

The following notation, similar to (23), can be introduced for a third order field

$$(skw_R \overset{(e)}{\mathcal{A}}_{\mathcal{K}}\, \tilde{\mathbf{u}})\tilde{\mathbf{v}} := \frac{1}{2}\big((\overset{(e)}{\mathcal{A}}_{\mathcal{K}}\, \tilde{\mathbf{u}})\tilde{\mathbf{v}} - (\overset{(e)}{\mathcal{A}}_{\mathcal{K}}\, \tilde{\mathbf{v}})\tilde{\mathbf{u}}\big). \tag{32}$$

We consider that it is possible to have $\mathbf{L}^p$, $\nabla_{\mathcal{K}}\mathbf{L}^p$ and $\delta\boldsymbol{\Lambda}$ vanishing, while the virtual velocities $\widetilde{\mathbf{L}}$, $\nabla_\chi \widetilde{\mathbf{L}}$ are arbitrarily given, then from (31) it follows that:

Proposition 3. *Constitutive equations for macroforces (physical forces) are expressed through free energy density function* $\psi_{\mathcal{K}}$,

$$\frac{1}{\rho_{\mathcal{K}}}\Pi_{\mathcal{K}} = 2\,\partial_{\mathbf{C}^e}\psi_{\mathcal{K}}, \quad or \quad \{\mathbf{T}\}^s = 2\rho\mathbf{F}^e\,\partial_{\mathbf{C}^e}\psi_{\mathcal{K}}(\mathbf{F}^e)^T, \quad \{\mathbf{T}\}^a = -\mathbf{T}^*,$$

$$sym_R\left\{\frac{1}{\rho_{\mathcal{K}}}\boldsymbol{\mu}_{\mathcal{K}}\right\} = 0, \tag{33}$$

under the supposition that $\mathbf{T}^*$ *is a skewsymmetric tensor.*

Here the (symmetric) Piola–Kirchhoff stress tensor with respect to the configuration with torsion was introduced

$$\frac{1}{\rho_{\mathcal{K}}}\Pi_{\mathcal{K}} = \frac{1}{\rho}\mathbf{F}^e\mathbf{T}(\mathbf{F}^e)^T. \tag{34}$$

Proposition 4. *When we introduce (33) into (31), the dissipation inequality is derived, i.e.*

$$\left\{\frac{1}{\rho_{\mathcal{K}}}\boldsymbol{\Upsilon}^p_{\mathcal{K}} + (\mathbf{F}^p)^{-T}\partial_{(\mathbf{F}^p)^{-1}}\psi_{\mathcal{K}}\right\}\cdot\widetilde{\mathbf{L}}^p + \left(\frac{1}{\rho_{\mathcal{K}}}\boldsymbol{\mu}^\lambda_{\mathcal{K}} - \partial_{\boldsymbol{\Lambda}\times\mathbf{I}}\psi_{\mathcal{K}}\right)\cdot(\delta\boldsymbol{\Lambda}\times\mathbf{I})$$

$$+\left(\frac{1}{\rho_{\mathcal{K}}}\boldsymbol{\mu}^p_{\mathcal{K}} - skw_R\left\{\frac{1}{\rho_{\mathcal{K}}}\boldsymbol{\mu}_{\mathcal{K}}\right\}\right)\cdot\nabla_{\mathcal{K}}\widetilde{\mathbf{L}}^p + 2\partial_{\tilde{\mathbf{S}}^e_{\mathcal{K}}}\psi_{\mathcal{K}}\cdot skw_R\{\nabla_{\mathcal{K}}\widetilde{\mathbf{L}}^p\}+$$

$$+2\partial_{\tilde{\mathbf{S}}^e_{\mathcal{K}}}\psi_{\mathcal{K}}\cdot skw_R\{\overset{(p)}{\mathcal{A}}_{\mathcal{K}}[\mathbf{I}, \widetilde{\mathbf{L}}^p] + \overset{(p)}{\mathcal{A}}_{\mathcal{K}}\widetilde{\mathbf{L}}^p - \widetilde{\mathbf{L}}^p\overset{(p)}{\mathcal{A}}_{\mathcal{K}}\}+$$

$$+\partial_{sym_R \overset{(p)}{\mathcal{A}}_{\mathcal{K}}} \psi_{\mathcal{K}} \cdot sym_R\{\nabla_{\mathcal{K}}\widetilde{\mathbf{L}}^p + \overset{(p)}{\mathcal{A}}_{\mathcal{K}}[\mathbf{I}, \widetilde{\mathbf{L}}^p] + \overset{(p)}{\mathcal{A}}_{\mathcal{K}}\widetilde{\mathbf{L}}^p - \widetilde{\mathbf{L}}^p \overset{(p)}{\mathcal{A}}_{\mathcal{K}}\} \geq 0. \tag{35}$$

Now we pass to the constitutive equations written with respect to the reference configuration k. We use the relationships between the partial derivatives of the free energy expressed relative to the initial configuration k, and to the configuration with torsion $\mathcal{K}$, which are derived by applying the chain rule to (26).

$$\partial_{\mathbf{C}}\bar{\psi}_k = (\mathbf{F}^p)^{-1}\partial_{\mathbf{C}^e}\psi_{\mathcal{K}}(\mathbf{F}^p)^{-T}$$

$$\partial_{sym_R \overset{(p)}{\mathcal{A}}_k}\bar{\psi}_k = -(\mathbf{F}^p)^T\partial_{sym_R \overset{(p)}{\mathcal{A}}_{\mathcal{K}}}\psi_{\mathcal{K}}[(\mathbf{F}^p)^{-T},(\mathbf{F}^p)^{-T}],$$

$$\partial_{\overset{(e)}{\tilde{\mathbf{S}}}_k}\bar{\psi}_k = -(\mathbf{F}^p)^{-T}\partial_{\overset{(e)}{\tilde{\mathbf{S}}}_{\mathcal{K}}}\psi_{\mathcal{K}}[(\mathbf{F}^p)^T,(\mathbf{F}^p)^T]. \quad (36)$$

In order to avoid certain indeterminacy looking at the dissipation inequality (35), we assume that

Ax. The stress momentum is derived from the free energy function via the constitutive relation

$$\frac{1}{\rho_{\mathcal{K}}}\boldsymbol{\mu}_{\mathcal{K}} = 2\partial_{\overset{(e)}{\tilde{\mathbf{S}}}_{\mathcal{K}}}\psi_{\mathcal{K}}. \quad (37)$$

Obviously, (33) holds.

As a consequence of (37) together with (36) we get from (14)

$$\frac{\boldsymbol{\mu}}{\rho} = -2\,\mathbf{F}^{-T}\partial_{\overset{(e)}{\tilde{\mathbf{S}}}_k}\bar{\psi}_k[\mathbf{F}^T,\mathbf{F}^T]. \quad (38)$$

Similar, from (33) together with (36), the constitutive equation to determine the Cauchy stress in terms of the potential yields

$$\frac{1}{\rho_{\mathcal{K}}}\Pi_{\mathcal{K}} = 2\,\partial_{\mathbf{C}^e}\psi_{\mathcal{K}}, \quad or \quad \{\mathbf{T}\}^s = 2\mathbf{F}\,\partial_{\mathbf{C}}\bar{\psi}_k\mathbf{F}^T. \quad (39)$$

The dissipation inequality (35) suggests the adoption of the constitutive visco-plastic relationships for the microforces.

Ax. Constitutive equations for *microforces* and *micromomenta* contain *dissipative* and non-dissipative parts, the so-called *energetic microforces*,

$$\boldsymbol{\Upsilon}^p_{\mathcal{K}} = 2\rho_{\mathcal{K}}\,\mathbf{C}^e\partial_{\mathbf{C}^e}\psi_{\mathcal{K}} + \rho_{\mathcal{K}}\partial_{\mathbf{F}^p}\bar{\psi}_k(\mathbf{F}^p)^T + Y_f\mathbf{L}^p$$

$$\boldsymbol{\mu}^p_{\mathcal{K}} = \rho_{\mathcal{K}}\partial_{sym_L \overset{(p)}{\mathcal{A}}_{\mathcal{K}}}\psi_{\mathcal{K}} + Y_m\nabla_{\mathcal{K}}\mathbf{L}^p,$$

$$\boldsymbol{\mu}^{\lambda}_{\mathcal{K}} = \rho_{\mathcal{K}}\partial_{\boldsymbol{\Lambda}\times\mathbf{I}}\psi_{\mathcal{K}} + Y_{\lambda}(\dot{\boldsymbol{\Lambda}}\times\mathbf{I}). \quad (40)$$

In (40) the partial derivatives of the potential $\psi_{\mathcal{K}}$ have to be replaced by those corresponding to $\bar{\psi}_k$, via the formulae (36). To ensure the positivity of the dissipated power we add another constitutive restriction.

Ax. The scalar constitutive functions Y_f, Y_m, Y_{λ} are defined in such a way to be compatible with the dissipation inequality

$$Y_f \mathbf{L}^p \cdot \mathbf{L}^p + Y_m \nabla_{\mathcal{K}} \mathbf{L}^p \cdot \nabla_{\mathcal{K}} \mathbf{L}^p + Y_\lambda (\dot{\mathbf{\Lambda}} \times \mathbf{I}) \cdot (\dot{\mathbf{\Lambda}} \times \mathbf{I}) \geq 0. \tag{41}$$

5 Conclusions

1. A non-Cosserat kinematics has been developed, but the stresses and couple stresses are involved, apart from Forest et al. [6], Teodosiu [15], and so on.
2. Not only physical forces are presented in our model, but also microforces and micromomenta, which should be consistent with their microbalance equations, like in the theories developed by Gurtin in [8], and Stumpf and Hackl in [14]. Here we distinguish between the microforces and the material forces in the sense well defined by Maugin, see [11].
3. The kinematics of the elasto-plastic processes involve non-local effects and rise from the decomposition rule of the second order pair of deformations, which involve connection with non-zero torsion and non-zero curvature, apart from the models with connection without torsion but having non-zero curvature, see [5].

Acknowledgement

The author acknowledges support from the Romanian Ministry of Education and Research through CEEX programm (Contract No. 11-12/25.07.2006).

References

1. Bilby, B.A., Continuous distribution of dislocations. In: Sneddon, I.N., Hill, R. (Eds.), *Progress in Solid Mechanics, I*, North-Holland, Amsterdam, 1960, pp. 329–398.
2. Cleja-Ţigoiu, S., Couple stresses and non-Riemannian plastic connection in finite elasto-plasticity. *ZAMP* **53**, 2002, 996–1013.
3. Cleja-Ţigoiu, S., Material forces in finite elasto-plasticity with continuously distributed dislocation. *Int. J. Fracture* **147**, 2007, 67–81.
4. Cross, J.J., Mixtures of fluids and isotropic solids. *Arch. Mech.* **25**, 1973, 1025–1039.
5. Epstein, M. and Maugin, G.A., Thermomechanics of volumetric growth in uniform bodies. *Int. J. Plast.* **16**, 2000, 951–978.
6. Forest, S., Cailletaud, G. and Sievert, R., A Cosserat theory for elastoviscoplastic single crystals at finite deformation. *Arch. Mech.* **49**, 1997, 705–736 .
7. Gurtin, M.E., A gradient theory of single-crystal viscoplasticity that accounts for geometricallynecessary dislocations. *J. Mech. Phys. Solids* **50**, 2002, 5–32.
8. Gurtin, M.E., A gradient theory of small-deformation isotropic plasticity that accounts for the Burgers and for dissipation due to plastic spin. J. Mech. Phys. Solids **52**, 2004, 2545–2568.
9. Kondo, K. and Yuki, M., On the current viewpoints of non-Riemannian plasticity theory. In: *RAAG Memoirs of the Unifying Study of Basic Problems in Engineering and Physical Sciences by Means of Geometry, II (D)*, Tokyo, 1958, pp. 202–226.
10. Liebe, T., Menzel, A. and Steinmann, P., Theory and numerics of geometrically non-linear gardient plasticity. *Int. J. Engng. Sci.* **41**, 2003, 1603–1629.

11. Maugin, G.A., Material forces: Concepts and applications. *Appl. Mech. Rev.* **48**, 1995, 213–245.
12. Noll, W., Materially uniform simple bodies with inhomogeneities. *Arch. Rat. Mech. Anal.* **27**, 1967, 1–32.
13. Schouten, J.A., *Ricci-Calculus*. Springer-Verlag, Berlin, 1954.
14. Stumpf, H. and Hackl, K., Micromechanical concept for analysis of damage evolution in thermo-viscoplastic and quasi-brittle materials. *Int. J. Solids Struct.* **40**, 2003, 1567–1584.
15. Teodosiu, C., Contribution to a continuum theory of dislocations and initial stresses. I, II, III, *Rev. Roum. Sci. Techn.- Mec.Appl.* **12**, 1967, 961–977, 1061–1077, 1291–1308.
16. Teodosiu, C., A dynamic theory of dislocations and its applications to the theory of the elastic-plastic continuum. In: Simmons, J.A., de Witt, R., Bullough, R. (Eds.), *Fundamental Aspects of Dislocation Theory*, Nat. Bur. Stand. (U.S.), Spec. Publ. 317, II, 1970, pp. 837–876.

A Variational Framework for Dual Solutions in the Physical and Material Space

Daniel Materna and Franz-Joseph Barthold

Abstract This contribution is concerned with duality techniques in the physical and material spaces. Variational formulations for the primal physical and primal material problem are derived and we introduce the corresponding dual problems. The dual solutions in the physical and material spaces can be used to compute the changes in a certain quantity of interest for arbitrary given changes in the physical and material residual, respectively.

1 Introduction

A variational framework is a common approach for deriving the equilibrium relations of the classical physical motion problem as well as for the corresponding material motion problem in elasticity, see e.g. the monographs [3, 8]. Besides the primal physical and the primal material problem we investigate duality techniques based on variational arguments for both problems. The concept of duality plays an important role in many fields, e.g. in structural mechanics, physics, optimization, control theory, computational methods and goal-oriented error estimation.

In the context of structural mechanics and mechanics in the physical space, duality relations are well-known as Betti's principle, also known as the reciprocity theorem. The corresponding *dual solutions* are Green's functions or influence functions, see e.g. [1].

In the context of configurational mechanics or mechanics in the material space, a reciprocity relation in the material space similar to the reciprocity relation for point loads in the physical space was proposed in [2]. We investigate a general variational approach for dual solutions in the physical and material space.

D. Materna · F.-J. Barthold
Chair of Numerical Methods and Information Processing, TU Dortmund,
August-Schmidt-Str. 8, D-44227 Dortmund, Germany; e-mail: daniel.materna@tu-dortmund.de; franz-joseph.barthold@tu-dortmund.de

P. Steinmann (ed.), IUTAM Symposium on Progress in the Theory and Numerics of Configurational Mechanics, 95–105.

We consider for motivation a linear problem. Let for instance $\boldsymbol{u} \in \mathcal{V}$ be the solution of the so-called *primal problem* in the variational form

$$a(\boldsymbol{u}, \boldsymbol{\eta}) = F(\boldsymbol{\eta}) \quad \forall \boldsymbol{\eta} \in \mathcal{V}, \tag{1}$$

where $a(\cdot,\cdot) : \mathcal{V} \times \mathcal{V} \to \mathbb{R}$ denotes a bilinear form corresponding to the considered differential operator and $F(\cdot) : \mathcal{V} \to \mathbb{R}$ is a linear functional associated with the given external loads in a given space $\mathcal{V}$.

In duality techniques we are concerned with a certain *quantity of interest* or *general output* or *cost functional* $J(\boldsymbol{u})$, which can be every functional value which corresponds in some sense to the solution $\boldsymbol{u}$. The quantities of interest can be represented as linear or nonlinear functionals $J(\cdot) : \mathcal{V} \to \mathbb{R}$ of the solution. This could be point values, e.g. a component u_i or the derivatives $\partial_k u_i$ at some given point $\boldsymbol{X}_p$, i.e.

$$J(\boldsymbol{u}) = u_i(\boldsymbol{X}_p) \quad \text{or} \quad J(\boldsymbol{u}) = \partial_k u_i(\boldsymbol{X}_p). \tag{2}$$

For a chosen quantity of interest $J(\boldsymbol{u})$, the corresponding *dual* or *adjoint solution* or *generalized Green's function* $\boldsymbol{z} \in \mathcal{V}$ is determined by the *dual* or *adjoint problem*

$$a(\boldsymbol{\eta}, \boldsymbol{z}) = J(\boldsymbol{\eta}) \quad \forall \boldsymbol{\eta} \in \mathcal{V}. \tag{3}$$

By using (1) and (3) as well as the symmetry of $a(\cdot,\cdot)$ follows

$$J(\boldsymbol{u}) = a(\boldsymbol{u}, \boldsymbol{z}) = F(\boldsymbol{z}). \tag{4}$$

If the dual solution $\boldsymbol{z}$ is known, the quantity of interest $J(\boldsymbol{u})$ can be computed for arbitrary functionals $F(\cdot)$, i.e. $J(\boldsymbol{u}) = F(\boldsymbol{z})$. In structural mechanics this relation is well-known as Betti's principle, see e.g. [1].

The relation (4) holds in the exact form only for linear problems. Nevertheless, the duality approach can be extended to general nonlinear problems. Furthermore, the concept is more general and not restricted to mechanical problems. In general, every quantity of interest can be associated with a functional $J(\cdot)$.

2 A General Optimal Control Approach

We consider an abstract optimal control approach, which yields a general framework for duality relations of variational problems, see e.g. [4]. Let $\boldsymbol{x} \in \mathcal{X}$ be the solution of a given (nonlinear) variational problem

$$B(\boldsymbol{x}; \boldsymbol{\eta}) := a(\boldsymbol{x}; \boldsymbol{\eta}) - F(\boldsymbol{\eta}) = 0 \quad \forall \boldsymbol{\eta} \in \mathcal{X} \tag{5}$$

in a space $\mathcal{X}$. Here, $a(\boldsymbol{x}; \cdot)$ is a semilinear form, i.e. only linear with respect to all arguments right from the semicolon and $F(\cdot)$ is again a linear functional. Furthermore, let $J(\boldsymbol{x}) : \mathcal{X} \to \mathbb{R}$ be the (possible nonlinear) quantity of interest with respect to the solution $\boldsymbol{x}$.

An optimal control approach ends in the following constraint optimization problem:

$$\min_{x \in \mathcal{X}} J(\boldsymbol{x}) \qquad \text{subject to} \qquad B(\boldsymbol{x}; \boldsymbol{\eta}) = 0 \quad \forall \boldsymbol{\eta} \in \mathcal{X}. \tag{6}$$

The constraint is just the variational problem (5). The corresponding Lagrangian functional reads

$$L(\boldsymbol{x}, \boldsymbol{g}) = J(\boldsymbol{x}) + F(\boldsymbol{g}) - a(\boldsymbol{x}; \boldsymbol{g}) = J(\boldsymbol{x}) - B(\boldsymbol{x}; \boldsymbol{g}) \tag{7}$$

and the first-order optimality condition follows in the form

$$L'(\boldsymbol{x}, \boldsymbol{g})(\boldsymbol{v}, \boldsymbol{\eta}) = \left\{ \begin{matrix} L'_x(\boldsymbol{x}, \boldsymbol{g})(\boldsymbol{v}) \\ L'_g(\boldsymbol{x}, \boldsymbol{g})(\boldsymbol{\eta}) \end{matrix} \right\} = \left\{ \begin{matrix} J'_x(\boldsymbol{x}; \boldsymbol{v}) - B'_x(\boldsymbol{x}; \boldsymbol{g}, \boldsymbol{v}) \\ - B(\boldsymbol{x}; \boldsymbol{\eta}) \end{matrix} \right\} = 0 \tag{8}$$

for all $\{\boldsymbol{v}, \boldsymbol{\eta}\} \in \mathcal{X} \times \mathcal{X}$. Here, $\boldsymbol{g} \in \mathcal{X}$ denotes the *dual* or *adjoint variable*. The first equation in (8) is the *dual* or *adjoint problem* and reads: Find $\boldsymbol{g} \in \mathcal{X}$ such that

$$B'_x(\boldsymbol{x}; \boldsymbol{g}, \boldsymbol{v}) = J'_x(\boldsymbol{x}; \boldsymbol{v}) \qquad \forall \boldsymbol{v} \in \mathcal{X}, \tag{9}$$

with the tangent operator $B'_x(\boldsymbol{x}; \cdot, \cdot)$ and the linearized functional $J'_x(\boldsymbol{x}; \cdot)$, i.e.

$$B'_x(\boldsymbol{x}; \boldsymbol{g}, \boldsymbol{v}) := \frac{d}{d\varepsilon} B(\boldsymbol{x} + \varepsilon\, \boldsymbol{v}; \boldsymbol{g}) \Big|_{\varepsilon=0}, \qquad J'_x(\boldsymbol{x}; \boldsymbol{v}) := \frac{d}{d\varepsilon} J(\boldsymbol{x} + \varepsilon\, \boldsymbol{v}) \Big|_{\varepsilon=0}. \tag{10}$$

The dual problem is a linear problem and is formulated at the current linearization point, i.e. at a given solution $\boldsymbol{x}$.

The second equation in (8) is just the variational equation of the primal problem (5). In the general nonlinear case, the solution of this equation within a Newton scheme requires the linearization $B(\boldsymbol{x}; \boldsymbol{\eta}) + D_x B(\boldsymbol{x}; \boldsymbol{\eta}) \cdot \Delta \boldsymbol{x} + \mathcal{O} = 0$. The tangent operator is given in (10) and denoted by $t(\boldsymbol{x}; \boldsymbol{\eta}, \Delta \boldsymbol{x}) := D_x B(\boldsymbol{x}; \boldsymbol{\eta}) \cdot \Delta \boldsymbol{x} = B'_x(\boldsymbol{x}; \boldsymbol{\eta}, \Delta \boldsymbol{x})$. The term $\mathcal{O}$ denotes a reminder of higher-order and can usually be neglected. Hence, the solution of the nonlinear primal problem (5) requires the solution of the linear equation

$$t(\boldsymbol{x}; \boldsymbol{\eta}, \Delta \boldsymbol{x}) = -B(\boldsymbol{x}; \boldsymbol{\eta}) \qquad \forall \boldsymbol{\eta} \in \mathcal{X} \tag{11}$$

in every Newton step in order to find the new increment $\Delta \boldsymbol{x}$.

By using (9) and (11) as well as the symmetry of the bilinear form $B'_x(\boldsymbol{x}; \cdot, \cdot) = t(\boldsymbol{x}; \cdot, \cdot)$, i.e. $B'_x(\boldsymbol{x}; \boldsymbol{\eta}, \Delta \boldsymbol{x}) = B'_x(\boldsymbol{x}; \Delta \boldsymbol{x}, \boldsymbol{\eta})$, we have

$$J'_x(\boldsymbol{x}; \Delta \boldsymbol{x}) = t(\boldsymbol{x}; \boldsymbol{g}, \Delta \boldsymbol{x}) = -B(\boldsymbol{x}; \boldsymbol{g}). \tag{12}$$

If the dual solution $\boldsymbol{g}$ is known, the change in the quantity of interest $J(\boldsymbol{x})$ can be computed for arbitrary functionals $B(\boldsymbol{x}; \cdot)$, i.e. $J'_x(\boldsymbol{x}; \Delta \boldsymbol{x}) = -B(\boldsymbol{x}; \boldsymbol{g})$.

In order to make this relation more transparent we consider a discrete formulation. Let $J(\boldsymbol{x}) = x_i$ be the quantity of interest and let $\boldsymbol{g} \in \mathbb{R}^n$ be the discrete dual solution according to x_i as well as $\boldsymbol{J}_x \in \mathbb{R}^n$ be a discrete vector which corresponds to $J'_x(\boldsymbol{x};\cdot)$. Furthermore, let $\boldsymbol{T} \in \mathbb{R}^{n\times n}$ be the tangent matrix and let $\boldsymbol{B} \in \mathbb{R}^n$ be a discrete vector corresponding to the bilinear form $t(\boldsymbol{x};\cdot,\cdot)$ and the functional $B(\boldsymbol{x};\cdot)$, respectively. With these, the discrete versions of (11) and (9), i.e. the discrete primal problem and the discrete dual problem are given as

$$\boldsymbol{T}\Delta\boldsymbol{x} = -\boldsymbol{B} \qquad \text{and} \qquad \boldsymbol{T}^T\boldsymbol{g} = \boldsymbol{J}_x, \tag{13}$$

respectively. Due to the symmetry of the tangent form $B'_x(\boldsymbol{x};\cdot,\cdot)$, we have $\boldsymbol{T} = \boldsymbol{T}^T$. Then, the change in the quantity of interest $J'_x(\boldsymbol{x};\Delta\boldsymbol{x}) = \Delta x_i$ at the current linearization point $\boldsymbol{x}$ is given from the discrete version of (12) in form of

$$\Delta x_i = \Delta\boldsymbol{x}^T\boldsymbol{J}_x = \Delta\boldsymbol{x}^T\boldsymbol{T}^T\boldsymbol{g} = \boldsymbol{g}^T\boldsymbol{T}\Delta\boldsymbol{x} = -\boldsymbol{g}^T\boldsymbol{B}. \tag{14}$$

3 Variational Balance Laws in the Physical and Material Spaces

Let $E(\boldsymbol{v},\boldsymbol{s})$ be the total potential energy of a hyperelastic body. The energy depends on the *generalized state function* $\boldsymbol{v} \in \mathcal{V}$ and on a *generalized design* or *control function* $\boldsymbol{s} \in \mathcal{S}$, which specifies in an abstract sense the current reference configuration Ω_R, i.e. $\Omega_R = \Omega_R(\boldsymbol{s})$. The space $\mathcal{V}$ denotes the usual Sobolev space and $\mathcal{S}$ the space with all admissible design functions. The total potential energy is given by

$$E(\boldsymbol{v},\boldsymbol{s}) := \int_{\Omega_R} W_R\, d\Omega - F(\boldsymbol{s};\boldsymbol{v}), \tag{15}$$

where W_R denotes the strain energy function and $F(\boldsymbol{s};\cdot)$ is a functional associated with the external potential.

The partial variation of E with respect to $\boldsymbol{v}$ yields the *primal physical residual*

$$R(\boldsymbol{v},\boldsymbol{s};\boldsymbol{\eta}) := E'_v(\boldsymbol{v},\boldsymbol{s};\boldsymbol{\eta}) = \int_{\Omega_R} \boldsymbol{P} : \operatorname{Grad}\boldsymbol{\eta}\, d\Omega - F(\boldsymbol{s};\boldsymbol{\eta}) = 0 \quad \forall\,\boldsymbol{\eta} \in \mathcal{V}. \tag{16}$$

In the same manner, variation with respect to $\boldsymbol{s}$ leads to the *primal material residual*

$$G(\boldsymbol{v},\boldsymbol{s};\boldsymbol{\psi}) := E'_s(\boldsymbol{v},\boldsymbol{s};\boldsymbol{\psi}) = \int_{\Omega_R} \boldsymbol{\Sigma} : \operatorname{Grad}\boldsymbol{\psi}\, d\Omega - F'_s(\boldsymbol{s};\boldsymbol{v},\boldsymbol{\psi}) = 0 \quad \forall\,\boldsymbol{\psi} \in \mathcal{S}. \tag{17}$$

Both residuals are written in terms of the reference configuration Ω_R in which $\boldsymbol{P}$ is the first Piola–Kirchhoff stress tensor and $\boldsymbol{\Sigma} = W_R\boldsymbol{I} - \operatorname{Grad}\boldsymbol{v}^T\boldsymbol{P}$ denotes the energy-momentum tensor. Using standard pull back and push forward operations we can transform all quantities into different configurations, see e.g. [8].

Let $\mathcal{Y} := \mathcal{V} \times \mathcal{S}$ be a product space and set $\boldsymbol{y} = \{\boldsymbol{v},\boldsymbol{s}\} \in \mathcal{Y}$, $\Delta\boldsymbol{y} = \{\Delta\boldsymbol{v},\Delta\boldsymbol{s}\} \in \mathcal{Y}$ as well as $\boldsymbol{\varphi} = \{\boldsymbol{\eta},\boldsymbol{\psi}\} \in \mathcal{Y}$. The simultaneous solution of the coupled physical (16) and material problem (17) ends in the linearized equation

$$\boldsymbol{H}(\boldsymbol{y};\boldsymbol{\varphi},\Delta\boldsymbol{y}) = -\boldsymbol{L}(\boldsymbol{y};\boldsymbol{\varphi}) \qquad \forall\,\boldsymbol{\varphi} \in \mathcal{Y}, \tag{18}$$

$$\boldsymbol{H}(\boldsymbol{y};\boldsymbol{\varphi},\Delta\boldsymbol{y}) := \left\{\begin{array}{l} k(\boldsymbol{v},\boldsymbol{s};\boldsymbol{\eta},\Delta\boldsymbol{v}) + p(\boldsymbol{v},\boldsymbol{s};\boldsymbol{\eta},\Delta\boldsymbol{s}) \\ p(\boldsymbol{v},\boldsymbol{s};\Delta\boldsymbol{v},\boldsymbol{\psi}) + d(\boldsymbol{v},\boldsymbol{s};\boldsymbol{\psi},\Delta\boldsymbol{s}) \end{array}\right\}, \boldsymbol{L}(\boldsymbol{y};\boldsymbol{\varphi}) := \left\{\begin{array}{l} R(\boldsymbol{v},\boldsymbol{s};\boldsymbol{\eta}) \\ G(\boldsymbol{v},\boldsymbol{s};\boldsymbol{\psi}) \end{array}\right\},$$

which has to be solved in each Newton step. The Hessian operator $\boldsymbol{H}(\boldsymbol{y};\boldsymbol{\varphi},\Delta\boldsymbol{y})$ contains the partial variations of R and G, respectively, i.e. the second and mixed variations of the energy functional. The variations are given as

$$k(\boldsymbol{v},\boldsymbol{s};\boldsymbol{\eta},\Delta\boldsymbol{v}) := R'_v(\boldsymbol{v},\boldsymbol{s};\boldsymbol{\eta},\Delta\boldsymbol{v}) = E''_{vv}, \tag{19}$$

$$p(\boldsymbol{v},\boldsymbol{s};\boldsymbol{\eta},\Delta\boldsymbol{s}) := R'_s(\boldsymbol{v},\boldsymbol{s};\boldsymbol{\eta},\Delta\boldsymbol{s}) = E''_{vs}, \tag{20}$$

$$d(\boldsymbol{v},\boldsymbol{s};\boldsymbol{\psi},\Delta\boldsymbol{s}) := G'_s(\boldsymbol{v},\boldsymbol{s};\boldsymbol{\psi},\Delta\boldsymbol{s}) = E''_{ss}, \tag{21}$$

$$p(\boldsymbol{v},\boldsymbol{s};\Delta\boldsymbol{v},\boldsymbol{\psi}) := G'_v(\boldsymbol{v},\boldsymbol{s};\boldsymbol{\psi},\Delta\boldsymbol{v}) = E''_{sv}, \tag{22}$$

where $k(\cdot;\cdot)$ is the *tangent physical stiffness operator* and $d(\cdot;\cdot)$ is the *tangent material stiffness operator*. Due to symmetry and permutableness of variations, the problem is coupled by the *tangent pseudo load operator* $p(\cdot;\cdot) := R'_s(\cdot;\cdot) = E''_{vs} = E''_{sv} = G'_v(\cdot;\cdot)$, see [5] for explicit formulations of the tangent operators.

4 Duality Techniques for the Physical Problem

4.1 The Primal Physical Problem

The variational equation of the primal problem is given in Eq. (16). The solution within a Newton scheme requires the linearization $R(\boldsymbol{v},\boldsymbol{s};\boldsymbol{\eta}) + D_v R(\boldsymbol{v},\boldsymbol{s};\boldsymbol{\eta})\cdot\Delta\boldsymbol{v} + \mathcal{O} = 0$. The tangent operator $D_v R(\boldsymbol{v},\boldsymbol{s};\boldsymbol{\eta})\cdot\Delta\boldsymbol{v} = R'_v(\boldsymbol{u},\boldsymbol{s};\boldsymbol{\eta},\Delta\boldsymbol{v}) = k(\boldsymbol{v},\boldsymbol{s};\boldsymbol{\eta},\Delta\boldsymbol{v})$ is the tangent physical stiffness operator (19). Hence, the solution of (16) requires the solution of the linear equation

$$k(\boldsymbol{v},\boldsymbol{s};\boldsymbol{\eta},\Delta\boldsymbol{v}) = -R(\boldsymbol{v},\boldsymbol{s};\boldsymbol{\eta}) \qquad \forall\,\boldsymbol{\eta}\in\mathcal{V} \tag{23}$$

in every Newton step in order to compute the increment $\Delta\boldsymbol{v}$.

4.2 The Dual Physical Problem

Let $J(\cdot):\mathcal{V}\to\mathbb{R}$ be a (possible nonlinear) quantity of interest for a given fixed $\boldsymbol{s}$. We use the optimal control approach from Section 2 with $\boldsymbol{x}=\boldsymbol{v}$ as the considered variable and we set $\mathcal{X}=\mathcal{V}$. Then, the corresponding *dual physical solution* $\boldsymbol{z}\in\mathcal{V}$ is determined by

$$k(\boldsymbol{v},\boldsymbol{s};\boldsymbol{z},\boldsymbol{\eta}) = J'_v(\boldsymbol{v};\boldsymbol{\eta}) \qquad \forall\,\boldsymbol{\eta}\in\mathcal{V}. \tag{24}$$

The variation of J according to (10) is given from

$$J'_v(\boldsymbol{v};\,\boldsymbol{\eta}) = \frac{d}{d\varepsilon} J(\boldsymbol{v} + \varepsilon\,\boldsymbol{\eta})\bigg|_{\varepsilon=0}.$$

The explicit form of $J'_v(\boldsymbol{v};\,\cdot)$ depends on the chosen quantity of interest J.

With (23) and (24) and by using the symmetry of $k(\boldsymbol{v}, \boldsymbol{s};\,\cdot,\cdot)$ we have

$$J'_v(\boldsymbol{v};\,\Delta\boldsymbol{v}) = k(\boldsymbol{v}, \boldsymbol{s};\,\boldsymbol{z}, \Delta\boldsymbol{v}) = -R(\boldsymbol{v}, \boldsymbol{s};\,\boldsymbol{z}). \tag{25}$$

Finally, if the dual solution $\boldsymbol{z}$ is known, the change in the quantity of interest can be computed for arbitrary functionals $R(\cdot)$.

We consider the discrete case. Let $J(\boldsymbol{v}_h) = v_i$ be the quantity of interest and let $\boldsymbol{z} \in \mathbb{R}^n$ be the vector with the discrete dual solution corresponding to v_i. Then, the discrete form of (25) for the change in the quantity of interest reads

$$\Delta v_i = -\boldsymbol{z}^T \boldsymbol{R}, \tag{26}$$

where $\boldsymbol{R} \in \mathbb{R}^n$ is the physical residual vector corresponding to $R(\boldsymbol{v}, \boldsymbol{s};\,\cdot)$.

4.3 Sensitivity Relation Using the Dual Solution

We consider a variational sensitivity relation for the change in the state $\boldsymbol{v}$ due to changes in the material configuration $\boldsymbol{s}$. This reads: Find $\delta\boldsymbol{v} \in \mathcal{V}$ for a given fixed design variation $\delta\hat{\boldsymbol{s}} \in \mathcal{S}$ such that

$$k(\boldsymbol{v}, \boldsymbol{s};\,\boldsymbol{\eta}, \delta\boldsymbol{v}) = -Q_p(\boldsymbol{v}, \boldsymbol{s};\,\boldsymbol{\eta}) \qquad \forall\,\boldsymbol{\eta} \in \mathcal{V}, \tag{27}$$

where $Q_p(\boldsymbol{v}, \boldsymbol{s};\,\cdot) := p(\boldsymbol{v}, \boldsymbol{s};\,\cdot, \delta\hat{\boldsymbol{s}}) = R'_s(\boldsymbol{v}, \boldsymbol{s};\,\cdot, \delta\hat{\boldsymbol{s}})$ is the *pseudo load* of the physical problem for the variation $\delta\hat{\boldsymbol{s}}$, see [5] for details. The variation in the state $\delta\boldsymbol{v} \in \mathcal{V}$ due to a variation in the design $\delta\boldsymbol{s} \in \mathcal{S}$ is given in (27). The dual solution $\boldsymbol{z}$ is given from (24). From this, we obtain a sensitivity relation for the change in the quantity of interest

$$J'_v(\boldsymbol{v};\,\delta\boldsymbol{v}) = k(\boldsymbol{v}, \boldsymbol{s};\,\boldsymbol{z}, \delta\boldsymbol{v}) = -Q_p(\boldsymbol{v}, \boldsymbol{s};\,\boldsymbol{z}). \tag{28}$$

If the dual solution $\boldsymbol{z}$ is known, Eq. (28) can be evaluated for arbitrary pseudo loads $Q_p(\boldsymbol{u}, \boldsymbol{s};\,\cdot)$, i.e. for arbitrary admissible design variations $\delta\boldsymbol{s}$ [7].

We consider the discrete case. Let again $J(\boldsymbol{v}_h) = v_i$ and $\boldsymbol{z} \in \mathbb{R}^n$ be the vector with the discrete dual solution. Furthermore, let $\delta\hat{\boldsymbol{s}} \in \mathbb{R}^m$ be a discrete vector with the changes in the design variables. Then, the discrete form of (28) is given as

$$\delta v_i = -\boldsymbol{z}^T \boldsymbol{Q}_p, \tag{29}$$

where $\boldsymbol{Q}_p = \boldsymbol{P}\delta\hat{\boldsymbol{s}} \in \mathbb{R}^n$ is the pseudo load vector and $\boldsymbol{P} \in \mathbb{R}^{n\times m}$ is the pseudo load operator matrix corresponding to $Q_p(\boldsymbol{v}, \boldsymbol{s};\,\cdot)$ and $p(\boldsymbol{v}, \boldsymbol{s};\,\cdot,\cdot)$, respectively.

4.4 The Material Residual of the Dual Problem

The dual solution $z \in \mathcal{V}$ is determined by the linear equation (24). We assume, that $J(\boldsymbol{v})$ is a regularized functional and well defined on $\mathcal{V}$. Then, the dual solution z is a minimizer of the corresponding energy functional of the dual problem

$$E^*(\boldsymbol{v})(z, \boldsymbol{s}) := \frac{1}{2} k(\boldsymbol{v}, \boldsymbol{s}; z, z) - J'_v(\boldsymbol{v}, \boldsymbol{s}; z). \tag{30}$$

By means of the tangent operator $k(\boldsymbol{v}, \boldsymbol{s}; \cdot, \cdot)$, we introduce the strain energy function of the dual problem $W_R^*(\boldsymbol{v}, z)$ such that the internal energy of the dual problem reads

$$\frac{1}{2} k(\boldsymbol{v}, \boldsymbol{s}; z, z) = \int_{\Omega_R} W_R^*(\boldsymbol{v}, z)\, d\Omega. \tag{31}$$

With this, the total potential energy of the dual problem can be written as

$$E^*(\boldsymbol{v})(z, \boldsymbol{s}) = \int_{\Omega_R} W_R^*(\boldsymbol{v}, z)\, d\Omega - J'_v(\boldsymbol{v}, \boldsymbol{s}; z). \tag{32}$$

The dual energy functional $E^*(\boldsymbol{v})(z, \boldsymbol{s})$ is constructed at a given deformed state $\boldsymbol{v}$ and describes an energy quantity due to the linear dual solution z at this point [7]. The variation of E^* with respect to z yields the *dual physical residual*

$$R^*(\boldsymbol{v}, \boldsymbol{s}; z, \boldsymbol{\eta}) := E_z^{*\prime}(\boldsymbol{v})(z, \boldsymbol{s}; \boldsymbol{\eta}) = k(\boldsymbol{v}, \boldsymbol{s}; z, \boldsymbol{\eta}) - J'_v(\boldsymbol{v}, \boldsymbol{s}; \boldsymbol{\eta}) = 0 \quad \forall \boldsymbol{\eta} \in \mathcal{V}. \tag{33}$$

This is just the dual problem (24). A variation of E^* with respect to $\boldsymbol{s}$ leads to the *dual material residual*

$$\begin{aligned} G^*(\boldsymbol{v}, \boldsymbol{s}; z; \boldsymbol{\psi}) &:= E_s^{*\prime}(\boldsymbol{v})(z, \boldsymbol{s}; \boldsymbol{\psi}) \\ &= \frac{1}{2} k'_s(\boldsymbol{v}, \boldsymbol{s}; z; \boldsymbol{\psi}) - J''_{vs}(\boldsymbol{v}, \boldsymbol{s}; z, \boldsymbol{\psi}) = 0 \quad \forall \boldsymbol{\psi} \in \mathcal{S}. \end{aligned} \tag{34}$$

The variation of the internal energy of the dual problem (31) is given by

$$\frac{1}{2} k'_s(\boldsymbol{v}, \boldsymbol{s}; z; \boldsymbol{\psi}) = \int_{\Omega_R} \boldsymbol{\Sigma}^*(\boldsymbol{v}, z) : \mathrm{Grad}\, \boldsymbol{\psi}\, d\Omega. \tag{35}$$

Here, we have introduced an *energy-momentum tensor of the dual problem*

$$\boldsymbol{\Sigma}^*(\boldsymbol{v}, z) := W_R^*(\boldsymbol{v}, z)\boldsymbol{I} - \mathrm{Grad}\, z^T \frac{\partial W_R^*(\boldsymbol{v}, z)}{\partial\, \mathrm{Grad}\, z} - \mathrm{Grad}\, \boldsymbol{v}^T \frac{\partial W_R^*(\boldsymbol{v}, z)}{\partial\, \mathrm{Grad}\, \boldsymbol{v}}. \tag{36}$$

The dual material residual G^* represents the sensitivity of the energy functional $E^*(\boldsymbol{v})(z, \boldsymbol{s})$ with respect to changes in the design $\boldsymbol{s}$, see [7] for details. It is well-known that the material residual of the primal problem G is an error indicator for a non-optimal finite element discretization of the primal solution $\boldsymbol{v}$ and used within global mesh optimization algorithms. The material residual of the dual problem

G^* is an error indicator for a non-optimal finite element discretization of the dual solution z and can be used within a goal-oriented r-adaptivity algorithm [6].

5 Duality Techniques for the Material Problem

5.1 The Primal Material Problem

The variational equation of the material problem is given in Eq. (17) and reads $G(\boldsymbol{v}, \boldsymbol{s}; \boldsymbol{\psi}) = 0$. In order to solve this nonlinear problem in a staggered way, we use the linearization $G(\boldsymbol{v}, \boldsymbol{s}; \boldsymbol{\psi}) + D_s G(\boldsymbol{v}, \boldsymbol{s}; \boldsymbol{\psi}) \cdot \Delta \boldsymbol{s} + \mathcal{O} = 0$, see [5] for details. The material tangent operator is introduced as $m(\boldsymbol{v}, \boldsymbol{s}; \boldsymbol{\psi}, \Delta \boldsymbol{s}) := D_s G(\boldsymbol{v}, \boldsymbol{s}; \boldsymbol{\psi}) \cdot \Delta \boldsymbol{s}$. This is a bilinear form $m : \mathcal{S} \times \mathcal{S} \to \mathbb{R}$, which has to be evaluated for the current $\{\boldsymbol{v}, \boldsymbol{s}\}$. Hence, we have to solve the linear equation

$$m(\boldsymbol{v}, \boldsymbol{s}; \boldsymbol{\psi}, \Delta \boldsymbol{s}) = -G(\boldsymbol{v}, \boldsymbol{s}; \boldsymbol{\psi}) \qquad \forall \boldsymbol{\psi} \in \mathcal{S} \tag{37}$$

in every Newton step in order to compute the increment $\Delta \boldsymbol{s}$.

5.2 The Dual Material Problem

Let $J(\cdot) : \mathcal{S} \to \mathbb{R}$ be a quantity of interest for a given fixed $\boldsymbol{v}$. We use the optimal control approach from Section 2 and we set $\boldsymbol{x} = \boldsymbol{s}$ and $\mathcal{X} = \mathcal{S}$. Then, the corresponding *dual material solution* $\boldsymbol{q} \in \mathcal{S}$ is determined by the linear equation

$$m(\boldsymbol{v}, \boldsymbol{s}; \boldsymbol{q}, \boldsymbol{\psi}) = J_s'(\boldsymbol{s}; \boldsymbol{\psi}) \qquad \forall \boldsymbol{\psi} \in \mathcal{S}. \tag{38}$$

The variation of J according to (10) is given from

$$J_s'(\boldsymbol{s}; \boldsymbol{\psi}) = \frac{d}{d\varepsilon} J(\boldsymbol{s} + \varepsilon \boldsymbol{\psi}) \bigg|_{\varepsilon=0}.$$

By using (37) and (38) as well as the symmetry of $m(\boldsymbol{v}, \boldsymbol{s}; \cdot, \cdot)$ we have

$$J_s'(\boldsymbol{s}; \Delta \boldsymbol{s}) = m(\boldsymbol{v}, \boldsymbol{s}; \boldsymbol{q}, \Delta \boldsymbol{s}) = -G(\boldsymbol{v}, \boldsymbol{s}; \boldsymbol{q}). \tag{39}$$

If the dual solution $\boldsymbol{q}$ is known $J_s'(\boldsymbol{s}; \Delta \boldsymbol{s})$ can be computed for any functionals $G(\cdot)$.

We consider the discrete case. Let $J(\boldsymbol{s}_h) = s_i$ be the quantity of interest and let $\boldsymbol{q} \in \mathbb{R}^m$ be the vector with the discrete dual solution corresponding to s_i. Then, the discrete form of (39) for the change in the quantity of interest reads

$$\Delta s_i = -\boldsymbol{q}^T \boldsymbol{G}, \tag{40}$$

where $\boldsymbol{G} \in \mathbb{R}^m$ is the material residual vector corresponding to $G(\boldsymbol{v}, \boldsymbol{s}; \cdot)$.

5.3 Duality Relation for the Coupled Problem

We consider the solution of the coupled physical and material problem given in Eq. (18). Let $J(\boldsymbol{v}, \boldsymbol{s}) : \mathcal{V} \times \mathcal{S} \to \mathbb{R}$ be a differentiable quantity of interest. The dual solution $\boldsymbol{y}^* := \{z_c, \boldsymbol{q}_c\} \in \mathcal{Y}$ is determined by the dual problem

$$\boldsymbol{H}(\boldsymbol{y}; \boldsymbol{y}^*, \boldsymbol{\varphi}) = \boldsymbol{J}'(\boldsymbol{y}; \boldsymbol{\varphi}) \qquad \forall\, \boldsymbol{\varphi} \in \mathcal{Y}, \tag{41}$$

$$\boldsymbol{H}(\boldsymbol{y}; \boldsymbol{y}^*, \boldsymbol{\varphi}) := \left\{ \begin{array}{l} k(\boldsymbol{v}, \boldsymbol{s}; z_c, \boldsymbol{\eta}) + p(\boldsymbol{v}, \boldsymbol{s}; \boldsymbol{\eta}, \boldsymbol{q}_c) \\ p(\boldsymbol{v}, \boldsymbol{s}; z_c, \boldsymbol{\psi}) + d(\boldsymbol{v}, \boldsymbol{s}; \boldsymbol{q}_c, \boldsymbol{\psi}) \end{array} \right\}, \quad \boldsymbol{J}'(\boldsymbol{y}; \boldsymbol{\varphi}) := \left\{ \begin{array}{l} J_v'(\boldsymbol{v}, \boldsymbol{s}; \boldsymbol{\eta}) \\ J_s'(\boldsymbol{v}, \boldsymbol{s}; \boldsymbol{\psi}) \end{array} \right\}.$$

Due to the symmetry properties of $\boldsymbol{H}$, i.e. $\boldsymbol{H}(\boldsymbol{y}; \boldsymbol{\varphi}, \Delta\boldsymbol{y}) = \boldsymbol{H}(\boldsymbol{y}; \Delta\boldsymbol{y}, \boldsymbol{\varphi})$, we have

$$\boldsymbol{J}'(\boldsymbol{y}; \Delta\boldsymbol{y}) = \boldsymbol{H}(\boldsymbol{y}; \boldsymbol{y}^*, \Delta\boldsymbol{y}) = -\boldsymbol{L}(\boldsymbol{y}; \boldsymbol{y}^*). \tag{42}$$

This equation can be evaluated for arbitrary residuals $\boldsymbol{L}(\boldsymbol{y}; \cdot)$ if the dual solutions $\boldsymbol{y}^* = \{z_c, \boldsymbol{q}_c\}$ are known. The result is the total change in the quantity of interest with respect to $\boldsymbol{v}$ and $\boldsymbol{s}$. The index c indicates that the dual solutions are obtained using the coupled system. Note, that these dual solutions differ in general from the results which are obtained with (24) and (38), respectively. Because the consecutive solution of (24) and (38) is part of a staggered solution scheme, i.e. a decoupled algorithm, and the increments $\{\Delta\boldsymbol{v}, \Delta\boldsymbol{s}\}$ differ from the results of (18).

6 Conclusions

We have considered duality techniques for the physical and material problem based on variational principles. A general optimal control approach yields the framework for duality relations of variational equations. The dual solution depends on the kind of the considered variational problem. For the physical and material problem the duality relations are summarized in the following table:

	linearized primal problem	dual problem
physical problem: $R = 0$	$k(\boldsymbol{\eta}, \Delta\boldsymbol{v}) = -R(\boldsymbol{\eta})$	$k(z, \boldsymbol{\eta}) = J_v'(\boldsymbol{\eta})$
material problem: $G = 0$	$m(\boldsymbol{\psi}, \Delta\boldsymbol{s}) = -G(\boldsymbol{\psi})$	$m(\boldsymbol{q}, \boldsymbol{\psi}) = J_s'(\boldsymbol{\psi})$

The dual problems in the physical and material spaces are linear problems and formulated at the current linearization points $\{\boldsymbol{v}, \boldsymbol{s}\}$, i.e. at a given deformed state. If the dual solutions are known, a quantity of interest can be computed for arbitrary physical and material residuals, respectively.

An example for a dual physical solution z and a dual material solution $\boldsymbol{q}$ is given in Fig. 1 (the vertical components). The approximate dual solutions z_h and $\boldsymbol{q}_h$ are

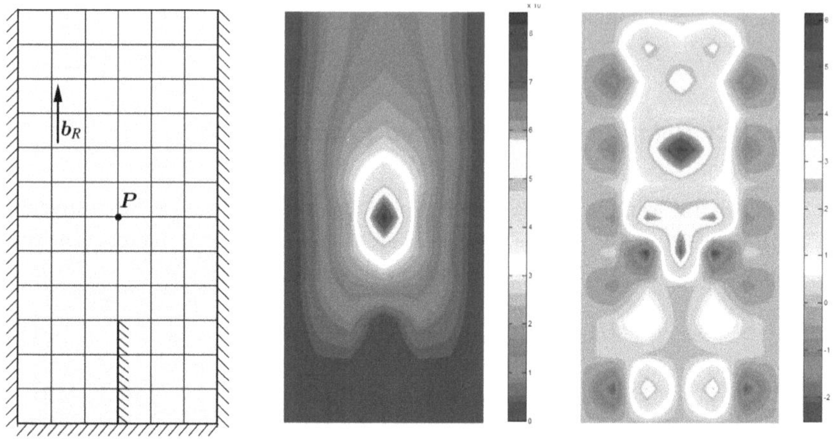

(a) primal physical problem (b) dual physical solution z_2 for $J(\boldsymbol{v}_h) = u_2(\boldsymbol{P})$ (c) dual material solution q_2 for $J(\boldsymbol{s}_h) = X_2(\boldsymbol{P})$

Fig. 1 Plate with a slit loaded by a body force $\boldsymbol{b}_R$: the dual physical solution z_h and the dual material solution $\boldsymbol{q}_h$ in vertical direction.

computed using the finite element method and the shown mesh by solving (24) and (38), respectively. The physical quantity of interest is the vertical component of the nodal displacement at a given point $\boldsymbol{P}$, i.e. $J(\boldsymbol{v}_h) = u_2(\boldsymbol{P})$. The material quantity of interest is the vertical component of the nodal coordinate at $\boldsymbol{P}$, i.e. $J(\boldsymbol{s}_h) = X_2(\boldsymbol{P})$. The dual solutions reflect the influence of the considered residual on the change in the physical and material quantity of interest, respectively. For instance, a large value of z in a certain domain Ω_i indicates that a physical residual in Ω_i causes a large change in u_2. In the same manner, a large value of $\boldsymbol{q}$ in Ω_i indicates that a material residual in Ω_i causes a large change in X_2. Hence, the dual solutions are the influence functions for the considered physical and material quantity, respectively.

Furthermore, we have investigated the material residual of the dual physical problem $G^*(\boldsymbol{v}, \boldsymbol{s}; z; \cdot)$. In the context of the finite element method, the residual $G^*(\boldsymbol{v}_h, \boldsymbol{s}_h; z_h; \cdot)$ is an error indicator for a non-optimal finite element mesh for the dual solution z_h and can be used within a goal-oriented r-adaptive mesh optimization algorithm.

References

1. Hartmann, F., *The Mathematical Foundation of Structural Mechanics*. Springer-Verlag, Berlin/Heidelberg, 1985.
2. Herrmann, G. and Kienzler, R., Reciprocity relations in Eshelbian mechanics. *Mech. Res. Commun.* **34**, 2007, 338–343.
3. Kienzler, R. and Herrmann, G., *Mechanics in Material Space*. Springer-Verlag, Berlin/Heidelberg/New York, 2000.

4. Lions, J.L., *Optimal Control of Systems Governed by Partial Differential Equations*. Springer-Verlag, Berlin/Heidelberg/New York, 1971.
5. Materna, D. and Barthold, F.J., Variational design sensitivity analysis in the context of structural optimization and configurational mechanics. *Int. J. Fract.* **147**(1–4), 2007, 133–155.
6. Materna, D. and Barthold, F.J., Goal-oriented r-adaptivity based on variational arguments in the physical and material spaces. *Comput. Methods Appl. Mech. Engrg.*, 2008, submitted.
7. Materna, D. and Barthold, F.J., Configurational variations for the primal and dual problem in elasticity. *Z. Angew. Math. Mech.*, 2009, DOI 10.1002/zamm.200800144, accepted for publication.
8. Maugin, G.A., *Material Inhomogeneities in Elasticity*. Chapman & Hall, London, 1993.

On the Nonlocal Symmetries, Group Invariant Solutions and Conservation Laws of the Equations of Nonlinear Dynamical Compressible Elasticity

G.W. Bluman, A.F. Cheviakov and J.F. Ganghoffer

Abstract A family of new PDE systems of one-dimensional nonlinear elastodynamics, which are nonlocally related to the classical Lagrange and Euler formulations, is derived. These new PDE systems provide alternative equivalent descriptions of the one-dimensional nonlinear elasticity model. In particular, nonlocally related systems are used to find nonlocal symmetries of the Euler system for various forms of constitutive and loading functions. Examples of new dynamical solutions arising as group invariant solutions with respect to such nonlocal symmetries are constructed. Another application of nonlocally related systems considered in this paper is the construction of nonlocal conservation laws. Examples of nonlocal conservation laws are derived for several classes of stress-strain relations and loading functions.

1 Introduction

Analytical studies of nonlinear dynamical elasticity models, and especially, the problem of finding exact solutions, have attracted significant attention of researchers in recent years. Lie symmetries are widely used in the analysis of contemporary nonlinear elasticity models, especially for the calculation of similarity (invariant) solutions, arising from symmetry reduction (see [1–4] and further references in [7]). Focusing on nonlinear elasticity, it is well known that very few closed-form solutions

G.W. Bluman
Department of Mathematics, University of British Columbia, Vancouver V6T 1Z2, Canada; e-mail: bluman@math.ubc.ca

A.F. Cheviakov
Department of Mathematics, University of Saskatchewan, Room 142, McLean Hall 106 Wiggins Road, Saskatoon SK S7N 5E6, Canada; e-mail: cheviakov@math.usask.ca

J.F. Ganghoffer
LEMTA-ENSEM Nancy Université, 2 Avenue de la Foret de Haye, BP 160, 54504 Vandoeuvre Cedex, France; e-mail: jean-francois.ganghoffer@ensem.inpl-nancy.fr

P. Steinmann (ed.), IUTAM Symposium on Progress in the Theory and Numerics of Configurational Mechanics, 107–120.

of BVPs for compressible elasticity have been obtained in the literature (contrary to incompressible elasticity), due to the absence of the kinematic incompressibility constraint, as pointed out in [8,9]. Lie group analysis is of further interest in setting up numerical schemes preserving the group properties of an initial boundary value problem (BVP) [5].

The problem of finding conservation laws (full divergence expressions) that hold for a system of partial differential equations (PDEs) is an important counterpart to symmetry analysis. In elasticity theory, one well-known application of conservation laws is the Eshelby energy-momentum tensor, and a related path-independent integral, which governs the energy release rate at a singularity [10]. Symmetries play moreover an important role in Eshelbian mechanics, since invariance of a suitable functional under translations in material space (corresponding to horizontal variations) highlights the Eshelby tensor in the resulting equilibrium equation [6]. In a series of papers, Olver studied conservation laws and related path-independent integrals in linear elastostatics within the framework of hyperelasticity (see [11] and references therein). Mathematically, conservation laws can be systematically calculated, both for variational problems (Noether's theorem) and for non-variational problems [13, 14].

The present contribution is organized as follows. In Section 2, the Euler and Lagrange PDE systems of one-dimensional systems of nonlinear elastodynamics are presented. In Section 3, a nonlocal relation between the Euler and Lagrange PDE systems is derived, and using conservation laws of the Euler system, a set of additional PDE systems, equivalent but nonlocally related to both the Euler and Lagrange PDE systems, is constructed. In Section 4, nonlocal symmetries of the Euler system are classified, arising as point symmetries of its nonlocally related systems. Such nonlocal symmetries are used in Section 5 to derive an example of an exact invariant solution of a nonlinear elastodynamics BVP, corresponding to a nonlinear stretching. In Section 6, one of the considered nonlocally related PDE systems is used to derive new nonlocal conservation laws of the Euler equations of nonlinear elastodynamics.

2 Nonlinear Elasticity: Boundary Value Problems in 1D

In the one-dimensional situation, since the transformation gradient is the ratio of initial to actual density, $F = q = \rho_0/\rho$, the Cauchy stress is given by $\sigma = \sigma(\rho)$. The 1D Euler system is given by [7]

$$\mathbf{E}\{x, t\,; v, \sigma, \rho\} : \quad \begin{cases} \rho_t + (\rho v)_x = 0, \\ \sigma_x + \rho f(x,t) = \rho(v_t + v v_x), \\ \sigma = K(\rho). \end{cases} \tag{1}$$

In this paper we will only consider conservative forces $f(x,t) = f(x)$.

The independent variables are the Eulerian coordinates (x, t), the dependent variables are (ρ, v, σ), and some freedom of choice is allowed for the possible forms of the load per unit mass $f = f(x)$ and the material's constitutive response, i.e. the function $K = K(\rho)$. Dimensionless variables are adopted in the sequel, following [7].

The relationship between the first Piola–Kirchhoff stress and the Cauchy stress leads in 1D to $\sigma = T$. Therefore, the 1D Lagrange system in dimensionless variables is given by

$$\mathbf{L}\{y, s\,;\, v, \sigma, q, x\} : \quad \begin{cases} q = x_y, \\ v = x_t, \\ v_t = \sigma_y + f(x), \\ \sigma = K(1/q). \end{cases} \tag{2}$$

In the Lagrange system, the independent variables are the Lagrangian coordinates (y, t), the dependent variables are (x, v, q, σ), and the free functions are $f = f(x)$ and $K_1(q) \equiv K(1/q)$.

3 Nonlocally Related Systems of 1D Nonlinear Elasticity

Consider a PDE system $\mathbf{R}\{x\,;\,u\}$ of N PDEs of order k with n independent variables $x = (x^1, \ldots, x^n)$ and m dependent variables $u(x) = (u^1(x), \ldots, u^m(x))$, given by

$$R^\sigma[u] \equiv R^\sigma(x, u, \partial u, \ldots, \partial^k u) = 0, \quad \sigma = 1, \ldots, N. \tag{3}$$

PDE systems *nonlocally related* to $\mathbf{R}\{x\,;\,u\}$ arise in the following two simple ways [12].

(a) If the system $\mathbf{R}\{x\,;\,u\}$ has a *conservation law*

$$D_i \Phi^i[u] = 0, \tag{4}$$

then one may accordingly introduce nonlocal (potential) variable(s) v, satisfying corresponding potential equations. The union of the set of equations of $\mathbf{R}\{x\,;\,u\}$ and the potential equations yields a *potential system* $\mathbf{S}\{x\,;\,u, v\}$.

(b) Exclusion of one of the dependent variables of $\mathbf{R}\{x\,;\,u\}$ by differential compatibility relations (e.g. $v_{xt} = v_{tx}$) yields a *nonlocally related subsystem.* For example, if u^1 can be excluded, the corresponding subsystem is denoted $\underline{\mathbf{R}}\{x\,;\,u^2, \ldots, u^n\}$.

Combinations of the above two constructions, including their use in combination with interchanges of dependent and independent variables, may be used to obtain further nonlocally related PDE systems.

Solution sets of nonlocally related PDE systems are *equivalent*, in the sense that the solution set of one such system can be found from the solution set of of any other one. Therefore any method of analysis (qualitative, perturbation, numerical, etc.) that fails to work for a given PDE system, especially a method that is not coordinate-dependent, could turn out to be successful when applied to such a nonlocally related PDE system. In particular, for a given PDE system, through Lie's algorithm applied to a nonlocally related system, one can systematically calculate *nonlocal symmetries* (which in turn are useful for obtaining new exact solutions from known ones), construct (further) invariant and nonclassical solutions, as well as obtain linearizations (see e.g. [7, 12]). One could also obtain *nonlocal conservation laws* of a given PDE system, through the application of a standard procedure for finding local conservation laws to a nonlocally related system (see Section 6).

We now construct a tree of nonlocally related systems for 1D nonlinear elastodynamics, starting with the Euler system $\mathbf{E}\{x,t\,;\,v,\sigma,\rho\}$ (1).

The first equation of the system $\mathbf{E}\{x,t\,;\,v,\sigma,\rho\}$ (1) is in the form of a conservation law (mass conservation) as it stands; hence a potential w can be introduced. The corresponding potential system takes the form

$$\mathbf{EW}\{x,t\,;\,v,\sigma,\rho,w\}:\quad \begin{cases} w_x=\rho,\\ w_t=-\rho v,\\ \sigma_x+\rho f(x)=\rho(v_t+vv_x),\\ \sigma=K(\rho). \end{cases} \tag{5}$$

It is remarkable that a *local* 1:1 point transformation (an interchange of a dependent and independent variable) of the system $\mathbf{EW}\{x,t\,;\,v,\sigma,\rho,w\}$ with $w=y$ and t treated as independent variables, and $x,v,\sigma,q=1/\rho$ as dependent variables, directly yields the Lagrange system $\mathbf{L}\{y,s\,;\,v,\sigma,q,x\}$ (2). Hence, the systems $\mathbf{EW}\{x,t\,;\,v,\sigma,\rho,w\}$ (5) and $\mathbf{L}\{y,s\,;\,v,\sigma,q,x\}$ (2) are *locally related to each other* (by a point transformation), but *nonlocally related to the Euler system* $\mathbf{E}\{x,t\,;\,v,\sigma,\rho\}$ (1). A similar connection exists in higher dimensions, expressed by the kinematic relation from configurational mechanics given by

$$y_{,t}+\mathbf{F}^{-1}.x_{,t}=0$$

In the Lagrange system $\mathbf{L}\{y,s\,;\,v,\sigma,q,x\}$ (or $\mathbf{EW}\{x,t\,;\,v,\sigma,\rho,w\}$), the independent variable $y=w=\int\rho(x,t)dx$ is a *mass coordinate.*

Note that in the case of *linear elastodynamics*, $\sigma=K(\rho)=\rho_0/\rho$ with linear loading $f(x)$, the system $\mathbf{EW}\{x,t\,;\,v,\sigma,\rho,w\}$ (5) is a nonlinear PDE system, whereas the locally equivalent system $\mathbf{L}\{y,s\,;\,v,\sigma,q,x\}$ (2) becomes *linear.*

To further extend the tree of nonlocally related systems of one-dimensional nonlinear elasticity equations, one can use additional conservation laws and consider potential systems of the PDE systems $\mathbf{E}\{x,t\,;\,v,\sigma,\rho\}$, $\mathbf{EW}\{x,t\,;\,v,\sigma,\rho,w\}$ and/or $\mathbf{L}\{y,s\,;\,v,\sigma,q,x\}$. In particular, the Euler system $\mathbf{E}\{x,t\,;\,v,\sigma,\rho\}$ (1) has the conservation law

$$D_t(v-f(x)t)+D_x\left(\frac{v^2}{2}-M(\rho)\right)=0, \tag{6}$$

where $M(\rho) = \int \frac{K'(\rho)}{\rho} d\rho$. Introducing a potential variable $r(x, t)$, one obtains the potential equations

$$r_x = v - f(x)t; \quad r_t = M(\rho) - \frac{v^2}{2} \tag{7}$$

The Euler system $\mathbf{E}\{x, t\,; v, \sigma, \rho\}$ (1) also has a conservation law corresponding to the conservation of energy:

$$D_t\left(\rho\frac{v^2}{2} - \int M(\rho)d\rho - \rho\int f(x)dx\right) + D_x\left(\rho v\left[\frac{v^2}{2} - M(\rho) - \int f(x)dx\right]\right) = 0, \tag{8}$$

which yields the potential equations

$$s_x = \rho\frac{v^2}{2} - \int M(\rho)d\rho - \rho\int f(x)dx; \quad s_t = -\rho v\left[\frac{v^2}{2} - M(\rho) - \int f(x)dx\right]. \tag{9}$$

The nonlocal variable

$$s(x, t) = \int\left(\rho\frac{v^2}{2} - \int M(\rho)d\rho - \rho\int f(x)dx\right)dx$$

is an "energy coordinate", analogous to the mass coordinate w and the "velocity coordinate" r. The three conservation laws (mass, average velocity and energy) yield the following seven distinct nonlocally related (potential) systems of the Euler system $\mathbf{E}\{x, t\,; v, \sigma, \rho\}$ [12]:

- Three *singlet* potential systems: $\mathbf{EW}\{x, t\,; v, \sigma, \rho, w\}$ (5),

$$\mathbf{ER}\{x, t\,; v, \sigma, \rho, r\}: \quad \begin{cases} \rho_t + (\rho v)_x = 0, \\ r_x = v - f(x)t, \\ r_t = M(\rho) - v^2/2, \\ \sigma = K(\rho). \end{cases} \tag{10}$$

  and

$$\mathbf{ES}\{x, t\,; v, \sigma, \rho, s\}: \quad \begin{cases} \rho_t + (\rho v)_x = 0, \\ s_x = \rho v^2/2 - \int M(\rho)d\rho - \rho\int f(x)dx, \\ s_t = -\rho v\left[v^2/2 - M(\rho) - \int f(x)dx\right], \\ \sigma_x + \rho f(x) = \rho(v_t + vv_x), \\ \sigma = K(\rho). \end{cases} \tag{11}$$

- Three *couplet* potential systems:

$$\mathbf{ERW}\{x,t\,;\,v,\sigma,\rho,r,w\}:\quad\begin{cases}\rho_t+(\rho v)_x=0,\\ r_x=v-f(x)t,\\ r_t=M(\rho)-v^2/2,\\ w_x=\rho,\\ w_t=-\rho v,\\ \sigma=K(\rho).\end{cases}\tag{12}$$

$$\mathbf{ESW}\{x,t\,;\,v,\sigma,\rho,s,w\}:\quad\begin{cases}\rho_t+(\rho v)_x=0,\\ s_x=\rho v^2/2-\displaystyle\int M(\rho)d\rho-\rho\int f(x)dx,\\ s_t=-\rho v\left[v^2/2-M(\rho)-\displaystyle\int f(x)dx\right],\\ w_x=\rho,\\ w_t=-\rho v,\\ \sigma_x+\rho f(x)=\rho(v_t+vv_x),\\ \sigma=K(\rho).\end{cases}\tag{13}$$

and

$$\mathbf{ERS}\{x,t\,;\,v,\sigma,\rho,r,s\}:\quad\begin{cases}\rho_t+(\rho v)_x=0,\\ r_x=v-f(x)t,\\ r_t=M(\rho)-v^2/2,\\ s_x=\rho v^2/2-\displaystyle\int M(\rho)d\rho-\rho\int f(x)dx,\\ s_t=-\rho v\left[v^2/2-M(\rho)-\displaystyle\int f(x)dx\right],\\ \sigma_x+\rho f(x)=\rho(v_t+vv_x),\\ \sigma=K(\rho).\end{cases}\tag{14}$$

- One *triplet* potential system

$$\mathbf{ERSW}\{x,t\,;\,v,\sigma,\rho,r,s,w\}:\quad\begin{cases}\rho_t+(\rho v)_x=0,\\ r_x=v-f(x)t,\\ r_t=M(\rho)-v^2/2,\\ s_x=\rho v^2/2-\displaystyle\int M(\rho)d\rho-\rho\int f(x)dx,\\ s_t=-\rho v\left[v^2/2-M(\rho)-\displaystyle\int f(x)dx\right],\\ w_x=\rho,\\ w_t=-\rho v,\\ \sigma_x+\rho f(x)=\rho(v_t+vv_x),\\ \sigma=K(\rho).\end{cases}\tag{15}$$

(For $f(x)=$ const, one can obtain additional nonlocally related PDE systems, as discussed in [7].) Hence, *for arbitrary forms of the constitutive functions $K(\rho)$ and $f(x)$, one has a tree of equivalent and nonlocally*

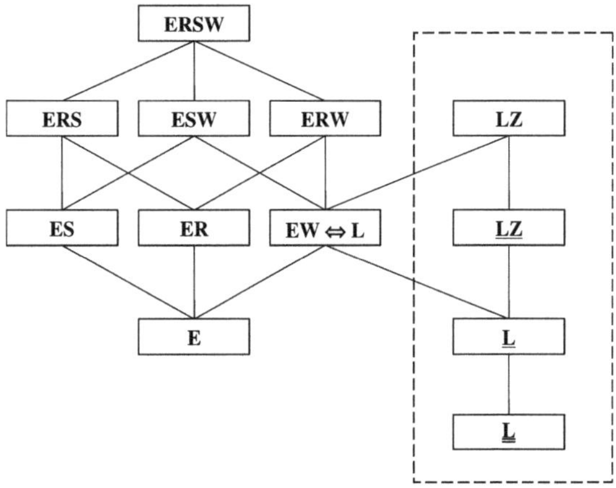

Fig. 1 Tree of nonlocally-related systems of nonlinear elasticity. (The dotted box corresponds to nonlocally related systems that arise for the case $f(x) = \text{const}$ [7].)

related systems of nonlinear elasticity, consisting of eight PDE systems $\mathbf{E}\{x, t\,;\, v, \sigma, \rho\}$, $\mathbf{L}\{y, s\,;\, v, \sigma, q, x\} \Leftrightarrow \mathbf{EW}\{x, t\,;\, v, \sigma, \rho, w\}$, $\mathbf{ER}\{x, t\,;\, v, \sigma, \rho, r\}$, $\mathbf{ES}\{x, t\,;\, v, \sigma, \rho, s\}$, $\mathbf{ERW}\{x, t\,;\, v, \sigma, \rho, r, w\}$, $\mathbf{ESW}\{x, t\,;\, v, \sigma, \rho, s, w\}$, $\mathbf{ERS}\{x, t\,;\, v, \sigma, \rho, r, s\}$, *and* $\mathbf{ERSW}\{x, t\,;\, v, \sigma, \rho, r, s, w\}$ (Figure 1). All nonlocally related PDE systems in the tree provide equivalent descriptions of nonlinear 1D elastodynamics, and thus naturally extend the traditional Lagrangian and Eulerian viewpoints.

4 Point and Nonlocal Symmetry Classification of the Lagrange System $\mathbf{EW}\{x, t\,;\, v, \sigma, \rho, w\} \Leftrightarrow \mathbf{L}\{y, s\,;\, v, \sigma, q, x\}$

A *symmetry* of a system of PDEs is any transformation of its solution manifold into itself (i.e., a symmetry transforms any solution to another solution of the same system).

Lie's algorithm is used to find one-parameter (ε) Lie groups of point transformations (point symmetries)

$$\begin{aligned} (x^*)^i &= f^i(x, u; \varepsilon), \quad i = 1, \dots, n, \\ (u^*)^j &= g^j(x, u; \varepsilon), \quad j = 1, \dots, m, \end{aligned} \tag{16}$$

that leave invariant a given system of N partial differential equations $\mathbf{R}\{x; u\}$ [1–4] such that $R^\rho[u^*] = 0$, $\rho = 1, \dots, N$, if and only if $R^\sigma[u] = 0$, $\sigma = 1, \dots, N$.

Global Lie transformation groups (16) are in one-to-one correspondence with local transformations

$$\begin{aligned} (x^*)^i &= x^i + \varepsilon \xi^i(x, u) + O(\varepsilon^2), \quad i = 1, \dots, n, \\ (u^*)^j &= u^j + \varepsilon \eta^j(x, u) + O(\varepsilon^2), \quad j = 1, \dots, m, \end{aligned} \tag{17}$$

where ξ^i, η^j are components of a vector field (infinitesimal generator)

$$X = \xi^i(x,u)\frac{\partial}{\partial x^i} + \eta^j(x,u)\frac{\partial}{\partial u^j} \tag{18}$$

tangent to the solution manifold of the given PDE system.

For a given PDE system, nonlocal symmetries called *potential symmetries* can arise naturally by applying Lie's algorithm to a related potential system. Such a symmetry is a nonlocal symmetry when at least one component of the symmetry generator has an essential dependence on a nonlocal variable. The complete point symmetry classification for the potential system **EW**$\{x,t\,;\,v,\sigma,\rho,w\}$ (5) in terms of its constitutive and loading functions was presented in [7]. In particular, it was found that in the cases

$$K(\rho) = \frac{1}{2}\left(\arctan\frac{1}{\rho} + \frac{\rho}{\rho^2+1}\right) \quad \text{or} \quad K(\rho) = \frac{1}{4}\ln\frac{\rho-1}{\rho+1} - \frac{1}{2}\frac{\rho}{\rho^2-1},$$

for a linear body force $f(x) = x$, the potential system **EW**$\{x,t\,;\,v,\sigma,\rho,w\}$ (5) has two point symmetries which are nonlocal symmetries of the Euler system **E**$\{x,t\,;\,v,\sigma,\rho\}$ (1). One of these nonlocal symmetries is used in the following section to construct a corresponding exact invariant solution of the Euler system **E**$\{x,t\,;\,v,\sigma,\rho\}$ (1).

5 Calculation of Group Invariant Solutions Arising from the Lagrange System EW$\{x,t\,;\,v,\sigma,\rho,w\}$ (5)

The general method for finding invariant solutions following from local symmetries is presented in detail in [1,4]. For invariant solutions arising from nonlocal (potential) symmetries, see also [15].

Let G be a one-parameter Lie group of point symmetries of the potential system **EW**$\{x,t\,;\,v,\sigma,\rho,w\}$ (5), with an infinitesimal generator

$$X = \xi\frac{\partial}{\partial x} + \tau\frac{\partial}{\partial t} + \eta^v\frac{\partial}{\partial v} + \eta^\rho\frac{\partial}{\partial \rho} + \eta^w\frac{\partial}{\partial w}. \tag{19}$$

Here $\xi, \tau, \eta^v, \eta^\rho$ and η^w are functions of x, t, v, σ, ρ and w. The corresponding invariant solutions

$$(v,\rho,w) \;=\; (V(x,t),\, R(x,t),\, W(x,t)) \tag{20}$$

of the potential system **EW**$\{x,t\,;\,v,\sigma,\rho,w\}$ satisfy

$$X\cdot\left.\begin{bmatrix} v - V(x,t) \\ \rho - R(x,t) \\ w - W(x,t)\end{bmatrix}\right|_{(v,\rho,w)\,=\,(V(x,t),R(x,t),W(x,t))} = 0 \tag{21}$$

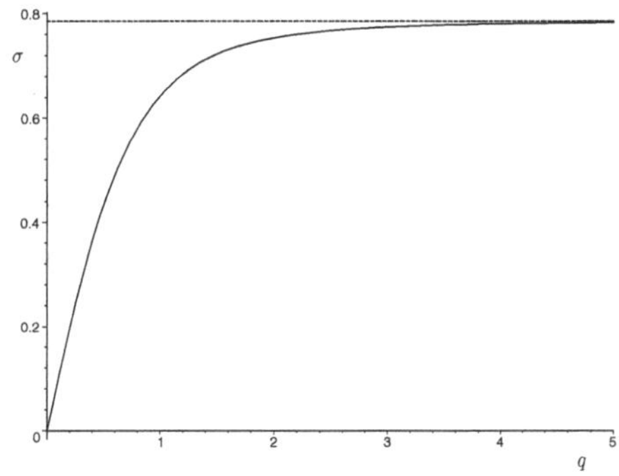

Fig. 2 Stress-strain curve for the constitutive relation $\sigma = K(\rho)$ given by (22). Here $q = 1/\rho$.

as well as the system $\mathbf{EW}\{x, t\,; v, \sigma, \rho, w\}$ (5). We now calculate specific invariant solutions of the potential system $\mathbf{EW}\{x, t\,; v, \sigma, \rho, w\}$ (5) arising as reductions from a point symmetry that is a potential symmetry of the Euler system $\mathbf{E}\{x, t\,; v, \sigma, \rho\}$ (1).

We choose the constitutive relation $\sigma = K(\rho)$ given by

$$\sigma = K(\rho) = \frac{1}{2}\arctan\frac{1}{\rho} + \frac{1}{2}\frac{\rho}{\rho^2+1}, \tag{22}$$

(see Figure 2), and a linear body force $f(x) = x$. In this case, the potential Euler system $\mathbf{EW}\{x, t\,; v, \sigma, \rho, w\}$ (5) has the point symmetry

$$Y_4 = \frac{e^t}{\rho}\left[\frac{\partial}{\partial t} + (v+\rho w)\frac{\partial}{\partial x} + (x+\rho w)\frac{\partial}{\partial v} - \rho(\rho^2+1)\frac{\partial}{\partial \rho} - \rho(x-v)\frac{\partial}{\partial w}\right],$$

which is clearly a nonlocal symmetry of the Euler system $\mathbf{E}\{x, t\,; v, \sigma, \rho\}$ (1), since its x- and v-components depend on the potential variable w.

The physical dependent variables ρ, v, w, σ are found as functions of x and t. The velocity $v(x, t)$ solves the implicit equation:

$$v(x,t) = e^t\left(\frac{C(U)}{A(U)} - \frac{U}{A^2(U)}\right) \tag{23}$$

where $A(U)$ and $C(U)$ are given by

$$A(U) = \sqrt{U^2+\alpha^2}, \quad C(U) = \frac{1}{2}\frac{\alpha U + (U^2+\alpha^2)\left(\beta - \arctan\frac{U}{\alpha}\right)}{\alpha(U^2+\alpha^2)}, \tag{24}$$

with α, β constants of integration, and the similarity variable given by $U = e^t(x - v(x, t))$.

For every value of x and t, the solution $v(x, t)$ of (23) can be found numerically. After $v(x, t)$ (and thus the similarity variable U) is determined, the density and the mass coordinate are obtained from the formulas

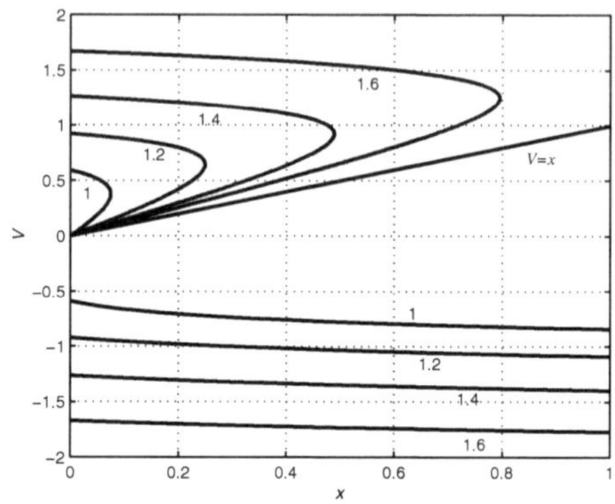

Fig. 3 Solution curves of the implicit equation (23) defining the material velocity $v(x,t)$ ($\alpha = 2$; $t = 1, 1.2, 1.4, 1.6$).

$$\rho(x,t) = \frac{\sqrt{U^2+\alpha^2}}{\sqrt{e^{2t}-U^2-\alpha^2}}, \qquad w(x,t) = \frac{v(x,t)-x}{\rho(x,t)}.$$

We seek a solution describing a nonlinear deformation of an elastic slab $x_0 < x < L(t)$, attached at $x = x_0$ (i.e., subject to the boundary condition $v\,(x_0,t) = 0$). We use the boundary conditions

$$v\,(x_0,t) = 0, \quad \rho\,(x_0,t) = R(t), \quad w(x_0,t) = 0. \tag{25}$$

The latter boundary condition is due to the definition of the mass coordinate (potential variable): $w(x,t) = \int_{x_0}^{x} \rho(s,t)ds$. When substituted into equation (23), this boundary condition yields $\beta = 0, \quad x_0 = 0$.

The velocity $v(x,t)$ following from equation (23) turns out to be a three-valued function for $0 \le x < x^*(t)$, where $x^*(t)$ is a bifurcation point. Sample curves of $v(x,t)$ for $\alpha = 2$ and times $t = 1, 1.2, 1.4, 1.6$ are shown in Figure 3.

From the three possible values of $v(x,t)$ that arise from the implicit equation (23), only one branch is physical. Indeed, one may check that only the middle branch (the one closest to $v(x,t) = x$) yields a real-valued density function [7].

It is also important to note that the geometrical velocity of the bifurcation point $v^*(t) = dx^*(t)/dt$ is always greater than the physical velocity $v(x^*(t),t)$ at the bifurcation point. The expression for the total mass between $x = 0$ and the bifurcation point $x^*(t)$ (per unit area of the slab cross-section) is given by

$$w(x^*(t),t) = 1 - \alpha e^{-t}, \tag{26}$$

which is an increasing function of time, in agreement with the previous remark. The invariant solutions are defined for $0 < x < x^*(t)$. If the initial length of the slab $L(t_0)$ is chosen ($0 < L(t_0) < x^*(t_0)$), then the solution is regular for all times.

The family of invariant solutions presented in this section describes the nonlinear deviation of a trivial "homogeneous stretching" solution $v(x,t) = x$, $\rho(x,t) = e^{-t}$

of the Euler system $\mathbf{E}\{x, t\,;\, v, \sigma, \rho\}$ (1). A specific numerical example of such an invariant solution was constructed in [7].

6 Conservation Laws of Dynamical Nonlinear Elasticity

For a PDE system $\mathbf{R}\{x\,;\, u\}$ (3), one can consider the problem of finding local conservation laws of the form (4). The fluxes $\Phi^i[u]$ may depend on x, u and derivatives of u up to an arbitrary order. In practice, conservation laws are used for direct physical interpretation, analysis, and development of efficient numerical methods.

The direct method of finding conservation laws involves considering a linear combination of equations of a given PDE system (3) with a set of *multipliers* $\{\Lambda_\sigma\}$, which may depend on independent and dependent variables and their derivatives. A linear combination yields a conservation law (4) if and only if

$$\Lambda_\sigma[U]R^\sigma[U] \equiv D_i\Phi^i[U] \tag{27}$$

for some fluxes $\{\Phi^i[U]\}$ (here U denotes a vector of arbitrary functions of x). Then the conservation law $D_i\Phi^i[u] = 0$ holds on the solutions $U = u(x)$ of the system (3).

In the direct method, the determining equations that yield sets of multipliers $\{\Lambda_\sigma[U]\}$ are found from the known fact: an expression is a divergence expression if and only if it is annihilated by Euler operators with respect to all dependent variables [2, 13, 14]:

$$E_{U^k}\left(\Lambda_\sigma[U]R^\sigma[U]\right) = 0, \quad k = 1, \ldots, m. \tag{28}$$

Here $U = (U^1(x), \ldots, U^m(x))$ is a set of arbitrary functions, and E_{U^k} is the *Euler operator* with respect to U^k, given by

$$E_{U^k} = \frac{\partial}{\partial U^k} - D_i\frac{\partial}{\partial U^k_i} + \cdots + (-1)^j D_{i_1}\cdots D_{i_j}\frac{\partial}{\partial U^k_{i_1\cdots i_j}} + \cdots.$$

Symbols $U^k_{i_1\cdots i_j}$ denote partial derivatives $\frac{\partial^j U^k}{\partial x^{i_1}\ldots\partial x^{i_j}}$.

Equations (28) are linear determining equations for the multipliers $\{\Lambda_\sigma[U]\}$. In practice, to perform a computation, one chooses the maximal order of derivatives $q \geq 0$ in the dependence of multipliers $\Lambda_\sigma[U]$. When the multipliers are determined, one finds the corresponding set of fluxes of the conservation law (4) either by solving (27) directly, or using integral homotopy operators [13, 14].

Note that the direct method does not require the PDE system to have a variational formulation, and does not use any version of Noether's theorem.

Now our goal is to construct examples of nonlocal conservation laws of the Euler system $\mathbf{E}\{x, t\,;\, v, \sigma, \rho\}$ (1) of nonlinear elastodynamics, which arise as local conservation laws of a potential system of $\mathbf{E}\{x, t\,;\, v, \sigma, \rho\}$. The following important theorem holds [16].

Theorem 1. *Let* $\mathbf{S}\{x\,;\, u, v\}$ *given by*

$$S^{\mu}[u, v] \equiv S^{\mu}(x, u, v, \partial u, \partial v, \ldots, \partial^k u, \partial^k v) = 0, \quad \mu = 1, \ldots, M \tag{29}$$

be a potential system of a PDE system $\mathbf{R}\{x\,;\, u\}$ (3)*, where* $v = (v^1, \ldots, v^l)$ *are nonlocal (potential) variables. A local conservation law*

$$\widetilde{\Lambda}_{\mu}[u, v] S^{\mu}[u, v] = D_i \Psi^i[u, v] = 0 \tag{30}$$

of the potential system $\mathbf{S}\{x\,;\, u, v\}$ (29) *yields a nonlocal conservation law of* $\mathbf{R}\{x\,;\, u\}$ (3) *if and only if the multipliers* $\widetilde{\Lambda}_{\mu}[u, v]$ *essentially depend on the nonlocal variable(s)* v.

We now seek local conservation laws of the potential system $\mathbf{EW}\{x, t\,;\, v, \sigma, \rho, w\}$ (5) that yield nonlocal conservation laws of the Euler system $\mathbf{E}\{x, t\,;\, v, \sigma, \rho\}$ (1), i.e., conservation laws of the potential system $\mathbf{EW}\{x, t\,;\, v, \sigma, \rho, w\}$ (5) arising from one or more of multipliers with an essential dependence on the potential variable w. One may write

$$\mathbf{EW}\{x, t\,;\, v, \sigma, \rho, w\} : \quad \begin{cases} w_x - \rho = 0, \\ w_t + \rho v = 0, \\ \dfrac{K'(\rho)}{\rho}\rho_x + f(x) - (v_t + v v_x) = 0. \end{cases} \tag{31}$$

For the conservation law multipliers, we use the ansatz

$$\widetilde{\Lambda}_{\mu} = \widetilde{\Lambda}_{\mu}(x, t, V, R, w, V_x, V_t), \quad \mu = 1, 2, 3,$$

and require that for arbitrary functions $V(x, t)$, $P(x, t)$, $W(x, t)$, one has

$$\widetilde{\Lambda}_1(x, t, V, P, W, V_x, V_t)(W_x - R) + \widetilde{\Lambda}_2(x, t, V, P, W, V_x, V_t)(W_t + RV) \\ + \widetilde{\Lambda}_3(x, t, V, P, W, V_x, V_t)\left(\frac{K'(R)}{R}R_x + f(x) - (V_t + V V_x)\right) \equiv D_i \Psi^i[U, V]. \tag{32}$$

Then on solutions $V = v$, $P = \rho$, $W = w$ of the PDE system $\mathbf{EW}\{x, t\,;\, v, \sigma, \rho, w\}$ (5), the expression (32) becomes a conservation law.

Subsequent application of the Euler operators with respect to V, P and W to the left-hand side of (32) yields determining equations for the multipliers $\widetilde{\Lambda}_{\mu}$, $\mu = 1, 2, 3$. From the determining equations, it follows that one or more of the multipliers essentially depend on W, and thus a nonlocal conservation law of the Euler system $\mathbf{E}\{x, t\,;\, v, \sigma, \rho\}$ (1) arises in the following cases.

Case 1: $f(x) = f_1 = \text{const}$, $K(\rho) = \frac{\rho^{1/3}}{(A\rho+B)^{1/3}} + C$, $A, B \neq 0$. In this case, the multipliers are given by

$$\Lambda_1 = V - f_1 t + V_x\left(f_1\frac{t^2}{2} - x - \frac{3AW}{B}\right),$$
$$\Lambda_2 = V^2 + xV_t + f_1\left[\frac{t^2}{2}(1-V_t) - x - tV\right] + \frac{3AW}{B}(V_t - f_1) + \frac{P^{-2/3}}{(AP+B)^{1/3}},$$
$$\Lambda_3 = P\left(f_1\frac{t^2}{2} - x - \frac{3AW}{B}\right) - 2W. \tag{33}$$

Case 2: $f(x) = f_0 x + f_1$, $K(\rho) = A\rho^{1/3} + B$, A, B, f_0, f_1 = const. The multipliers are given by

$$\begin{aligned}
\Lambda_1 &= f_0(V - xV_x) - f_1 V_x,\\
\Lambda_2 &= -f_0^2x^2 + f_0(x(V_t - 2f_1) + V^2) + f_1(V_t - f_1) + Af_0P^{-2/3},\\
\Lambda_3 &= -f_0(xP + 2W) - f_1 P.
\end{aligned} \tag{34}$$

Case 3: $f(x) = f_1$, $K(P) = A\rho^{1/3} + B$. The multipliers are given by

$$\begin{aligned}
\Lambda_1 &= -f_1 t + V - V_x\left(x - f_1\frac{t^2}{2}\right),\\
\Lambda_2 &= f_1^2 - f_1\left(x + tV + \frac{t^2}{2}V_t\right) + V^2 + xV_t + AP^{-2/3},\\
\Lambda_3 &= -\left(x - f_1\frac{t^2}{2}\right)P - 2W.
\end{aligned} \tag{35}$$

Case 4: $f(x) = f_0 x + f_1$, $K(\rho) = \rho_0/\rho$, ρ_0 = const. This case corresponds to linear elasticity with linear loading. Here one finds that the potential system $\mathbf{EW}\{x, t\,;\, v, \sigma, \rho, w\}$ (5) has an infinite number of conservation laws corresponding to nonlocal conservation laws of the Euler equations $\mathbf{E}\{x, t\,;\, v, \sigma, \rho\}$. This reflects the fact that for linear elasticity with linear loading, the system $\mathbf{EW}\{x, t\,;\, v, \sigma, \rho, w\}$ (5) can be linearized by a point transformation [17]. Indeed, this transformation is the interchange of dependent and independent variables that transforms the system $\mathbf{EW}\{x, t\,;\, v, \sigma, \rho, w\}$ (5) to the Lagrange system $\mathbf{L}\{y, s\,;\, v, \sigma, q, x\}$ (2) (see Section 3).

7 Conclusions

In this paper, we presented the complete set of dynamic nonlinear elasticity equations in Lagrangian and Eulerian formulations, as well as in several other equivalent formulations. The corresponding nonlocally related systems were used for the classification of nonlocal symmetries and construction of examples of invariant solutions of the Euler system $\mathbf{E}\{x, t\,;\, v, \sigma, \rho\}$ (1). Moreover, we demonstrated how nonlocal conservation laws can be obtained for the Euler system through consideration of local conservation laws of its potential system.

Future work will include flux computation, interpretation and applications of the nonlocal conservation laws of the Euler system $\mathbf{E}\{x, t\,;\, v, \sigma, \rho\}$ (1) obtained in this paper, and also the study of conservation laws of two-dimensional models of nonlinear elasticity.

References

1. Bluman, G.W. and Anco, S.C., *Symmetry and Integration Methods for Differential Equations*. Springer, New York, 2002.
2. Olver, P.J., *Application of Lie Groups to Differential Equations*. Springer, New York, 1993.
3. Ovsiannikov, L.V., *Group Analysis of Differential Equations*. Academic Press, New York, 1982.
4. Bluman, G.W. and Kumei, S., *Symmetries and Differential Equations*. Springer, New York, 1989.
5. Dorodnitsyn, V. and Winternitz, P., Lie point symmetry preserving discretizations for variable coefficient Korteweg–de Vries equations. Modern group analysis. *Nonlin. Dynam.* **22**, 2000, 49–59.
6. Maugin, G.A., *Material Inhomogeneities in Elasticity*. Cambridge University Press, 1993.
7. Bluman, G., Cheviakov, A. and Ganghoffer, J.F., Nonlocally related PDE systems for one-dimensional nonlinear elastoydnamics. *Int. J. Engng Math.*, 2008, doi: 10.1007/s10665-008-9221-7.
8. Horgan, C.O. and Murphy, J.H., Lie group analysis and plane strain bending of cylindrical sectors for compressible nonlinearly elastic materials. *IMA J. Appl. Math.* **70**, 2005, 80–91.
9. Horgan, C.O. and Murphy, J.H., A Lie group analysis of the axisymmetric equations of finite elastostatics for compressible materials. *Math. Mech. Sol.* **10**, 2005, 311–333.
10. Budiansky, B. and Rice, J.R., Conservation laws and energy release rates. *J. Appl. Mech.* **40**, 1968, 201–203.
11. Hatfield, G.A. and Olver P.J., Canonical forms and conservation laws in linear elastostatics. *Arch. Mech.* **50**, 1998, 389–404.
12. Bluman, G., Cheviakov, A.F. and Ivanova, N.M., Framework for nonlocally related PDE systems and nonlocal symmetries: extension, simplification, and examples. *J. Math. Phys.* **47**, 2006, 113505.
13. Anco, S. and Bluman, G., Direct construction method for conservation laws of partial differential equations. Part I: Examples of conservation law classfications. *Eur. J. Appl. Math.* **13**, 2002, 545–566.
14. Anco, S. and Bluman, G., Direct construction method for conservation laws of partial differential equations. Part II: General treatment. *Eur. J. Appl. Math.* **13**, 2002, 567–585.
15. Cheviakov, A.F., An extended procedure for finding exact solutions of partial differential equations arising from potential symmetries. Applications to gas dynamics. *J. Math. Phys.* **49**, 2008, 083502.
16. Kunzinger, M. and Popovych, R.O., *Potential Conservation Laws*. arXiv:0803.1156v2 [math-ph], 2008.
17. Anco, S., Bluman, G. and Wolf, T., Invertible mappings of nonlinear pdes through admitted conservation laws, *Acta Appl. Math.* **101**, 2008, 21–38.

Configurational Forces in the Theory of Two-Phase Plates

Victor A. Eremeyev and Holm Altenbach

Abstract Within the framework of the direct approach to the theory of plates the configurational forces are introduced taking into account the transverse shear stiffness. Their applications to plates made of non-homogeneous materials such as laminates, functionally graded materials, two-phase materials, etc., are discussed. The basic items of the applied theory of plates are related to the formulation of all balances for a deformable directed surface and to the specification of the constitutive equations. The variational principle of the total energy is established for plates assuming the presence of a variable surface singular curve. The surface singular curve separates the parts of the plate with different material properties. The continuity conditions at the singular curve are obtained. The energy-momentum tensor is constructed. Equilibrium and quasistatic motion of the surface singular curve are investigated on the base of the proposed kinetic equation. As an example of application of configurational forces we present the model of the deformation of a plate made of a material undergoing phase transitions (e.g., a shape memory alloy).

1 Introduction

The concept of the configurational force is useful in continuum mechanics as was shown in [13, 18, 20, 26, 29] among others. Configuration forces arise from the consideration of motion of defects in the continuous medium. Generally, the nature of defects may be different. Here we consider a defect as a singular curve which separates the parts of a plate consisting of different phases that means, we consider the equilibrium of plates made of materials undergoing phase transitions.

Victor A. Eremeyev
South Scientific Center of RASci & South Federal University, Milchakova str. 8a, 344090 Rostov on Don, Russia; e-mail: eremeyev.victor@gmail.com

Holm Altenbach
Martin-Luther-Universität Halle-Wittenberg, D-06099 Halle (Saale), Germany; e-mail: holm.altenbach@iw.uni-halle.de

P. Steinmann (ed.), IUTAM Symposium on Progress in the Theory and Numerics of Configurational Mechanics, 121–130.

The phase transitions (PT) of martensitic type are very important in the mechanics of materials. In particular, PT are responsible for the shape memory effect in some materials, see, e.g., [8, 23]. Mechanics of solids with PT has been developed in a number of papers and books, see for instance [1, 7, 8, 16–18, 23, 24].

The growing interest to understanding of the behavior of such two-dimensional structures as thin films, plates and shells made of shape memory alloys and other materials undergoing PT is based on the perspective applications of these structures for design of microelectromechanical systems (MEMS). The mechanics of martensititic thin films was discussed, for example, in [8, 9, 19]. Let us note that experiments on shape memory alloys are usually performed with thin-walled samples, for example rectangular plates [8, 12, 27].

Within the dynamically and kinematically exact theory of shells presented in [11, 25], the equilibrium conditions as well as the continuity conditions for quasi-static deformations of elastic thin-walled structures (plates and shells) undergoing PT of the martensitic type were formulated in [14, 15, 28]. By analogy to the three-dimensional case, in the two-dimensional theory of shells a singular surface curve was taken as the phase boundary in the shell. For the two-phase shells the Eshelby's (energy-momentum) tensor was introduced in [14]. From the point of view of configurational mechanics this tensor represents the configurational (or driving) force acting on the phase interface. Thus, the tensor is responsible for its motion. The configurational forces were investigated considering the theory of elastic beams in [20–22] while for the elastic plates in [10, 20], respectively.

Following [14, 28] we obtain here the equilibrium conditions at the phase interface applying the 5-parametric theory of plates presented in [4–6]. The two-phase plate is considered as some material surface consisting of two material phases divided by a sufficiently smooth surface curve. Below the following items are discussed. Considering [4–6, 30, 31] in Section 2 we recall the governing equations of the linear theory of plates based on the direct approach. Here we also introduce both the static and the kinematic compatibility conditions at the curvilinear phase boundary. In Section 3 we formulate the variational principle of stationarity of the total energy functional taking into account the variable surface singular curve. From the variational principle we deduce the additional thermodynamic equilibrium continuity condition satisfied at the phase interface curve. This condition is written using the energy-momentum tensor $\boldsymbol{\mu}$. In Section 4 we assume the kinetic equation describing the motion of the phase interface curve during the quasistatic deformation processes of the two-phase plate. As an example the tension of a two-phase rectangular plate is briefly discussed in Section 5.

2 Basic Equations of the Direct Theory of Plates

Let us assume the geometrically and physically linear plate theory based on the so-called direct approach. In this case one states a two-dimensional deformable surface. On each part of this deformable surface forces and moments are acting – they are

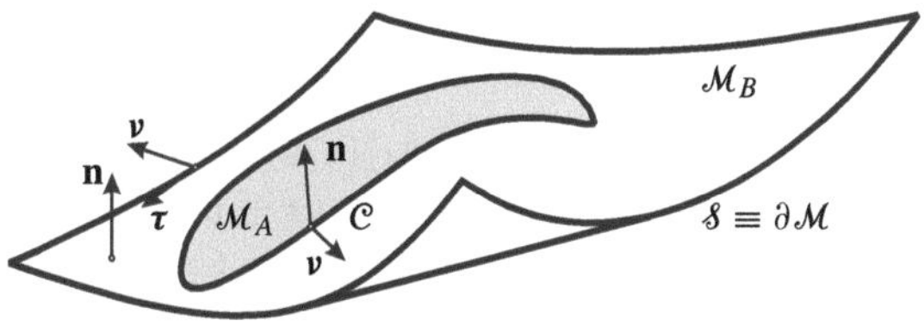

Fig. 1 Two-phase plate

the primary variables. The next step is the introduction of the deformation measures. Finally, it is necessary to interlink the forces and the moments with the deformation variables (constitutive equations). Such a plate theory is formulated by a more natural way in comparison with the other approaches because it is so strong and so exact as the three-dimensional continuum mechanics. But the identification of the stiffness and other parameters is a non-trivial problem and must be realized for each class of plates individually.

In the considered theory of plates we make two basic assumptions:

Assumption 1: The plate (homogeneous or inhomogeneous in transverse direction) can be represented by a deformable surface $\mathcal{M}$ (Figure 1).
Assumption 2: Each material point is an infinitesimal rigid body with 5 degrees of freedom (3 translations and 2 rotations).

In addition, the theory presented here is limited by small displacements and rotations and the quadratic strain energy density assumptions.

The equilibrium equations and the kinematic equations are given by the relations [4–6, 30, 31]

$$\nabla \cdot \mathsf{T} + \mathbf{q} = \mathbf{0}, \qquad \nabla \cdot \mathsf{M} + \mathsf{T}_{\times} + \mathbf{m} = \mathbf{0}, \tag{1}$$

$$\boldsymbol{\epsilon} = \frac{1}{2}\left[\nabla \mathbf{v} + (\nabla \mathbf{v})^{\mathrm{T}}\right], \qquad \boldsymbol{\gamma} = \nabla w + \mathsf{c} \cdot \boldsymbol{\varphi}, \qquad \boldsymbol{\kappa} = \nabla \boldsymbol{\varphi}. \tag{2}$$

Here T, M are the tensors of forces and moments, $\mathbf{q}$, $\mathbf{m}$ are the surface load vectors (forces and moments), $\mathsf{T}_{\times}$ is the vector invariant of the force tensor, ∇ is the nabla operator, $\mathbf{v} = \mathbf{u} \cdot \mathsf{a}$, $w = \mathbf{u} \cdot \mathbf{n}$, $\mathbf{u}$, $\boldsymbol{\varphi}$ are the vectors of displacements and rotations, and $(\ldots)^{\mathrm{T}}$ denotes transposed. a is the first metric tensor, $\mathbf{n}$ is the unit normal vector, $\mathsf{c} = -\mathsf{a} \times \mathbf{n}$ is the discriminant tensor, $\boldsymbol{\epsilon}$, $\boldsymbol{\gamma}$ and $\boldsymbol{\kappa}$ are the tensor of in-plane strains, the vector of transverse shear strains and the tensor of the out-of-plane strains, respectively. In the case of an orthotropic material behavior and a plane mid-surface we assume the following strain energy and constitutive equations

$$\begin{aligned} W(\boldsymbol{\epsilon}, \boldsymbol{\gamma}, \boldsymbol{\kappa}) = {} & \frac{1}{2}\boldsymbol{\epsilon} \cdot\cdot \mathsf{A} \cdot\cdot \boldsymbol{\epsilon} + \boldsymbol{\epsilon} \cdot\cdot \mathsf{B} \cdot\cdot \boldsymbol{\kappa} + \frac{1}{2}\boldsymbol{\kappa} \cdot\cdot \mathsf{C} \cdot\cdot \boldsymbol{\kappa} + \frac{1}{2}\boldsymbol{\gamma} \cdot \Gamma \cdot \boldsymbol{\gamma} \\ & + \mathsf{N}_0 \cdot\cdot \boldsymbol{\varepsilon} + \mathsf{M}_0^{\mathrm{T}} \cdot\cdot \boldsymbol{\kappa} + \mathbf{Q}_0 \cdot \boldsymbol{\gamma} + W_0, \end{aligned} \tag{3}$$

$$\mathsf{N} \equiv \mathsf{T} \cdot \mathsf{a} = \frac{\partial W}{\partial \boldsymbol{\epsilon}}, \quad \mathbf{Q} \equiv \mathsf{T} \cdot \mathsf{n} = \frac{\partial W}{\partial \boldsymbol{\gamma}}, \quad \mathsf{M}^{\mathrm{T}} = \frac{\partial W}{\partial \boldsymbol{\kappa}}. \tag{4}$$

A, B, C are 4th rank tensors, Γ is a 2nd rank tensor expressing the effective stiffness properties. They depend on the material properties and the cross-section geometry and given by the relations [31]

$$\begin{aligned}
\mathsf{A} &= A_{11}\mathsf{a}_1\mathsf{a}_1 + A_{12}(\mathsf{a}_1\mathsf{a}_2 + \mathsf{a}_2\mathsf{a}_1) + A_{22}\mathsf{a}_2\mathsf{a}_2 + A_{44}\mathsf{a}_4\mathsf{a}_4,\\
\mathsf{B} &= B_{13}\mathsf{a}_1\mathsf{a}_3 + B_{14}\mathsf{a}_1\mathsf{a}_4 + B_{23}\mathsf{a}_2\mathsf{a}_3 + B_{24}\mathsf{a}_2\mathsf{a}_4 + B_{42}\mathsf{a}_4\mathsf{a}_2,\\
\mathsf{C} &= C_{22}\mathsf{a}_2\mathsf{a}_2 + C_{33}\mathsf{a}_3\mathsf{a}_3 + C_{34}(\mathsf{a}_3\mathsf{a}_4 + \mathsf{a}_4\mathsf{a}_3) + C_{44}\mathsf{a}_4\mathsf{a}_4,\\
\mathsf{\Gamma} &= \Gamma_1\mathsf{a}_1 + \Gamma_2\mathsf{a}_2,
\end{aligned} \tag{5}$$

where $\mathsf{a}_1 = \mathsf{a} = \mathbf{e}_1\mathbf{e}_1 + \mathbf{e}_2\mathbf{e}_2$, $\mathsf{a}_2 = \mathbf{e}_1\mathbf{e}_1 - \mathbf{e}_2\mathbf{e}_2$, $\mathsf{a}_3 = \mathsf{c} = \mathbf{e}_1\mathbf{e}_2 - \mathbf{e}_2\mathbf{e}_1$, $\mathsf{a}_4 = \mathbf{e}_1\mathbf{e}_2 + \mathbf{e}_2\mathbf{e}_1$, and $\mathbf{e}_1$, $\mathbf{e}_2$ are unit basis vectors of an orthonormal coordinate system. In addition, one obtains the orthogonality condition for the a_i $(i = 1, 2, 3, 4)$

$$\frac{1}{2}\mathsf{a}_i \cdot\cdot \mathsf{a}_j = \delta_{ij},$$

where $\cdot\cdot$ is the double inner (dot) product. N_0, M_0 are the tensors of the initial in-plane forces and moments, $\mathbf{Q}_0$ is the vector of the initial shear forces, while W_0 is the initial value of the strain energy.

The identification of the effective stiffness tensors A, B, C and $\mathsf{\Gamma}$ should be performed on the base of the properties of the real material. Let us assume the Hooke's law with material properties which depend on the normal coordinate z. The identification of the effective properties can be performed with the help of static boundary value problems (two-dimensional, three-dimensional) and the comparison of the forces and moments (in the sense of averaged stresses or stress resultants). Finally, we get the following expressions for the classical stiffness tensor components [2–4]

$$\begin{aligned}
(A_{11};\, -B_{13};\, C_{33}) &= \frac{1}{4}\left\langle \frac{E_1 + E_2 + 2E_1\nu_{21}}{1 - \nu_{12}\nu_{21}}(1;\, z;\, z^2)\right\rangle,\\
(A_{22};\, B_{24};\, C_{44}) &= \frac{1}{4}\left\langle \frac{E_1 + E_2 - 2E_1\nu_{21}}{1 - \nu_{12}\nu_{21}}(1;\, z;\, z^2)\right\rangle,\\
(A_{12};\, -B_{23} = B_{14};\, -C_{34}) &= \frac{1}{4}\left\langle \frac{E_1 - E_2}{1 - \nu_{12}\nu_{21}} z(1;\, z;\, z^2)\right\rangle,\\
(A_{44};\, -B_{42};\, C_{22}) &= \langle G_{12}(1;\, z;\, z^2)\rangle,
\end{aligned} \tag{6}$$

where $\langle \ldots \rangle$ is the integral over the plate thickness h, while E_1, E_2, ν_{12}, ν_{21}, G_{12} are the elastic moduli of the orthotropic bulk material. In addition, two non-classical stiffness are obtained

$$\Gamma_1 = \frac{1}{2}(\lambda^2 + \eta^2)\frac{A_{44}C_{22} - B_{42}^2}{A_{44}}, \quad \Gamma_2 = \frac{1}{2}(\eta^2 - \lambda^2)\frac{A_{44}C_{22} - B_{42}^2}{A_{44}}. \tag{7}$$

Here η^2 and λ^2 are the smallest non-zero eigen-values of Sturm–Liouville problems

$$\frac{\mathrm{d}}{\mathrm{d}z}\left(G_{1n}\frac{\mathrm{d}Z}{\mathrm{d}z}\right) + \eta^2 G_{12}Z = 0, \quad \frac{\mathrm{d}}{\mathrm{d}z}\left(G_{2n}\frac{\mathrm{d}Z}{\mathrm{d}z}\right) + \lambda^2 G_{12}Z = 0, \quad \left.\frac{\mathrm{d}Z}{\mathrm{d}z}\right|_{|z|=h/2} = 0.$$

The boundary conditions are given by

$$\boldsymbol{\nu} \cdot \mathsf{T} = \mathbf{f}, \quad \boldsymbol{\nu} \cdot \mathsf{M} = \mathbf{l}, \quad (\mathbf{l} \cdot \mathbf{n} = 0) \quad \text{along } \mathcal{S}_f, \tag{8}$$

and

$$\mathbf{u} = \mathbf{u}^0, \quad \boldsymbol{\varphi} = \boldsymbol{\varphi}^0 \quad \text{along } \mathcal{S}_u. \tag{9}$$

Here $\mathbf{f}$ and $\mathbf{l}$ are external force and couple vectors acting along the part S_f of the boundary of the plate $\mathcal{S} = \mathcal{S}_f \cup \mathcal{S}_u \equiv \partial \mathcal{M}$, while $\mathbf{u}^0$ and $\boldsymbol{\varphi}^0$ are given functions describing the displacements and rotation of the plate boundary $\mathcal{S}_u$, respectively. $\boldsymbol{\nu}$ is the unit normal vector to $\mathcal{S}$ $(\boldsymbol{\nu} \cdot \mathbf{n} = 0)$. The relations (8) and (9) are the static and kinematic boundary conditions, respectively. Other mixed types of boundary conditions are possible. For example, the simple support boundary conditions corresponding to a hinge are given by

$$\boldsymbol{\nu} \cdot \mathsf{M} \cdot \boldsymbol{\tau} = \mathbf{0}, \quad \mathbf{u} = \mathbf{0}, \quad \boldsymbol{\varphi} \cdot \boldsymbol{\tau} = 0.$$

Here $\boldsymbol{\tau}$ is the unit tangent vector to $\mathcal{S}$ $(\boldsymbol{\tau} \cdot \mathbf{n} = \boldsymbol{\tau} \cdot \boldsymbol{\nu} = 0)$.

Let us consider the two-phase plate consisting of a material undergoing the phase transformations. In this case the plate is modeled by the surface $\mathcal{M} = \mathcal{M}_A \cup \mathcal{M}_B$, where $\mathcal{M}_{A,B}$ are surfaces consisting of phases A and B, respectively. The phase interface is a smooth curve $\mathcal{C}$ separating the surfaces $\mathcal{M}_{A,B}$ (Figure 1).

The curvilinear phase interfaces in plates can be either coherent or incoherent in rotations, see [14]. For the *coherent interface* both fields $\mathbf{u}$ and $\boldsymbol{\varphi}$ are supposed to be continuous at $\mathcal{C}$

$$[\![\mathbf{u}]\!] = \mathbf{0}, \quad [\![\boldsymbol{\varphi}]\!] = \mathbf{0}, \tag{10}$$

where double square brackets denote a jump of discontinuity across $\mathcal{C}$: $[\![(\ldots)]\!] = (\ldots)_B - (\ldots)_A$.

The phase interface is called *incoherent in rotations* if only $\mathbf{u}$ is continuous at $\mathcal{C}$ but the continuity of $\boldsymbol{\varphi}$ may be violated. In this case the condition $(10)_1$ is still satisfied, but $(10)_2$ may be violated.

3 Variation Principle of Total Energy

Following [14] we obtain the phase equilibrium conditions along the phase interface curve $\mathcal{C}$ using the variational principle

$$\delta E = 0, \quad E = \int_{\mathcal{M}_A} W_A \, da + \int_{\mathcal{M}_B} W_B \, da - A^{\text{ext}}, \tag{11}$$

where E is the functional of the total energy, $W_{A,B}$ are the strain energies corresponding to the phases A, B, and A^{ext} is the functional of external loads. We assume that δA^{ext} is given by the relation

$$\delta A^{\text{ext}} = \int_{\mathcal{M} \setminus \mathcal{C}} (\mathbf{q} \cdot \delta \mathbf{u} + \mathbf{m} \cdot \delta \boldsymbol{\varphi}) \, da + \int_{\mathcal{S}_f} (\mathbf{f} \cdot \delta \mathbf{u} + \mathbf{l} \cdot \delta \boldsymbol{\varphi}) \, ds, \tag{12}$$

where $\delta\mathbf{u}$ and $\delta\boldsymbol{\varphi}$ are variations of $\mathbf{u}$ and $\boldsymbol{\varphi}$, respectively. Let us note that $\delta\mathbf{u}$ and $\delta\boldsymbol{\varphi}$ satisfy the kinematic relations

$$\delta\mathbf{u} = \mathbf{0}, \quad \delta\boldsymbol{\varphi} = \mathbf{0} \quad \text{along } \mathcal{S}_u, \tag{13}$$

as well as the continuity conditions along $\mathcal{C}$. For the coherent interface the continuity conditions are

$$[\![\delta\mathbf{u}]\!] + V[\![\boldsymbol{\nu}\cdot\nabla\mathbf{u}]\!] = \mathbf{0}, \quad [\![\delta\boldsymbol{\varphi}]\!] + V[\![\boldsymbol{\nu}\cdot\nabla\boldsymbol{\varphi}]\!] = \mathbf{0}. \tag{14}$$

where V is the virtual velocity of $\mathcal{C}$, see [14], while for the phase interfaces incoherent in rotations the continuity conditions reduce to the relation $(14)_1$.

The variation of E is given by

$$\delta E = \int_{\mathcal{M}\setminus\mathcal{C}} \delta W\, da - \int_{\mathcal{C}} V\,[\![W]\!]\, ds - \delta A^{\mathrm{ext}}, \tag{15}$$

where

$$\delta W = \frac{\partial W}{\partial \boldsymbol{\epsilon}} \cdot\cdot\, \delta\boldsymbol{\epsilon}^{\mathrm{T}} + \frac{\partial W}{\partial \boldsymbol{\gamma}}\cdot\delta\boldsymbol{\gamma} + \frac{\partial W}{\partial \boldsymbol{\kappa}} \cdot\cdot\, \delta\boldsymbol{\kappa}^{\mathrm{T}}.$$

Using Eqs (13), (14) from Eq. $(11)_1$ one obtains the equilibrium equations (1), the static boundary conditions (8) as well as the following general continuity condition to be satisfied at the interface $\mathcal{C}$:

$$\int_{\mathcal{C}} \{V[\![W]\!] + [\![\boldsymbol{\nu}\cdot\mathsf{T}\cdot\delta\mathbf{u}]\!] + [\![\boldsymbol{\nu}\cdot\mathsf{M}\cdot\delta\boldsymbol{\varphi}]\!]\}\, ds = 0. \tag{16}$$

The balance equation on the phase interface $\mathcal{C}$ following from Eq. (16) consist of both the static balance equations and the so-called thermodynamical balance equation which is necessary for the determination of $\mathcal{C}$. For the coherent interface the static balance equations are

$$\boldsymbol{\nu}\cdot[\![\mathsf{T}]\!] = \mathbf{0}, \quad \boldsymbol{\nu}\cdot[\![\mathsf{M}]\!] = \mathbf{0}, \tag{17}$$

while the thermodynamical balance equation is given by

$$\boldsymbol{\nu}\cdot[\![\boldsymbol{\mu}_{\mathrm{C}}]\!]\cdot\boldsymbol{\nu} = 0. \tag{18}$$

For the phase interface incoherent in rotations from the variational principle we obtain the following continuity conditions along $\mathcal{C}$:

$$\boldsymbol{\nu}\cdot[\![\mathsf{T}]\!] = \mathbf{0}, \quad \boldsymbol{\nu}\cdot\mathsf{M}_{A,B} = \mathbf{0}, \quad \boldsymbol{\nu}\cdot[\![\boldsymbol{\mu}_{\mathrm{I}}]\!]\cdot\boldsymbol{\nu} = 0$$

Here

$$\boldsymbol{\mu}_{\mathrm{C}} = W\mathsf{a} - \mathsf{N}\cdot(\nabla\mathbf{v})^{\mathrm{T}} - \mathbf{Q}\nabla w - \mathsf{M}^{\mathrm{T}}\cdot(\nabla\boldsymbol{\varphi})^{\mathrm{T}}, \quad \boldsymbol{\mu}_{\mathrm{I}} = W\mathsf{a} - \mathsf{N}\cdot(\nabla\mathbf{v})^{\mathrm{T}} - \mathbf{Q}\nabla w.$$

$\boldsymbol{\mu}_{\mathrm{C}}$ and $\boldsymbol{\mu}_{\mathrm{I}}$ are the Eshelby's tensors (the energy-momentum tensors) for the coherent interface and the incoherent one in the two-phase plate, respectively. The condition $\boldsymbol{\nu}\cdot[\![\boldsymbol{\mu}]\!]\cdot\boldsymbol{\nu} = 0$ should be satisfied, where $\boldsymbol{\mu} = \boldsymbol{\mu}_{\mathrm{C}}$ for the coherent phase interface and $\boldsymbol{\mu} = \boldsymbol{\mu}_{\mathrm{I}}$ for the incoherent in rotations one. From the point of view of configurational mechanics the quantity $\boldsymbol{\nu}\cdot[\![\boldsymbol{\mu}]\!]\cdot\boldsymbol{\nu}$ represents the configurational (or driving) force acting on $\mathfrak{C}$ and responsible for its motion for non-equilibrium deformations.

4 Kinetic Equation

Let us consider quasistatic deformation process, where the equilibrium conditions (1), (8) and (17) are satisfied while the $\boldsymbol{\nu}\cdot[\![\boldsymbol{\mu}]\!]\cdot\boldsymbol{\nu} = 0$ is not. For the quasi-static process we obtain the formula

$$\frac{\mathrm{d}E}{\mathrm{d}t} = -\int_{\mathfrak{C}} VF\,ds, \qquad F = \boldsymbol{\nu}\cdot[\![\boldsymbol{\mu}]\!]\cdot\boldsymbol{\nu}. \tag{19}$$

Here F is the *configurational force* acting on the phase boundary, while V is the velocity of $\mathfrak{C}$. By analogy to three-dimensional case [1] Eq. (19) leads to the kinetic equation describing the propagation of $\mathfrak{C}$

$$V = k(F)$$

with a non-negative definite kinetic function k. In the theory of elasticity the kinetic equations of the type (19) were discussed in a number of papers and monographs, see for example [1, 7, 8, 17, 18, 24]. The equation (19) can also be regarded as a constitutive relation consistent with the thermodynamic requirement $\mathrm{d}E/\mathrm{d}t \le 0$.

Following [1], let us assume $k(\varsigma)$ in the form

$$k(F) = \begin{cases} K(F - F_0), & F \ge F_0, \\ 0, & -F_0 < F < F_0, \\ K(F + F_0), & F \le -F_0. \end{cases} \tag{20}$$

Here F_0 describes effects associated with the nucleation of the new phase and action of the surface tension, see [1], and K is a positive kinetic factor. If $F_0 = 0$ then the function (20) reduces to the linear kinetic function $k(F) = KF$.

5 Tension of Two-Phase Rectangular Plate

As an example let us consider tension of an isotropic rectangular plate undergoing PT. The forces p uniformly distributed at the left and right plate boundaries are applied (Figure 2). We assume that in the undeformed state the plate consist of the phase B. Under such loading there is an plane deformation state with

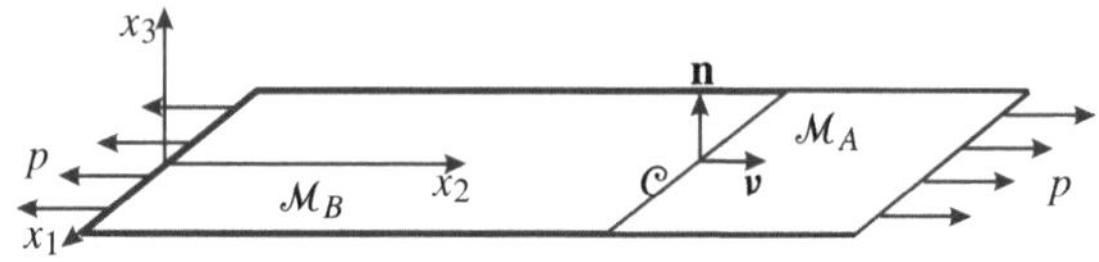

Fig. 2 Tension of a two-phase rectangular plate

$\mathbf{u} = u_1\mathbf{i}_1 + u_2\mathbf{i}_2$, $\boldsymbol{\varphi} = \mathbf{0}$, where $\mathbf{i}_1$, $\mathbf{i}_2$, $\mathbf{i}_2$ are the Cartesian base vectors. According to the assumption of the plane deformation, we search for the phase interface $\mathcal{C}$ in the form of the line, which position on $\mathcal{M}$ is given by the equation $x_1 = \ell(t)$, $0 \le \ell(t) \le L$. Hence, $V = d\ell(t)/dt$.

For the one-phase plate there is the following solution

$$u_1 = \frac{p}{Eh}x_1 + \text{const},$$

where E and h are the Young's modulus and the plate thickness respectively. In the two-phase plate such solution is possible only when $\nu_A = \nu_B = 0$, where ν_A and ν_B are the Poisson ratios for the bulk material. For the sake of simplicity we follow [15] and assume that $\nu_A = \nu_B = 0$. In this case the problem reduces to an one-dimensional problem which is similar to the considered one in [1] or [15].

For $F_0 = 0$, the equilibrium deformation process is described by the path $OABC$, Figure 3(a). The segment AB describes two-pase state of the plate corresponding the force $p^\star$. If we take into account the kinetic equation then in the process of loading the plate deforms according to the path $OAB'C$ while for the unloading the plate deforms according to the path $CBA'O$. The size of the hysteresis loop $AB'A'$ depends on K and the loading/unloading velocity. When $K \to \infty$ the hysteresis loop reduces to the segment AB.

When $F_0 \neq 0$, the plate deforms according to the path OA_+B_+C while for unloading we have the path CB_-A_-O, Figure 3(b). The size of hysteresis loop becomes larger with the growing value of F_0. Here growing values of K also lead to decreasing area of the hysteresis loop. For $K \to \infty$ the limiting paths reduce to

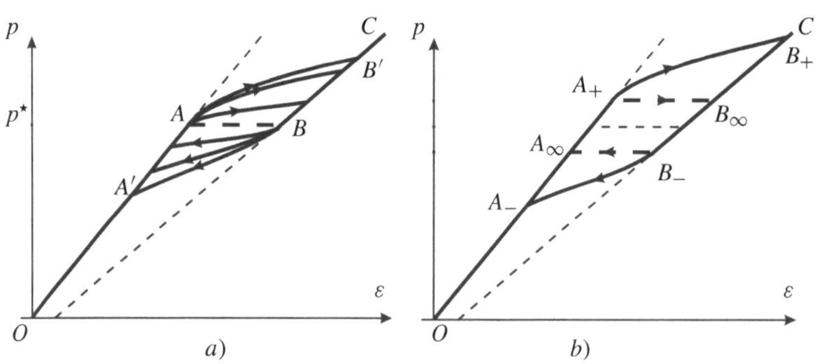

Fig. 3 $p - \varepsilon$ curves: a) $F_0 = 0$, b) $F_0 \neq 0$.

two different respective segments A_+B_∞ and B_-A_∞. Such limit case describes the so-called rate-independent phase transitions [1, 23].

Let us note that in the general case $\nu_A \neq \nu_B$ the deformation of the two-pase plate is described by two-dimensional problem. In this case some boundary-layer type solutions may appeared.

6 Conclusion

Here we introduced the configurational forces for the plate undergoing phase transformations. These expressions may be also used in the global analysis of a plate with a singular curves of another nature. For example, these approach may be used for modeling of delamination process in a sandwich plate, crack propagation, etc. Indeed, we postulate here the kinetic equation describing the quasistatic motion of singular curve on the base of the consideration of the total energy rate. For example, if one consider a sandwich plate consisting of two parts, delaminated part and undamaged one, then one can introduce a singular curve which separates these parts. Using the concept of the configurational force one may assume the kinetic equation describing the propagation of the delamination curve in the plate.

Acknowledgements

The authors acknowledge Professors R. Kienzler and A. Berezovski for fruitful discussions and involvement in the field of configurational mechanics.

The second author was supported with the grant of RFBR (07-01-00525) and by the DAAD program "Forschungsaufenthalte für Hochschullehrer und Wissenschaftler" in 2008.

References

1. Abeyaratne, R. and Knowles, J.K., *Evolution of Phase Transitions. A Continuum Theory.* Cambridge University Press, Cambridge, 2006.
2. Altenbach, H., An alternative determination of transverse shear stiffnesses for sandwich and laminated plates. *Int. J. Solids Struct.* **37**(25), 2000, 3503–3520.
3. Altenbach, H., On the determination of transverse shear stiffnesses of orthotropic plates. *ZAMP* **51**, 2000, 629–649.
4. Altenbach, H. and Eremeyev, V.A., Direct approach based analysis of plates composed of functionally graded materials. *Arch. Appl. Mech.* **78**(10), 2007, 775–794.
5. Altenbach, H. and Zhilin, P., A general theory of elastic simple shells. *Usp. Mekh.* **11**(4), 1988, 107–148 [in Russian].
6. Altenbach, H. and Zhilin, P.A., The theory of simple elastic shells. In: R. Kienzler, H. Altenbach and I. Ott (Eds.), *Critical Review of the Theories of Plates and Shells and New Applications*, Lect. Notes Appl. Comp. Mech., Vol. 16. Springer, Berlin, 2004, pp. 1–12.

7. Berezovski, A., Engelbrecht, J. and Maugin, G.A., *Numerical Simulation of Waves and Fronts in Inhomogeneous Solids*. World Scientific, New Jersey, 2008.
8. Bhattacharya, K., *Microstructure of Martensite: Why It Forms and How It Gives Rise to the Shape-Memory Effect*. Oxford University Press, Oxford, 2003.
9. Bhattacharya, K. and James, R.D., A theory of thin films of martensitic materials with applications to microactuators. *J. Mech. Phys. Solids* **36**, 1999, 531–576.
10. Bose, D.K. and Kienzler, R., On material conservation laws for a consistent plate theory. *Arch. Appl. Mech.* **75**, 2006, 607–617.
11. Chróścielewski, J., Makowski, J. and Pietraszkiewicz, W., *Statics and Dynamics of Multifold Shells. Non-Linear Theory and Finite Element Method*. Wydawnictwo IPPT PAN, Warszawa, 2004 [in Polish].
12. Daly, S., Ravichandran, G. and Bhattacharya, K., Stress-induced martensitic phase transformation in thin sheets of nitinol. *Acta Mater.* **55**, 2007, 3593–3600.
13. Dascalu, C., Maugin, G.A. and Stolz, C. (Eds.), *Defect and Material Mechanics*. Springer, Dordrecht, 2008.
14. Eremeyev, V. and Pietraszkiewicz, W., The non-linear theory of elastic shells with phase transitions. *J. Elasticity* **74**(1), 2004, 67–86.
15. Eremeyev, V.A. and Pietraszkiewicz, W., Phase transitions in thermoelastic and thermoviscoelastic shells. *Arch. Mech.* **61**(1), 2008, 41–67.
16. Grinfeld, M., *Thermodynamics Methods in the Theory of Heterogeneous Systems*. Longman, Harlow, 1991.
17. Gurtin, M.E., *Thermomechanics of Evolving Phase Boundaries in the Plane*. Clarendon-Press, Oxford, 1993.
18. Gurtin, M.E., *Configurational Forces as Basic Concepts of Continuum Physics*. Springer-Verlag, Berlin, 2000.
19. James, R.D. and Rizzoni, R., Pressurized shape memory thin films. *J. Elasticity* **59**, 2000, 399–436.
20. Kienzler, R. and Herrman, G., *Mechanics in Material Space with Applications to Defect and Fracure Mechanics*. Springer-Verlag, Berlin, 2000.
21. Kienzler, R. and Herrmann, G., An elementary theory of defective beams. *Acta Mech.* **62**, 1986, 37–46.
22. Kienzler, R. and Herrmann, G., On material forces in elementary beam theory. *J. Appl. Mech.* **53**, 1986, 561–564.
23. Lagoudas, D.C. (Ed.), *Shape Memory Alloys. Modeling and Engineering Applications*. Springer, Berlin, 2008.
24. Le, K.C., On kinetics of hysteresis. *Continuum Mech. Thermodyn.* **18**, 2007, 335–342.
25. Libai, A. and Simmonds, J.G., *The Nonlinear Theory of Elastic Shells*, 2nd edn. Cambridge University Press, Cambridge, 1998.
26. Maugin, G.A., *Material Inhomogeneities in Elasticity*. Chapman Hall, London, 1993.
27. Pieczyska, E.A., Gadaj, S.P., Nowacki, W.K. and Tobushi, H., Phase-transformation fronts evolution for stress- and strain-controlled tension tests in TiNi shape memory alloy. *Exp. Mech.* **46**, 2006, 531–542.
28. Pietraszkiewicz, W., Eremeyev, V.A. and Konopińska, V., Extended non-linear relations of elastic shells undergoing phase transitions. *ZAMM* **87**(2), 2007, 150–159.
29. Steinmann, P. and Maugin, G.A. (Eds.), *Mechanics of Material Forces*. Advances in Mechanics and Mathematics, Vol. 11. Springer, Berlin, 2006.
30. Zhilin, P.A., Mechanics of deformable directed surfaces. *Int. J. Solids Struct.* **12**, 1976, 635–648.
31. Zhilin, P.A., *Applied Mechanics. Foundations of the Theory of Shells*. St. Petersburg State Polytechnical University, 2006 [in Russian].

On Configurational Formulations in the Director Theory of Rods

Hans Irschik

Abstract It is shown how the configurational laws of balance and jump for the director theory of rods can be derived as consequences of the fundamental rod laws of balance of linear, director and angular momentum, and of energy. The configurational relations so derived are independent from constitutive relations.

1 Introduction

Recently, a material momentum balance law (a relation of balance of pseudomomentum) and a corresponding configurational jump condition have been presented by O'Reilly [1] in the framework of a director theory of elastic rods. The configurational formulations presented in [1] have been applied by Majidi [2], who has discussed adhering of an elastic rod to a flat rigid surface, the rod being composed of a non-contacting and contacting portion, where singular supply and production terms have to be taken into account at the cross-section between the two rod portions. A non-conventional aspect of this problem consists in finding an additional relation for computing the place of the interface. In particular, Majidi [2] has shown that applying the configurational jump condition presented by O'Reilly [1] to the singular cross-section is equivalent to alternative solution methods, see [2]. Not only that the practical applicability of the configurational formulations given in [1] has become evident from the study given in [2], further theoretical advantages have been pointed out by O'Reilly [1]. However, a believe has been expressed in [1] that the configurational relation of jump for the elastic director theory of rods would represent an extra law. In the present paper, we demonstrate that both, the configurational laws of balance and jump for rods can be derived as consequences of the rod laws of local balance and jump of linear, director and angular momentum, and of total

Hans Irschik
Institut für Technische Mechanik, Johannes Kepler Universität Linz, A-4040 Linz-Auhof, Austria; e-mail: irschik@mechatronik.uni-linz.ac.at

P. Steinmann (ed.), IUTAM Symposium on Progress in the Theory and Numerics of Configurational Mechanics, 131–138.

energy, simply by performing straight-forward mathematical manipulations. Thus, the configurational formulations have a role similar to the derived relations of local balance and jump of moment of momentum, kinetic energy and free energy. In showing this, a systematic strategy developed and discussed by the present author in [3] for three-dimensional problems of continuum mechanics is applied and extended to the director theory of rods. We particularly show how the configurational local rod laws of balance and jump can be derived without reference to elastic constitutive relations, which should be of interest in case of inelastic rod problems, and which extends the relations presented by O'Reilly in [1]. Last not least, we point out that that assigning a constitutive relation to the singular supply of material momentum at a singular cross-section is equivalent to assign constitutive relations to the singular supply of total energy.

2 Fundamental Relations of Balance and Jump

The present contribution is based on the Green and Naghdi director theory of rods, see [1] for a recent presentation. The local forms of the equations of balance of linear momentum, director momentum, angular momentum and total energy for this theory read:

$$\dot{J} = \rho_0 f + n', \tag{1}$$

$$\dot{J}^{\alpha} = \rho_0\, l^{\alpha} - k^{\alpha} + m^{\alpha'}, \tag{2}$$

$$\dot{H} = r \times \rho_0 f + d_{\alpha} \times \rho_0\, l^{\alpha} + (r \times n + d_{\alpha} \times m^{\alpha})', \tag{3}$$

$$\dot{E} = \dot{r} \cdot \rho_0\, f + \dot{d}_{\alpha} \cdot \rho_0\, l^{\alpha} + (\dot{r} \cdot n + \dot{d}_{\alpha} \cdot m^{\alpha})'\,. \tag{4}$$

The material time derivative of some entity g is indicated by a superimposed dot:

$$\dot{g} = \frac{\partial}{\partial t} g(\xi, t). \tag{5}$$

A prime stands for the spatial derivative with respect to the coordinate ξ of the (possibly curved) rod axis in the undeformed reference configuration

$$g' = \frac{\partial}{\partial \xi} g(\xi, t). \tag{6}$$

Linear momentum, director momentum, angular momentum and total energy are

$$J = \rho_0\, \dot{r} + \rho_0\, y^{0\alpha}\, \dot{d}_{\alpha}, \tag{7}$$

$$J^{\alpha} = \rho_0\, y^{0\alpha}\, \dot{r} + \rho_0 y^{\alpha\beta}\, \dot{d}_{\beta}, \tag{8}$$

$$H = r \times J + d_{\alpha} \times J^{\alpha}, \tag{9}$$

$$E = \rho_0 \psi + T. \tag{10}$$

The reference mass density is ρ_0, the position vector of a point of the rod axis in the deformed configuration is denoted by r, and d_α, $\alpha = 1, 2$, are the direction vectors, which span the rod cross-sections in the deformed configuration. Cross-sectional moments are denoted by $y^{0\alpha}$ and $y^{\alpha\beta}$, respectively, where

$$y^{\alpha\beta} = y^{\beta\alpha}. \tag{11}$$

Summation upon properly repeated indices is understood. The theory under consideration is considered as a purely mechanical one, ψ in Equation (10) thus being the free energy per unit mass. The kinetic energy T is given by

$$2T = \dot{r} \cdot J + \dot{d}_\alpha \cdot J^\alpha. \tag{12}$$

In Equations (1–4), f is the external force per unit mass, and l^α are the external director forces per unit mass. These supplies are given in advance. The so-called influx terms, which appear as spatial derivatives in Equations (1–4), are associated with the axial contact force n, and with the contact director forces m^α. For the latter forces, constitutive relations are needed in order to close the problem. When a singular cross-section is present within the rod, across which some of the balanced entities do suffer jumps, the following relations of jump are to be stisfied:

$$[\![J]\!]\dot{\gamma} = -[\![n]\!] - F, \tag{13}$$

$$[\![J^\alpha]\!]\dot{\gamma} = -[\![m^\alpha]\!] - L^\alpha, \tag{14}$$

$$[\![H]\!]\dot{\gamma} = -[\![r \times n + d_\alpha \times m^\alpha]\!] - M, \tag{15}$$

$$[\![E]\!]\dot{\gamma} = -[\![\dot{r} \cdot n + \dot{d}_\alpha \cdot m^\alpha]\!] - \Phi\,. \tag{16}$$

In Equations (13–16), the singular supplies of linear momentum, director momentum, angular momentum and total energy are denoted as F, L^α, M and Φ, respectively. The jump of some entity g, say, is indicated by

$$[\![g]\!] = g^+ - g^-\,. \tag{17}$$

The corresponding mean value across the singular cross-section follows from

$$2\langle g\rangle = g^+ + g^-. \tag{18}$$

The general decomposition

$$[\![g \circ h]\!] = \langle g\rangle \circ [\![h]\!] + [\![g]\!] \circ \langle h\rangle \tag{19}$$

is used frequently in the sequel. In addition to Equations (13) and (14), the following conditions of compatibility are required to hold across a singular cross-section:

$$[\![r]\!] = 0, \tag{20}$$

$$[\![d_\alpha]\!] = 0, \tag{21}$$

$$\nu_\gamma = \dot{r}^+ + r'^+ \dot{\gamma} = \dot{r}^- + r'^- \dot{\gamma}, \tag{22}$$

$$\nu_{\gamma\alpha} = \dot{d}_\alpha^+ + d_\alpha'^+ \dot{\gamma} = \dot{d}_\alpha^- + d_\alpha'^- \dot{\gamma}, \tag{23}$$

such that

$$[\![r']\!]\dot{\gamma} = -[\![\dot{r}]\!], \tag{24}$$

$$[\![d_\alpha']\!]\dot{\gamma} = -[\![\dot{d}_\alpha]\!]. \tag{25}$$

The coordinate of the place of the singular cross-section in the reference configuration is denoted by γ, and ν_γ is the velocity of the singular cross-section in the deformed configuration, with an analogous meaning of $\nu_{\gamma\alpha}$. The above relations of jump do indicate, which of the entities under consideration are supposed to suffer jumps across a singular cross-section. We subsequently assume that mass and cross-sectional moments are continuous along the rod.

3 Classical Derived Relations of Balance and Jump

The local relations of balance of linear momentum, director momentum, angular momentum and total energy, Equations (1–4), are considered as fundamental relations, since no terms do appear, which are formed by the product of an entity which may suffer a jump and the derivative of an entity that also may suffer a jump at the same singular cross-section. We thus talk about integrable local relations in the following. Indeed, the fundamental relations of balance are the consequence of corresponding integral statements over some part of the reference configuration, which are assumed to hold also when a singular cross-section is present within that part, which requires integrability. The fundamental jump relations, Equations (13–16), are then obtained by what nowadays is called a pill-box procedure, namely by letting the considered part of the reference configuration shrink down to the singular cross-section, see [4]. One also may state that any one of the above four fundamental balance equations does include new physically meaningful entities, or at least some new information.

The above fundamental relations of balance and jump however may be mathematically manipulated in order to obtain derived forms. Particularly, noting that

$$\dot{r} \times J + \dot{d}_\alpha \times J^\alpha = 0, \tag{26}$$

see Equations (7) and (8), we can derive the relation of balance of moment of momentum as a consequence of proper vectorial multiplications of the relations of linear and director momentum balance, Equations (1) and (2):

$$\dot{H} = r \times \rho_0\, f + d_\alpha \times (\rho_0\, l^\alpha - k^\alpha) - r' \times n - d_\alpha' \times m^\alpha + (r \times n + d_\alpha \times m^\alpha)'. \tag{27}$$

Note that in the present theory the notions of angular momentum and moment of momentum do coincide. Subtracting Equation (27) from Equation (3), we obtain

$$d_\alpha \times k^\alpha + r' \times n + d'_\alpha \times m^\alpha = 0, \tag{28}$$

which can be used to replace the relation of balance of angular momentum, Equation (3), see [1, equation (16.3)]. Hence, the intrinsic director forces k^α in Equation (2) are related to the influxes for n and m^α. Analogously, crossing Equations (13) and (14) by r and d_α, respectively, and using Equations (19–21), yields the following relation of jump of moment of momentum

$$[\![H]\!]\dot{\gamma} = -r \times [\![n]\!] - d_\alpha \times [\![m^\alpha]\!] - r \times F - d_\alpha \times L^\alpha, \tag{29}$$

such that that the relation of jump of angular momentum, Equation (15), proves that

$$r \times F + d_\alpha \times L^\alpha - M = 0. \tag{30}$$

This can be used to replace the original relation of jump of angular momentum, Equations (15), see [1, equation (18.3)].

We also may operate on the relations of balance and jump of linear and director momentum by scalar multiplications. Particularly, noting that

$$\ddot{r} \cdot J + \ddot{d}_\alpha \cdot J^\alpha = \dot{r} \cdot \dot{J} + \dot{d}_\alpha \cdot \dot{J}^\alpha, \tag{31}$$

we obtain the relation of balance of kinetic energy, see Equations (7), (8) and (12):

$$\dot{T} = \dot{r} \cdot \rho_0\, f + \dot{d}_\alpha \cdot \rho_0\, l^\alpha - \dot{d}_\alpha \cdot k^\alpha - \dot{r}' \cdot n - \dot{d}'_\alpha \cdot m^\alpha + (\dot{r} \cdot n + \dot{d}_\alpha \cdot m^\alpha)', \tag{32}$$

which is a consequence of proper scalar multiplications of the relations of linear and director momentum, Equations (1) and (2). Substituting into the relation of balance of total energy, Equation (4), gives the relation of balance of free energy, see also [1, equation (16.4)]:

$$\rho_0 \dot{\Psi} = \dot{d}_\alpha \cdot k^\alpha - \dot{r} \cdot n' - \dot{d}_\alpha \cdot m^{\alpha'} + (\dot{r} \cdot n + \dot{d}_\alpha \cdot m^\alpha)'. \tag{33}$$

In our present derivation, the relations of balance of kinetic and free energy, Equations (32) and (33), have been derived from the relations of balance of linear and director momentum and total energy, Equations (1), (2) and (4). In contrast to the latter three relations, however, Equations (32) and (33) cannot be considered as being fundamental, in the sense of representing integrable relations. Nevertheless, corresponding jump conditions can be derived from the relations of balance of linear momentum by a scalar multiplication with the mean values and $\langle \dot{r} \rangle$ and $\langle \dot{d}_\alpha \rangle$. From the definitions in Equations (7) and (8), and using (12) and (19), we find that

$$[\![T]\!] = \langle J \rangle \cdot [\![\dot{r}]\!] + \langle J^\alpha \rangle \cdot [\![\dot{d}_\alpha]\!] = \langle \dot{r} \rangle \cdot [\![J]\!] + \langle \dot{d}_\alpha \rangle \cdot [\![J^\alpha]\!]. \tag{34}$$

Multiplication by $\dot{\gamma}$ and substituting Equations (13) and (14) results in the relation of jump of kinetic energy

$$[\![T]\!]\dot{\gamma} = -\langle \dot{r} \rangle \cdot ([\![n]\!] + F) - \langle \dot{d}_\alpha \rangle \cdot \left([\![m^\alpha]\!] + L^\alpha\right) \tag{35}$$

from which Equations (10, 16) and (19) yield the jump relation of internal energy as

$$[\![\rho_0 \Psi]\!]\dot{\gamma} = -[\![\dot{r}]\!] \cdot \langle n \rangle - [\![\dot{d}_\alpha]\!] \cdot \langle m^\alpha \rangle + \langle \dot{r} \rangle \cdot F + \langle \dot{d}_\alpha \rangle \cdot L^\alpha - \Phi. \qquad (36)$$

4 Derivation of the Configurational Formulations of Balance and Jump

Based partially on the works of Maugin [5], Gurtin [6] and their coworkers on material and configurational forces in mechanics, an additional balance law denoted as balance of material momentum has been presented by O'Reilly [1]. It is the scope of the present section to demonstrate that this relation of balance of material momentum and the corresponding jump relation can be derived as a consequence of the fundamental local relations of balance and jump, and thus have a role similar to the above derived relations of moment of momentum, kinetic energy and free energy. For corresponding derivation for three-dimensional problems of continuum mechanics, see [3]. We again perform scalar multiplications of the linear and director momentum, this time however involving the spatial derivatives r' and d'_α. We start with definition of the material momentum for director rods that has been introduced in [1, equation (20)]:

$$P = -r' \cdot J - d'_\alpha \cdot J^\alpha. \qquad (37)$$

In the present contribution, the latter is understood as an entitity the balance of which can be derived by a mathematical manipulation of linear and director momentum, Equations (7) and (8). From the latter, and from the definition of the kinetic energy, Equation (12), we have

$$\dot{r}' \cdot J + \dot{d}'_\alpha \cdot J^\alpha = \dot{r} \cdot J' + \dot{d}_\alpha \cdot J^{\alpha'} = T'. \qquad (38)$$

We thus may write

$$\dot{P} = -r' \cdot \dot{J} - d'_\alpha \cdot \dot{J}^\alpha - T' + (\rho_0 \Psi)' - (\rho_0 \Psi)'\,. \qquad (39)$$

where we have performed an identical expansion by means of the term $(\rho_0 \Psi)'$. Substituting Equations (1) and (2), we eventually obtain an equation of balance material momentum of the form

$$\dot{P} = b + C'. \qquad (40)$$

In (40), the so-called material force b turns out to be

$$b = -r' \cdot \rho_0\, f - d'_\alpha \cdot \rho_0\, l^\alpha + r'' \cdot n + d'_\alpha \cdot k^\alpha + d''_\alpha \cdot m^\alpha - (\rho_0 \Psi)'. \qquad (41)$$

and the so-called material contact force, an influx term, becomes

$$C = \rho_0 \Psi - T - r' \cdot n - d'_\alpha \cdot m^\alpha . \tag{42}$$

The material contact force is identical to the one stated in [1, equation (19)]. Several remarks seem to be in order. First, while the balance of material momentum, Equation (40), was introduced by O'Reilly [1] as an additional law of balance, our derivation shows that it represents a consequence of mathematical manipulations of the fundamental laws of balance of linear and director momentum, Equations (1) and (2), and of an identical expansion by the term $(\rho_0 \psi)'$. Second, Equation (40) cannot be considered as a fundamental statement of balance, since, due to b in Equation (41), it cannot be considered as integrable without further considerations. The integral form of Equation (40), which was introduced in [1, equation (21)], and which involves a spatial integral of b, thus must be considered as doubtful in general. Third, the above formulation of balance of material momentum is independent from constitutive modeling. However, when one uses the special constitutive relations introduced in [1, equation (17)], then Equation (41) can be specialized to

$$b = -r' \cdot \rho_0 \, f - d'_\alpha \cdot \rho_0 \, l^\alpha - \rho_0 \, \frac{\partial \Psi}{\partial \xi_{\mathrm{expl}}}, \tag{43}$$

the index "expl" indicting the explicit dependence of the free energy on the place ξ, see [1, equation (25)]. This special form somewhat hides the above addressed question of integrability. Likewise to the jump relations of kinetic energy and free energy, see Equations (36) and (37), a jump relation of material momentum can be however derived from the jump relations of linear and director momentum, Equations (13) and (14), without involving constitutive modeling as follows. From Equations (24, 25) and (34), we find that

$$[\![T]\!] = [\![\dot{r}]\!] \cdot \langle J \rangle + [\![\dot{d}_\alpha]\!] \cdot \langle J^\alpha \rangle = -\left([\![r']\!] \cdot \langle J \rangle + [\![d'_\alpha]\!] \cdot \langle J^\alpha \rangle \right) \dot{\gamma} \, . \tag{44}$$

Hence, using Equation (19) in connection with Equation (37) and performing an identical expansion with the jump of $\rho_0 \psi$, one obtains

$$[\![P]\!] \dot{\gamma} = -\langle r' \rangle \cdot [\![J]\!] \dot{\gamma} - \langle d'_\alpha \rangle \cdot [\![J^\alpha]\!] \dot{\gamma} + [\![T]\!] + [\![\rho_0 \psi]\!] - [\![\rho_0 \Psi]\!] \tag{45}$$

We thus may set

$$[\![P]\!] \dot{\gamma} = -[\![C]\!] - B. \tag{46}$$

From the material contact force, Equation (42), and using the jump relations of linear and director momentum, Equations (13) and (14), the singular supply of material momentum becomes

$$B = -[\![\rho_0 \Psi]\!] - \langle r' \rangle \cdot F + [\![r']\!] \cdot \langle n \rangle - \langle d'_\alpha \rangle \cdot L^\alpha + [\![d'_\alpha]\!] \cdot \langle m^\alpha \rangle . \tag{47}$$

which coincides with [1, equation (32.1)]. Multiplying with $\dot{\gamma}$ and substituting the relation of jump of free energy, Equation (36), as well as Equations (22–24) and (25), we arrive at the relation

$$B\dot{\gamma} = \Phi - \nu_\gamma \cdot F - \nu_{\gamma\alpha} \cdot L^\alpha, \tag{48}$$

which is in coincidence with [1, equation (34)], and which can be checked by subtracting equation (31.1) from equation (31.2) of the latter reference. In case of $F = 0$ and $L^\alpha = 0$ this reduces to

$$B\dot{\gamma} = \Phi, \tag{49}$$

which clearly demonstrates that, when one assigns a constitutive relation to the singular supply of material momentum B in order to obtain an additional equation for determining the place of the singular-cross section γ, this is equivalent to assign constitutive relations to the singular supply of total energy Φ.

Acknowledgement

Support of this work in the framework of the Comet-K2 Austrian Center of Competence in Mechatronics (ACCM) is gratefully acknowleged.

References

1. O'Reilly, O.M., A material momentum balance law for rods. *J. Elasticity* **86**, 2007, 155–172.
2. Majidi, C., Remarks on formulating an adhesion problem using Euler's elastica. *Mech. Res. Commun.* **34**, 2007, 85–90.
3. Irschik, H., On rational treatments of the general laws of balance and jump, with emphasis on configurational formulations. *Acta Mech.* **194**, 2007, 11–32.
4. Truesdell, C.A. and Toupin, R., The classical field theories. In: *Handbuch der Physik, Band III/1, Prinzipien der Klassischen Mechanik und Feldtheorie*, S. Fluegge (Ed.). Springer, Berlin, 1960, pp. 226–793.
5. Maugin, G.A., Material forces: Concepts and applications. *Appl. Mech. Rev.* **48**, 1995, 213–245.
6. Gurtin, M.E., *Configurational Forces as Basic Concepts in Continuum Physics*. Springer , New York, 2000.

Macroscopic Elasticity of Nanoporous Silicon: Bulk and Surface Effects

H. Magoariec and A. Danescu

Abstract Nanoporous silicon is an elastic material for which the length scale of the pores renders inappropriate the use of the tools from continuum homogenization theories to obtain overall mechanical behavior. To encompass this difficulty we propose a model based on discrete interactions at the microscale and we use discrete homogenization to compute the bulk macroscopic material response for various sizes and shapes of pores. When pore sizes are only several nm, the specific surface of porous silicon is very important and surface effects, like oxidation in dry environment, have a significant impact on the macroscopic bulk properties. We extend the proposed model to include also structural changes that model oxidation near the surface. We discuss the size effect due to the presence of bulk and surface energy and compare our theoretical predictions with the available experimental data.

1 Introduction

Porous silicon is obtained by electrochemical etching of crystalline silicon wafers in a solution of ethanolic hydrofluoric acid. The size of pores created is controlled by tuning the electrochemical parameters, i.e. the modulation of the current density. Using this process one can obtain porosities up to 95% and nanopores up to several nm of diameter over a thickness of a few microns. As a consequence porous silicon possess a very large specific surface, which may attain orders of 100–1000 m^2/g.

Among the various applications of porous silicon we cite: fabrication of chemical sensors (for detection of toxins, volatile organic compounds, explosives, DNA and

H. Magoariec
LTDS – Ecole Centrale de Lyon, 36 Avenue Guy de Collongue, 69134 Ecully, France; e-mail: helene.magoariec@ec-lyon.fr

A. Danescu
LTDS – Ecole Centrale de Lyon, 36 Avenue Guy de Collongue, 69134 Ecully, France; e-mail: danescu@ec-lyon.fr

P. Steinmann (ed.), IUTAM Symposium on Progress in the Theory and Numerics of Configurational Mechanics, 139–148.

proteins) and its potential role as a hydrogen reservoir. This work is a part of a larger project whose goal is the study of the role of nanoporous silicon as an active part of a deformable substrate for molecular beam epitaxy. To this aim, we intend to use the deformation induced by oxidation of the nanopore surface in order to stretch a crystalline silicon film and use it as a substrate with tunable lattice parameters [3].

2 Elasticity of Nanoporous Silicon: Bulk Effects

In a fundamental paper dedicated to the structure and energy of covalent systems [13], Tersoff note that the models able to predict macroscopic properties from interatomic interactions fall in two groups: the group of pair potentials like Lennard-Jones [7] and Morse [9] and those intended to accurately describe the small distorsions (phonons and elastic interactions) from the ground state. More elaborate models for interatomic interactions are actually available (Stillinger–Weber [12], Tersoff [13]) but they are constructed to cover a wide range of bonding geometry and coordination. As we are interested here in the elastic properties of porous silicon, we shall use the simplest model in the second group, i.e., the Keating model [5]. This is the simplest model that goes beyond pair interactions and account for near-neighbor (NN) and next-to-near-neighbor (NNN) interactions.

2.1 Keating Model for Bulk Silicon

Widely used in semiconductors technology, crystalline silicon has a diamond-like structure, which is a non-primitive lattice with lattice parameter $a = 5.43$ Å. The Keating model [5] assumes that in the harmonic regime near a ground state, the total elastic energy of a covalent system can be decomposed as

$$W = \sum_{(j,k)} w_{jk} + \sum_{(j,k,l)} w_{jkl}, \tag{1}$$

where the NN and NNN interaction energies are expressed as

$$w_{jk} = \frac{A}{2}\left[(\boldsymbol{u}^k - \boldsymbol{u}^j)\cdot \boldsymbol{n}^{kj}\right]^2 \tag{2}$$

and

$$w_{jkl} = \frac{B}{2}\left[(\boldsymbol{u}^k - \boldsymbol{u}^j)\cdot \boldsymbol{n}^{lj} + (\boldsymbol{u}^l - \boldsymbol{u}^j)\cdot \boldsymbol{n}^{kj}\right]^2. \tag{3}$$

For bulk crystalline silicon the macroscopic energy can be explicitly computed as

$$\hat{W}(\boldsymbol{E}) = \min_{\{\tilde{\boldsymbol{u}}\}} W(\boldsymbol{E}\boldsymbol{x} + \tilde{\boldsymbol{u}}), \tag{4}$$

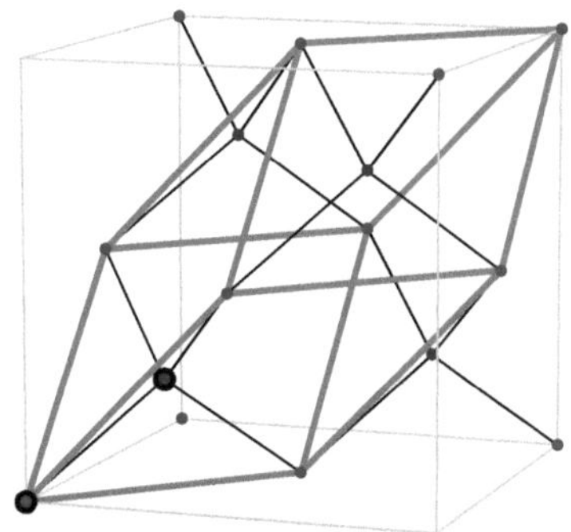

Fig. 1 The crystal structure of Si and the unit rhombohedron cell containing one atom in each lattice.

where minimisation is needed to account for internal displacement since diamond-like structure is a non-primitive lattice.[1] The macroscopic stress can therefore be expressed as

$$\mathbf{\Sigma} = \frac{1}{vol(Y)} \frac{\partial \hat{W}}{\partial \boldsymbol{E}} \tag{5}$$

and using as unit cell Y a rhombohedron containing one atom in each lattice (see Figure 1) and periodic boundary conditions we obtain, using Voigt notation,

$$C_{11} = \frac{A + 12B}{12a}, \qquad C_{12} = \frac{A - 4B}{12a}, \qquad C_{44} = \frac{4AB}{a(A + 4B)}. \tag{6}$$

This result (Keating [5]) shows that if the previous assumptions hold then the three macroscopic elasticites of bulk silicon are not independent and they obey the nonlinear relation

$$2C_{44}(C_{11} + C_{12}) = (C_{11} - C_{12})(C_{11} + 3C_{12}). \tag{7}$$

This is verified remarkably well since the values

$$A = 1.45 \cdot 10^{-8}\ \mathrm{N/Å} \qquad B = 1.03 \cdot 10^{-9}\ \mathrm{N/Å} \tag{8}$$

provide the macroscopic constants $C_{11} = 166$ GPa, $C_{12} = 64$ GPa and $C_{44} = 79$ GPa, in very good agreement (within 1%) with the experimental data.

2.2 *Extension to Porous Silicon*

We shall investigate the modeling of the macroscopic elastic response of porous silicon using the Keating model. Under suitable assumptions about the uniformity

[1] This procedure is the simplest discrete version of what is commonly called today *periodic homogenization theory*.

of shapes and sizes of pores, and accounting only for microscopic NN and NNN interactions in a sufficiently large representative volume element, we shall show that the use of (4) and (5) will still provide the macroscopic stress-strain relation. However, in this case, several important remarks are needed:

1. Excepting some special cases, explicit relations like (6) are not available. From a numerical perspective, minimization in (4) still leads to a linear problem since the Keating energy is a quadratic form with respect to displacements.
2. For pores of significant size the computation using discrete homogenization becomes inappropriate but, in these situations, one can switch to continuum models.
3. In general, the cubic symmetry of the bulk crystalline silicon is lost and the symmetry properties of the porous silicon will depend strongly on the shape and size of the pores.

In order to verify the validity of the extension of the Keating model for porous silicon we have explored different porosities using homogeneous imposed macroscopic strains and periodic boundary conditions. We start with the crystal structure on a large domain Y which represents N^3 copies of a unit cell and fix a simple geometry of the pore, compatible with the cubic symmetry and inter-connectivity between the pores.[2] If we denote by $L = N \cdot 5.43$ Å the lateral size of the reference volume Y, a typical choice for the pore volume is

$$\mathcal{P}_\alpha = \mathcal{C}_x^\alpha \cup \mathcal{C}_y^\alpha \cup \mathcal{C}_z^\alpha \tag{9}$$

where $\mathcal{C}_x^\alpha$ denotes the volume in the cylinder

$$\mathcal{C}_x^\alpha = \{(x, y, z) \in Y \text{ such that } |y - L/2| < \alpha L/2 \text{ and } |z - L/2| < \alpha L/2\} \tag{10}$$

for $0 < \alpha < 1$. The geometry of the pore described in (9) is illustrated in Figure 2 and two other possible geometries leading respectively to cubic and transversely isotropic elastic behavior are shown in Figure 3.

In the continuum case this choice will lead to a porosity given by

$$p(\alpha) = 1 - 3\alpha^2 + 2\alpha^3, \tag{11}$$

but the actual discrete setting will cover only a finite number of values for α.

In spite of the numerical complexity which increases rapidly with N, as already noted, for small porosity and large N the continuum theory can provide accurate results.

We have performed the numerical implementation of the above procedure for arbitrary geometries and arbitrary size of pores. The particular geometry described

---

[2] The inter-connectivity restriction, which is a well-founded physical assumption cannot be satisfied in a two-dimensional situation. The cubic symmetry of the pore was chosen for simplicity, since otherwise, the mechanical interpretation of the obtained result and its comparison with available experimental data become more intricate. The indirect methods used in experiments provide only partial information about the Young modulus in the crystallographic direction [100] at some fixed porosities.

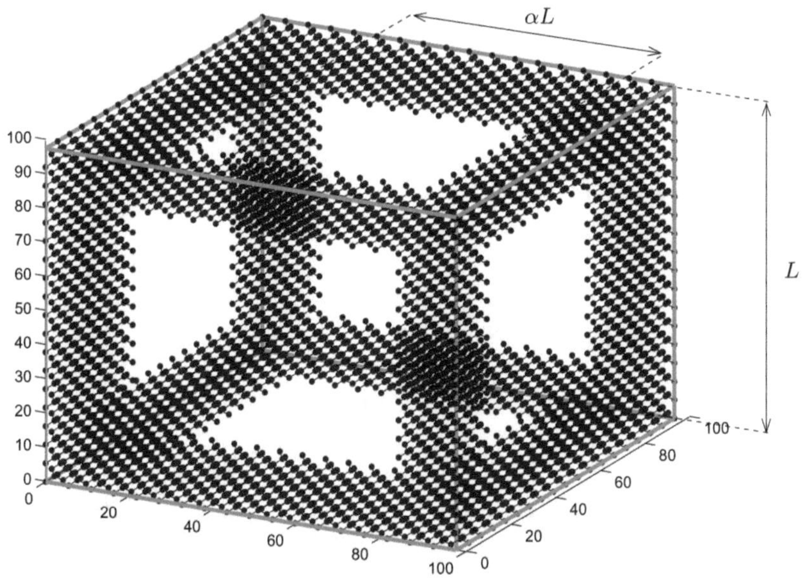

Fig. 2 A unit cell containing a pore whose geometry is described in (9).

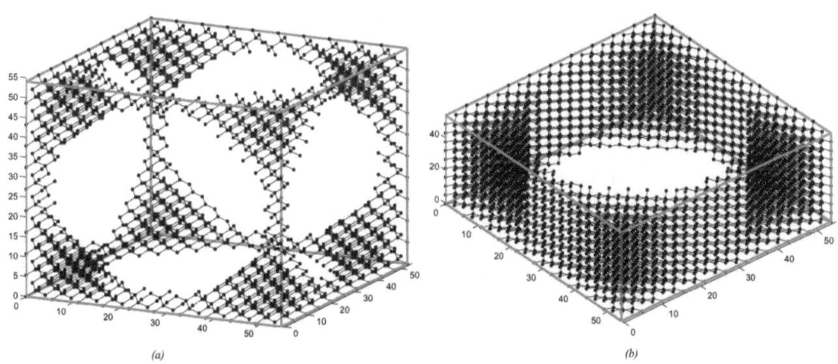

Fig. 3 A unit cell leading to cubic macroscopic behavior (a) and a second unit cell (b) leading to transversely isotropic macroscopic response.

in (9), which leads to cubic macroscopic response, allows the computation of elastic constants C_{11}, C_{12} and C_{44}. The Young modulus in the crystallographic direction [100] is therefore defined as

$$E_{[100]} = C_{11}(1-2R)(1+R)/(1-R) \quad \text{where} \quad R = C_{12}/C_{11}. \tag{12}$$

The comparison of the numerical results and the experimental data obtained using X-ray diffraction, nano-indentation, Brillouin spectroscopy and acoustic measurements is presented in Figure 4. We conclude that, using discrete homogenization,

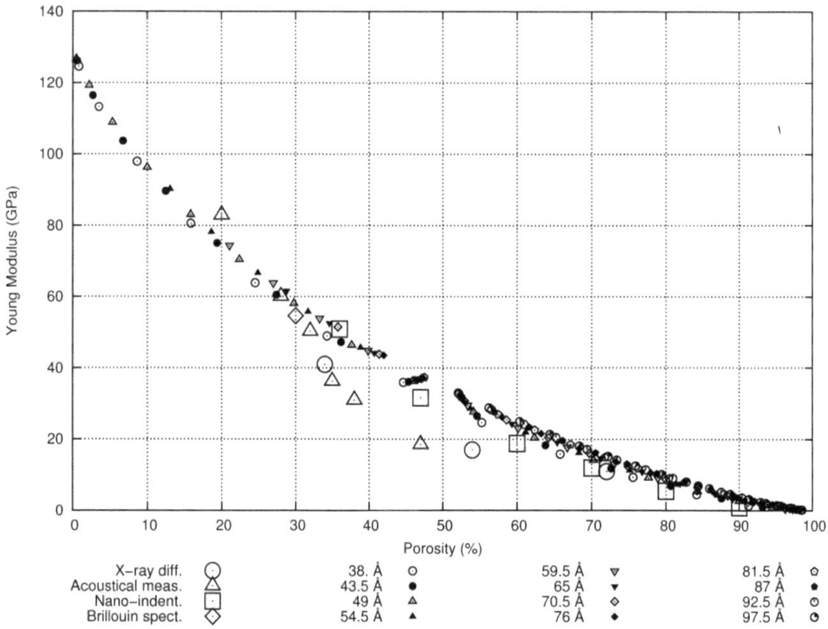

Fig. 4 Young modulus in the crystallographic direction [100] obtained using discrete homogenization and the Keating model and available experimental data from [1].

the extension to porous silicon of the Keating model is able to predict accurately the elastic macroscopic behavior.

A particular feature of porous silicon is its extremely high specific area induced by the presence of nanopores. At this scale there is a complex interplay between bulk and surface elastic energy, which in turn induce a size effect.

3 Oxidation of Porous Silicon: Bulk and Surface Effects

In this section we extend the previous results to account for the macroscopic effects of the nanopores free-surface oxidation in porous silicon.

3.1 Experimental Evidence

From the macroscopic point of view, the oxidation of porous silicon has a two-fold effect: firstly, it induces microscopic morphological changes which change the *reference* for a part of the structure. Secondly, oxidation induces a shift of the macroscopic response, i.e., the Young modulus. The second effect is less important since the porous silicon has already a weak Young modulus. The first effect is more important since it provides a way to stretch the bulk structure acting only on the nanopores surface. The more accurate, indirect observation of the first effect can be measured using Raman spectroscopy [4, 6, 10]. The shift of a Raman peak is commonly interpreted [2] as a strong physical evidence of *an increase of the lattice parameter.*[3]

Experimental results in [8, 10] show that in dry or wet environment the main characteristic feature in the oxidation of porous silicon is the back-bond oxidation of the Si-H surface bonds. This results in either splitting of the bond and creation of SiH and SiOH or creation of Si-O-Si [14]. Both processes induce strain at the free-surface of the nanopores. Due to the singificant number of nanopores, this induced strain will modify the macroscopic response of the material.

3.2 Surface Effects

As known in continuum mechanics, accounting for surface and bulk effects leads to a *size effect*. In this case it is important to account properly for as much as possible realistic features of the oxidation process at the appropriate scale from the available experimental data. Experiments reported in [11] were conducted on nanoporous silicon at 55% porosity. At fixed geometry proposed in (9), the porosity fixes $\alpha = 0.47$ and accounting for the order of magnitude of the specific surface of the nanoporous material which is larger than 100 m^2/g, one can deduce[4] the magnitude of L which is about 10 nm.

The equilibrium distance of a Si-Si bond in crystalline silicon equals 2.35 Å. Available data for silica, which is an amorphous material, show that the equilibrium distance between silicon and oxygen atoms in Si-O-Si structures is 1.61 Å and the minimum Si-O-Si bond angle is about 100°. This means that, during the back-bond oxidation the equilibrium distance of the Si-Si bond needs to be strained by, at least, 10%. In order to adjust the previous discrete model to account for the modification of structural changes at the free-surface of the nanopores we have modified physical characteristics of all Si-Si bonds which contain at least one Si at the surface to include:

1. a different reference distance induced by surface oxidation;
2. a different rigidity A' which is two times lower than that of crystalline silicon. This is a simple pertinent choice since crystalline silicon is about two times stiffer that the amorphous silica.

Thus, for NN interactions for which at least one Si atom is situated on the nanopore surface we modify the expression in (2) to

[3] Raman spectroscopy provides an average lattice parameter, at a length scale much larger than that of the pore.

[4] The mass density of crystalline silicon is 2.329 g/cm^3.

$$w_{jk}^{(s)} = \frac{A'}{2}\left[(\boldsymbol{u}^k - \boldsymbol{u}^j) \cdot \boldsymbol{n}^{kj} - l_0\right]^2. \tag{13}$$

There are two reasons for which the NNN interaction energy was not modified in this part of our study. The first one concerns the significant computational complexity needed to account for both the amorphous structure induced by back-bond oxidation, for instance using a combined random network at the surface of the nanopores, and a crystalline model in the bulk. The second reason rests on the construction of the Keating model, where the NNN interactions are considered only as a *second order term* and confirmed by the numerical values of A and B in (8).

We shall show in the following that, under the previous assumptions, the numerical results obtained using discrete homogenization lead to estimations for the strain induced by surface oxidation of the nanopores in good agreement with the experimental results.

3.3 Numerical Results and Size Effects

The introduction of a different reference for surface NN interactions, and the use of (4) and (5) lead to a macroscopic law in the form $\boldsymbol{\Sigma} = \boldsymbol{C}[\boldsymbol{E} + \boldsymbol{E}_0]$ where both the elasticities $\boldsymbol{C}$ and the pre-strain $\boldsymbol{E}_0$ will depend on the values of A, B, A', l_0, the size and the geometry of the pore. For simplicity, we have considered here the same geometry as in the previous section, choice which leads to a pre-strain which is a uniform extension, i.e. $\boldsymbol{E} = \varepsilon_0 \boldsymbol{I}$ as a consequence of the symmetry of the nanopore. An illustration of a 6nm structure containing a pore is provided in Figure 5 where for convenience all Si-Si bonds starting from one Si at the surface were shown in gray.

It is obvious that at fixed geometry, when the size of the unit cell is very large, the surface oxidation process do not affect the macroscopic behavior of the bulk. This is due to the fact that the ratio between the number of interactions modified by surface oxidation and that of interactions not affected by surface oxidation tends to 0.

We have explicitly compute the macroscopic elasticity tensor $\boldsymbol{C}$ and the pre-strain ε_0, for a wide range of values of L and α. The resulting values of the pre-strain are represented in Figure 6. We note that at fixed porosity, depending on the pore size, the pre-strain induced by surface oxidation varies with an order of magnitude. This is the effect of the interplay between the bulk and surface energy which has a significant impact on the final result only for nanoporous materials. The porosity of interest for the actual application is about 55%, in which case for a cell with $L \simeq 10$ nm the computed value of the pre-strain is 1.5%. This value is in good agreement with the experimental results [11] based on Raman spectroscopy. For nanopores with sizes of about 10 nm the surface oxidation has a very small impact on the pre-strain for porosities less than 50%, and a drastic increase is observed above this value.

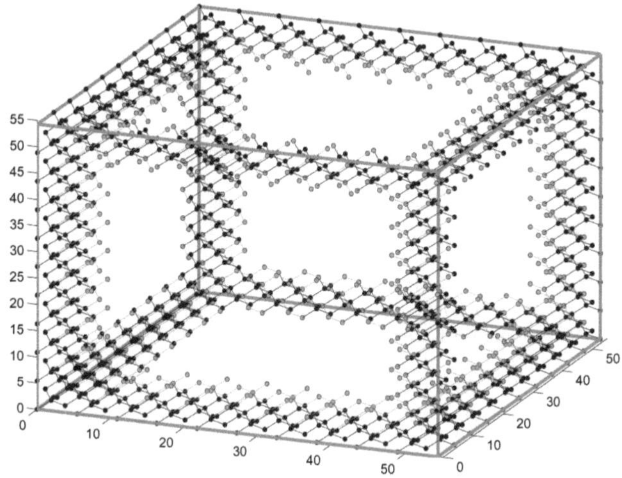

Fig. 5 A 6 nm structure containing a pore; for convenience, the NN interactions starting from one Si at the pore surface are represented in gray.

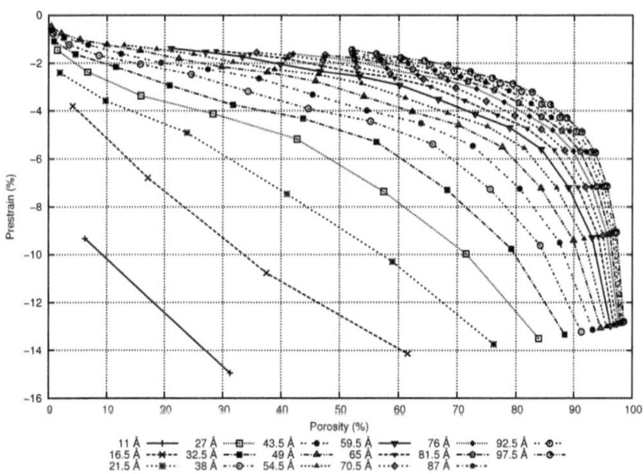

Fig. 6 Variation of the computed pre-strain ε_0 as a function of porosity for different sizes of the unit cell.

Acknowledgements

This work is part of the research project ANR-06-SPAME-0274 funded by the French National Research Agency. We acknowledge helpful discussions with G. Grenet, V. Lysenko, O. Marty, J-M. Bluet and A. Boucherif.

References

1. Bellet, D., in *Properties of Porous Silicon*, L. Canham (Ed.), EMIS Data Reviews 18, 1997.
2. Gouadec, G. and Colomban, P., Raman spectroscopy of nanomaterials: How spectra relate to disorder, particle size and mechanical properties. *Prog. Cryst. Growth Charact. Mat.* **53**, 2007, 1–56.
3. Kim, J. and Xie, Y.A., Fabrication of dislocation-free tensile strained Si thin film using controllably oxidized porous Si substrate. *Appl. Phys. Lett.* **89**, 2006, 152117.
4. Kang, Y., Qiu, Y., Lei, Z. et al., An application of Raman spectroscopy on the measurement of residual stress in porous silicon. *Optics and Lasers in Engineering* **43**, 2005, 847–855.
5. Keating, P.N., Effect of invariance requirements on the elastic strain energy of crystals with application to the diamond structure. *Phys. Rev.* **145**, 1966, 637–645.
6. Lei, Z., Kang, Y., Cen, H. et al., Variability on Raman shift to stress coefficient of porous silicon. *Chem. Phys. Lett.* **23**, 2006, 1623–1626.
7. Lennard-Jones, J., Cohesion. *Proc. Phys. Soc.* **43**, 1931, 461–482.
8. Mawhinney, D.B., Glass, J.A. and Yates, J.T., FTIR study of the oxidation of porous silicon. *J. Phys. Chem. B* **101**, 1996, 1202–1206.
9. Morse, R.M., Diatomic molecules according to the wave mechanics. *Phys. Rev.* **34**, 1929, 57–64.
10. Ogata, Y.H., Tsuboi, T., Sakka, T. et al., Oxidation of porous silicon in dry and wet environments under mild temperature conditions. *J. Porous Mater.* **7**, 2000, 63–66.
11. Ostapenko, D., Lysenko, V., Regreny, P. et al., Deformation of thin crystalline silicon films by stress generated in bulk porous silicon substrate. In *Proceedings of Porous Semiconductors Science and Technology Conference*, Mallorca, 2008, in press.
12. Stillinger, F. and Weber, T., *Phys. Rev. B* **31**, 1985, 5262–5271.
13. Tersoff, J., New empirical approach for the structure and energy of covalent systems. *Phys. Rev. B* **37**, 1988, 6991–7000.
14. Theiß, W., Optical properties of porous silicon. *Surface Sci. Rep.* **29**, 1997, 91–192.

Internal Variables and Generalized Continuum Theories

Arkadi Berezovski, Jüri Engelbrecht and Gérard A. Maugin

Abstract The canonical thermomechanics on the material manifold is enriched by the introduction of dual weakly non-local internal variables and extra entropy fluxes. In addition to the dissipative reaction-diffusion equation for a single internal variable of state, a hyperbolic evolution equation for the internal degree of freedom can be also recovered in the non-dissipative case. It is demonstrated that the Mindlin micromorphic theory can be represented in terms of dual internal variables in a natural way in the framework of the canonical thermomechanics.

1 Motivation

The description of any phenomenon depends on how many details we take into account. Any description can be improved, e.g., by the transition to finer space and time scales. Though such a transition may be desirable for the understanding of a process at microscopic or quantum level, it is hardly acceptable from the practical point of view. Fortunately, there exists a possibility to include the influence of microstructural effects into the description of a phenomenon without changing of space and time scales. This is the introduction of internal variables.

The use of internal variables in the description of the behavior of materials with microstructure has a long tradition [1–11], and nowadays it is practically commonly accepted. However, there are two clearly distinctive types of internal variables: internal degrees of freedom and internal variables of state [5, 12]. By definition, internal variables of state must have no inertia, and they produce no external work.

Arkadi Berezovski · Jüri Engelbrecht
Centre for Nonlinear Studies, Institute of Cybernetics at Tallinn University of Technology, Akadeemia tee 21, 12618 Tallinn, Estonia; e-mail: arkadi.berezovski@cs.ioc.ee

Gérard A. Maugin
Institut Jean Le Rond d'Alembert, Université Pierre et Marie Curie, UMR 7190, Case 162, 4 Place Jussieu, 75252, Paris Cédex 05, France; e-mail: gerard.maugin@upmc.fr

P. Steinmann (ed.), IUTAM Symposium on Progress in the Theory and Numerics of Configurational Mechanics, 149–158.

The internal variables of state are not governed by a field equation, i.e., by their own balance law; the power expended by internal variables will be only of the dissipated type. From another side, internal degrees of freedom are endowed with both inertia and flux, where the latter is not necessarily purely dissipative (on the contrary, it could be purely non-dissipative) [5, 12].

Starting the modeling of dynamics of a microstructured material, we do not know definitely *a priori* what kind of internal variables is more suitable in the particular case. It is useful therefore to have a procedure which formalizes the choice. The main idea of such a formalization can be illustrated on the simple example of linear elasticity in one dimension.

One-dimensional elastodynamics is described by a Lagrangian density $\mathcal{L}$ that depends on displacement $u(x,t)$ and its first derivatives, which we denote by u_t for the time derivative and by u_x for the spatial derivative. This leads to the Euler–Lagrange equation of motion,

$$\frac{\partial}{\partial t}\left(\frac{\partial \mathcal{L}}{\partial u_t}\right)+\frac{\partial}{\partial x}\left(\frac{\partial \mathcal{L}}{\partial u_x}\right)-\frac{\partial \mathcal{L}}{\partial u}=0. \tag{1}$$

In the linear case, the Lagrangian density has the form

$$\mathcal{L}(u,u_x,u_t)=\frac{1}{2}\left(\rho_0 {u_t}^2-Eu_x^2\right), \tag{2}$$

and we obtain the second-order wave equation for the single field variable u

$$\frac{\partial^2 u}{\partial t^2}-c^2\frac{\partial^2 u}{\partial x^2}=0, \tag{3}$$

where $c=\sqrt{E/\rho_0}$, E is the Young's modulus, and ρ_0 is the density.

Introducing velocity and strain by

$$v=u_t, \qquad \varepsilon=u_x, \tag{4}$$

we can represent the wave equation as the system of two first-order equations for the two field variables

$$\frac{\partial \varepsilon}{\partial t}=\frac{\partial v}{\partial x}, \tag{5}$$

$$\rho_0\frac{\partial v}{\partial t}=E\frac{\partial \varepsilon}{\partial x}. \tag{6}$$

The two variables, v and ε, are dual ones in the sense that the evolution of one of them is governed by another and vice versa. Just the same underlying idea is used for the introduction of dual internal variables, thermodynamics of which is given in [13].

In what follows, we will introduce two internal variables (which may have distinct tensorial nature) in the material formulation of thermomechanics and analyze the conditions that are necessary to classify the internal variables as internal degrees of freedom or internal variables of state. As an example, the micromorphic elasticity theory is presented as a particular case of the obtained formalization.

2 Canonical Thermomechanics on the Material Manifold

First, we need to recall certain basic definitions. A motion of a body is considered as a time-parametrized sequence of mappings χ between the reference configuration and the actual configuration: $\mathbf{x} = \chi(\mathbf{X}, t)$, where t is time, $\mathbf{X}$ represents the position of a material point in the reference configuration, and $\mathbf{x}$ is its position in the actual configuration. The deformation gradient is defined by

$$\mathbf{F} = \left.\frac{\partial \chi}{\partial \mathbf{X}}\right|_t = \nabla_R \chi. \tag{7}$$

If the constitutive relation for free energy has the form $W = \overline{W}(\mathbf{F}, \ldots, \mathbf{X}, t)$, then the first Piola–Kirchhoff stress tensor $\mathbf{T}$ is defined by

$$\mathbf{T} = \frac{\partial \overline{W}}{\partial \mathbf{F}}. \tag{8}$$

The local balance laws for sufficiently smooth fields at any regular material point $\mathbf{X}$ in the body read (cf. [14]):

$$\left.\frac{\partial \rho_0}{\partial t}\right|_{\mathbf{X}} = 0, \tag{9}$$

$$\left.\frac{\partial (\rho_0 \mathbf{v})}{\partial t}\right|_{\mathbf{X}} - \mathrm{Div}_R \mathbf{T} = \mathbf{f}_0, \tag{10}$$

$$\left.\frac{\partial (K + E)}{\partial t}\right|_{\mathbf{X}} - \nabla_R \cdot (\mathbf{T} \cdot \mathbf{v} - \mathbf{Q}) = \mathbf{f}_0 \cdot \mathbf{v}, \tag{11}$$

where ρ_0 is the mass density in the reference configuration, $\mathbf{v} = \partial \chi / \partial t|_X$ is the physical velocity, $\mathbf{f}_0$ is a body force per unit reference volume, $K = \rho_0 \mathbf{v}^2/2$ is the kinetic energy, E is the internal energy per unit reference volume, $\mathbf{Q}$ is the material heat flux, $d/dt = \partial/\partial t|_X$ or a superimposed dot denotes the material time derivative.

The second law of thermodynamics is written as

$$\left.\frac{\partial S}{\partial t}\right|_{\mathbf{X}} + \nabla_R \cdot \mathbf{S} \geq 0, \quad \mathbf{S} = (\mathbf{Q}/\theta) + \mathbf{K}, \tag{12}$$

where S is the entropy density per unit reference volume, θ is the absolute temperature, $\mathbf{S}$ is the entropy flux, and the "extra entropy flux" $\mathbf{K}$ vanishes in most cases, but this is not a basic requirement.

The canonical form of the energy conservation has the form [12, 15]

$$\left.\frac{\partial (S\theta)}{\partial t}\right|_{\mathbf{X}} + \nabla_R \cdot \mathbf{Q} = h^{\text{int}}, \quad h^{\text{int}} := \mathbf{T} : \dot{\mathbf{F}} - \left.\frac{\partial \overline{W}}{\partial t}\right|_{\mathbf{X}}, \tag{13}$$

where the right-hand side of Eq. $(13)_1$ is formally an internal heat source.

Then the second law can be represented in the form

$$S\dot{\theta} + \mathbf{S} \cdot \nabla_R \theta \leq h^{\text{int}} + \nabla_R \cdot (\theta \mathbf{K}). \tag{14}$$

Correspondingly, the canonical (material) momentum conservation equation is obtained as [12, 15]

$$\left.\frac{\partial \mathbf{P}}{\partial t}\right|_{\mathbf{X}} - \text{Div}_R \mathbf{b} = \mathbf{f}^{\text{int}} + \mathbf{f}^{\text{ext}} + \mathbf{f}^{\text{inh}}, \tag{15}$$

where the *material momentum* $\mathbf{P}$, the material *Eshelby stress* $\mathbf{b}$, the material *inhomogeneity force* $\mathbf{f}^{\text{inh}}$, the material *external* (or body) force $\mathbf{f}^{\text{ext}}$, and the material *internal* force $\mathbf{f}^{\text{int}}$ are defined by

$$\mathbf{P} := -\rho_0 \mathbf{v} \cdot \mathbf{F}, \quad \mathbf{b} = -\left(L\mathbf{I}_R + \mathbf{T} \cdot \mathbf{F}\right), \quad L = K - \overline{W}, \tag{16}$$

$$\mathbf{f}^{\text{inh}} := \left.\frac{\partial L}{\partial \mathbf{X}}\right|_{\text{expl}} \equiv \left.\frac{\partial L}{\partial \mathbf{X}}\right|_{\text{fixed fields}} = \left(\frac{1}{2}\mathbf{v}^2\right) \nabla_R \rho_0 - \left.\frac{\partial \overline{W}}{\partial \mathbf{X}}\right|_{\text{expl}}, \tag{17}$$

$$\mathbf{f}^{\text{ext}} := -\mathbf{f}_0 \cdot \mathbf{F}, \quad \mathbf{f}^{\text{int}} = \mathbf{T} : (\nabla_R \mathbf{F})^T - \nabla_R \overline{W}\big|_{\text{impl}}. \tag{18}$$

Here the subscript notations *expl* and *impl* mean, respectively, the material gradient keeping the fields fixed (and thus extracting the explicit dependence on $\mathbf{X}$), and taking the material gradient only through the fields present in the function.

3 Dual Internal Variables

Our goal is to show how the dual internal variables can be introduced in canonical thermomechanics. The corresponding theory with a single internal variable was recently presented in [12, 15]. The generalization of the internal variable theory to the case of two internal variables is straightforward. Let us consider the free energy W as a function of two internal variables, $\boldsymbol{\alpha}$ and $\boldsymbol{\beta}$, each of which is a second-order tensor

$$W = \overline{W}(\mathbf{F}, \theta, \boldsymbol{\alpha}, \nabla_R \boldsymbol{\alpha}, \boldsymbol{\beta}, \nabla_R \boldsymbol{\beta}). \tag{19}$$

In this case, the equations of state are given by

$$\mathbf{T} = \frac{\partial \overline{W}}{\partial \mathbf{F}}, \quad S = -\frac{\partial \overline{W}}{\partial \theta}, \quad \mathbf{A} := -\frac{\partial \overline{W}}{\partial \boldsymbol{\alpha}}, \quad \mathcal{A} := -\frac{\partial \overline{W}}{\partial \nabla_R \boldsymbol{\alpha}}, \tag{20}$$

$$\mathbf{B} := -\frac{\partial \overline{W}}{\partial \boldsymbol{\beta}}, \quad \mathcal{B} := -\frac{\partial \overline{W}}{\partial \nabla_R \boldsymbol{\beta}}. \tag{21}$$

We include into consideration the non-zero extra entropy flux according to the case of the single internal variable [12, 15]

$$\mathbf{K} = -\theta^{-1}\mathcal{A} : \boldsymbol{\alpha} - \theta^{-1}\mathcal{B} : \boldsymbol{\beta}. \tag{22}$$

The canonical equations of momentum and energy keep their form

$$\frac{\partial \mathbf{P}}{\partial t} - \mathrm{Div}_R\widetilde{\mathbf{b}} = \mathbf{f}^{th} + \widetilde{\mathbf{f}}^{\mathrm{intr}}, \qquad \frac{\partial (S\theta)}{\partial t} + \nabla_R \cdot \widetilde{\mathbf{Q}} = h^{th} + \widetilde{h}^{\mathrm{intr}}, \tag{23}$$

with the modified Eshelby stress tensor

$$\widetilde{\mathbf{b}} = -(L\mathbf{1}_R + \mathbf{T}\cdot\mathbf{F} - \mathcal{A} : (\nabla_R\boldsymbol{\alpha})^T - \mathcal{B} : (\nabla_R\boldsymbol{\beta})^T), \tag{24}$$

and intrinsic source terms

$$\widetilde{\mathbf{f}}^{\mathrm{intr}} := \widetilde{\mathcal{A}} : \nabla_R\boldsymbol{\alpha} + \widetilde{\mathcal{B}} : \nabla_R\boldsymbol{\beta}, \quad \widetilde{h}^{\mathrm{intr}} := \widetilde{\mathcal{A}} : \dot{\boldsymbol{\alpha}} + \widetilde{\mathcal{B}} : \dot{\boldsymbol{\beta}}. \tag{25}$$

In the above equations the following definitions are used

$$\widetilde{\mathcal{A}} \equiv -\frac{\delta \overline{W}}{\delta \boldsymbol{\alpha}} := -\left(\frac{\partial \overline{W}}{\partial \boldsymbol{\alpha}} - \mathrm{Div}_R\frac{\partial \overline{W}}{\partial(\nabla_R\boldsymbol{\alpha})}\right) = \mathbf{A} - \mathrm{Div}_R\mathcal{A}, \tag{26}$$

$$\widetilde{\mathcal{B}} \equiv -\frac{\delta \overline{W}}{\delta \boldsymbol{\beta}} := -\left(\frac{\partial \overline{W}}{\partial \boldsymbol{\beta}} - \mathrm{Div}_R\frac{\partial \overline{W}}{\partial(\nabla_R\boldsymbol{\beta})}\right) = \mathbf{B} - \mathrm{Div}_R\mathcal{B}, \tag{27}$$

$$\widetilde{\mathbf{S}} = \theta^{-1}\widetilde{\mathbf{Q}}, \quad \widetilde{\mathbf{Q}} = \mathbf{Q} - \mathcal{A} : \dot{\boldsymbol{\alpha}} - \mathcal{B} : \dot{\boldsymbol{\beta}}, \tag{28}$$

$$\mathbf{f}^{th} = S\nabla_R\theta, \quad h^{th} = S\dot{\theta}, \tag{29}$$

which are similar to those in the case of the single internal variable [12, 15].

The corresponding dissipation inequality

$$\Phi = \widetilde{h}^{\mathrm{intr}} - \widetilde{\mathbf{S}}\nabla_R\theta \geq 0, \tag{30}$$

is reduced in the isothermal case to

$$\widetilde{h}^{\mathrm{intr}} := \widetilde{\mathcal{A}} : \dot{\boldsymbol{\alpha}} + \widetilde{\mathcal{B}} : \dot{\boldsymbol{\beta}} \geq 0. \tag{31}$$

The introduction of the second internal variable results in a more general form of evolution equations for the internal variables $\boldsymbol{\alpha}$ and $\boldsymbol{\beta}$ than in the case of a single internal variable [12,15]. In accordance with (31) these evolution equations are chosen as

$$\begin{pmatrix}\dot{\boldsymbol{\alpha}}\\ \dot{\boldsymbol{\beta}}\end{pmatrix} = \mathbf{L}\begin{pmatrix}\widetilde{\mathcal{A}}\\ \widetilde{\mathcal{B}}\end{pmatrix}, \quad \text{or} \quad \begin{pmatrix}\dot{\boldsymbol{\alpha}}\\ \dot{\boldsymbol{\beta}}\end{pmatrix} = \begin{pmatrix}\mathbf{L}^{11} & \mathbf{L}^{12}\\ \mathbf{L}^{21} & \mathbf{L}^{22}\end{pmatrix}\begin{pmatrix}\widetilde{\mathcal{A}}\\ \widetilde{\mathcal{B}}\end{pmatrix}, \tag{32}$$

where components $\mathbf{L}^{11}, \ldots, \mathbf{L}^{22}$ of the linear operator $\mathbf{L}$ are dependent on state variables. Representing the linear operator $\mathbf{L}$ as the sum of symmetric and skew-symmetric components $\mathbf{L} = (\mathbf{L} + \mathbf{L}^T)/2 + (\mathbf{L} - \mathbf{L}^T)/2$, i.e.

$$\begin{pmatrix} \dot{\boldsymbol{\alpha}} \\ \dot{\boldsymbol{\beta}} \end{pmatrix} = \begin{pmatrix} \mathbf{L}^{11} & (\mathbf{L}^{12} + \mathbf{L}^{21})/2 \\ (\mathbf{L}^{21} + \mathbf{L}^{12})/2 & \mathbf{L}^{22} \end{pmatrix} \begin{pmatrix} \widetilde{\mathscr{A}} \\ \widetilde{\mathscr{B}} \end{pmatrix} + \\ + \begin{pmatrix} 0 & (\mathbf{L}^{12} - \mathbf{L}^{21})/2 \\ (\mathbf{L}^{21} - \mathbf{L}^{12})/2 & 0 \end{pmatrix} \begin{pmatrix} \widetilde{\mathscr{A}} \\ \widetilde{\mathscr{B}} \end{pmatrix}, \tag{33}$$

we can see that the symmetry of the linear operator $\mathbf{L}$, which is equivalent to the Onsagerian reciprocity relations $\mathbf{L}^{12} = \mathbf{L}^{21}$, leads to the elimination of the antisymmetric part of the linear operator $\mathbf{L}$. However, we have no reasons to assume the symmetry of the linear operator $\mathbf{L}$ in the case of arbitrary internal variables.

To provide the satisfaction of the dissipation inequality

$$\widetilde{h}^{\text{intr}} := \widetilde{\mathscr{A}} : \dot{\boldsymbol{\alpha}} + \widetilde{\mathscr{B}} : \dot{\boldsymbol{\beta}} = \widetilde{\mathscr{A}} : (\mathbf{L}^{11} \cdot \widetilde{\mathscr{A}} + \mathbf{L}^{12} \cdot \widetilde{\mathscr{B}}) + \widetilde{\mathscr{B}} : (\mathbf{L}^{21} \cdot \widetilde{\mathscr{A}} + \mathbf{L}^{22} \cdot \widetilde{\mathscr{B}}) \geq 0, \tag{34}$$

we may require that

$$\widetilde{\mathscr{A}} : (\mathbf{L}^{12} \cdot \widetilde{\mathscr{B}}) = -\widetilde{\mathscr{B}} : (\mathbf{L}^{21} \cdot \widetilde{\mathscr{A}}). \tag{35}$$

If $\widetilde{\mathscr{A}} \cdot \widetilde{\mathscr{B}}^T$ is symmetric, the latter relation is reduced to the Casimir reciprocity relations

$$\mathbf{L}^{12} = -\mathbf{L}^{21}. \tag{36}$$

On account of the relation (35), we arrive at the decomposition of evolution equations into dissipative and non-dissipative parts

$$\begin{pmatrix} \dot{\boldsymbol{\alpha}} \\ \dot{\boldsymbol{\beta}} \end{pmatrix} = \begin{pmatrix} \mathbf{L}^{11} & 0 \\ 0 & \mathbf{L}^{22} \end{pmatrix} \begin{pmatrix} \widetilde{\mathscr{A}} \\ \widetilde{\mathscr{B}} \end{pmatrix} + \begin{pmatrix} 0 & \mathbf{L}^{12} \\ -\mathbf{L}^{12} & 0 \end{pmatrix} \begin{pmatrix} \widetilde{\mathscr{A}} \\ \widetilde{\mathscr{B}} \end{pmatrix}, \tag{37}$$

and the dissipation inequality is reduced to

$$\widetilde{h}^{\text{intr}} = \widetilde{\mathscr{A}} : (\mathbf{L}^{11} \cdot \widetilde{\mathscr{A}}) + \widetilde{\mathscr{B}} : (\mathbf{L}^{22} \cdot \widetilde{\mathscr{B}}) = (\widetilde{\mathscr{A}} \cdot \widetilde{\mathscr{A}}^T) : \mathbf{L}^{11} + (\widetilde{\mathscr{B}} \cdot \widetilde{\mathscr{B}}^T) : \mathbf{L}^{22} \geq 0. \tag{38}$$

As it is seen, the form of evolution equations is determined by components of the linear operator $\mathbf{L}$. To analyze the possible forms of the evolution equations, we consider two limiting cases, corresponding to pure symmetric and pure skew-symmetric linear operator $\mathbf{L}$.

The most remarkable feature of the considered approach is its applicability to nondissipative processes. It is clear that in the skew-symmetric case ($\mathbf{L}^{11} = \mathbf{L}^{22} = \mathbf{0}$) the dissipation $\widetilde{h}^{\text{intr}}$ vanishes, while evolution equations for the two internal variables are fully coupled

$$\dot{\boldsymbol{\alpha}} = \mathbf{L}^{12} \cdot \widetilde{\mathscr{B}}, \quad \dot{\boldsymbol{\beta}} = -\mathbf{L}^{12} \cdot \widetilde{\mathscr{A}}. \tag{39}$$

In this case, the evolution of one internal variable is driven by another one that means the duality between the internal variables.

To be more specific, let us consider a simple case with $\mathscr{B} = \mathbf{0}$. In this case, the free energy function $\overline{W}$ is independent of $\nabla_R \boldsymbol{\beta}$, and the kinetic relations (39) are reduced to

$$\dot{\boldsymbol{\alpha}} = \mathbf{L}^{12} \cdot \mathbf{B}, \quad \dot{\boldsymbol{\beta}} = -\mathbf{L}^{12} \cdot \widetilde{\mathscr{A}}. \tag{40}$$

Assuming further a quadratic dependence of the free energy function with respect to the internal variable $\boldsymbol{\beta}$

$$\mathbf{B} := -\frac{\partial \overline{W}}{\partial \boldsymbol{\beta}} = -\boldsymbol{\beta}, \tag{41}$$

we reduce Eq. $(40)_1$ to

$$\dot{\boldsymbol{\alpha}} = -\mathbf{L}^{12} \cdot \boldsymbol{\beta}, \tag{42}$$

while Eq. $(40)_2$ is not changed

$$\dot{\boldsymbol{\beta}} = -\mathbf{L}^{12} \cdot \widetilde{\mathscr{A}}. \tag{43}$$

Substituting Eq. (42) into Eq. (43), we obtain a hyperbolic evolution equation for the internal variable $\boldsymbol{\alpha}$:

$$\ddot{\boldsymbol{\alpha}} = (\mathbf{L}^{12} \cdot \mathbf{L}^{12}) \cdot \widetilde{\mathscr{A}}. \tag{44}$$

This means that the introduced internal variable $\boldsymbol{\alpha}$ now is practically an internal degree of freedom, and the structure of Eqs. (42), (43) and (44) is similar to that in the case of elasticity.

If, vice versa, $\mathbf{L}^{11} \neq \mathbf{0}, \mathbf{L}^{22} \neq \mathbf{0}$, while $\mathbf{L}^{12} = \mathbf{0}$, we return to the classical situation, where internal variables are fully independent:

$$\dot{\boldsymbol{\alpha}} = \mathbf{L}^{11} \cdot \widetilde{\mathscr{A}}, \quad \dot{\boldsymbol{\beta}} = \mathbf{L}^{22} \cdot \widetilde{\mathscr{B}}. \tag{45}$$

Therefore, the classical internal variable theory implicitly includes the Onsagerian reciprocity relations. In the fully dissipative case we are dealing with true internal variables of state.

4 Example: Micromorphic Linear Elasticity

In the framework of the Mindlin micromorphic theory [16], each material point is endowed with three translational degrees of freedom u_i and a full microdeformation tensor ψ_{ij} with nine independent components. Three strain tensors are deduced: the classical strain tensor ε_{ij}

$$\varepsilon_{ij} \equiv \frac{1}{2}\left(\partial_i u_j + \partial_j u_i\right), \tag{46}$$

the relative deformation tensor γ_{ij}

$$\gamma_{ij} \equiv \partial_i u_j - \psi_{ij}, \tag{47}$$

and the microdeformation gradient $\varkappa_{ijk}$ defined by

$$\varkappa_{ijk} \equiv \partial_i \psi_{jk}. \tag{48}$$

The free energy density $\overline{W}$ is supposed to be a homogeneous, quadratic function of 42 variables $\varepsilon_{ij}, \gamma_{ij}, \varkappa_{ijk}$ [16]

$$\overline{W} = \frac{1}{2} c_{ijkl} \varepsilon_{ij} \varepsilon_{kl} + \frac{1}{2} b_{ijkl} \gamma_{ij} \gamma_{kl} + \frac{1}{2} a_{ijklmn} \varkappa_{ijk} \varkappa_{lmn} + \\ + d_{ijklm} \gamma_{ij} \varkappa_{klm} + f_{ijklm} \varkappa_{ijk} \varepsilon_{lm} + g_{ijkl} \gamma_{ij} \varepsilon_{kl}. \tag{49}$$

As it was emphasized, only 903 of the 1764 coefficients in the former equation are independent. In the case of centrosymmetric, isotropic materials the number of independent coefficients is greatly reduced [16]

$$\overline{W} = \frac{1}{2} \lambda \varepsilon_{ii} \varepsilon_{jj} + \mu \varepsilon_{ij} \varepsilon_{ij} + \frac{1}{2} b_1 \gamma_{ii} \gamma_{jj} + \frac{1}{2} b_2 \gamma_{ij} \gamma_{ij} + \\ + \frac{1}{2} b_3 \gamma_{ij} \gamma_{ji} + g_1 \gamma_{ii} \varepsilon_{jj} + g_2 \left(\gamma_{ij} + \gamma_{ji} \right) \varepsilon_{ij} + \\ + a_1 \varkappa_{iik} \varkappa_{kjj} + a_2 \varkappa_{iik} \varkappa_{jkj} + \frac{1}{2} a_3 \varkappa_{iik} \varkappa_{jjk} + \frac{1}{2} a_4 \varkappa_{ijj} \varkappa_{ikk} + \\ + a_5 \varkappa_{ijj} \varkappa_{kik} + \frac{1}{2} a_8 \varkappa_{iji} \varkappa_{kjk} + \frac{1}{2} a_{10} \varkappa_{ijk} \varkappa_{ijk} + a_{11} \varkappa_{ijk} \varkappa_{jki} + \\ + \frac{1}{2} a_{13} \varkappa_{ijk} \varkappa_{ikj} + \frac{1}{2} a_{14} \varkappa_{ijk} \varkappa_{jik} + \frac{1}{2} a_{15} \varkappa_{ijk} \varkappa_{kij}. \tag{50}$$

The corresponding stress tensors are the following ones [16]:
Cauchy stress

$$\sigma_{ij} \equiv \frac{\partial \overline{W}}{\partial \varepsilon_{ij}} = \sigma_{ji} = \lambda \delta_{ij} \varepsilon_{kk} + 2\mu \varepsilon_{ij} + g_1 \delta_{ij} \gamma_{kk} + g_2 (\gamma_{ij} + \gamma_{ji}), \tag{51}$$

relative stress

$$\tau_{ij} \equiv \frac{\partial \overline{W}}{\partial \gamma_{ij}} = g_1 \delta_{ij} \varepsilon_{kk} + 2 g_2 \varepsilon_{ij} + b_1 \delta_{ij} \gamma_{kk} + b_2 \gamma_{ij} + b_3 \gamma_{ji}, \tag{52}$$

and double stress

$$\mu_{ijk} \equiv \frac{\partial \overline{W}}{\partial \varkappa_{ijk}}. \tag{53}$$

The equations of motion in terms of stresses have the form (no body force) [16]

$$\rho \ddot{u}_j = \partial_i \left(\sigma_{ij} + \tau_{ij} \right), \tag{54}$$

$$\frac{1}{3} \rho' d_{ji}^2 \ddot{\psi}_{ik} = \partial_i \mu_{ijk} + \tau_{jk} + \Phi_{jk}, \tag{55}$$

where $\rho' d^2_{ji}$ is a microinertia tensor, Φ_{jk} is a volume double force.

In order to apply the internal variable theory, we need to represent the constitutive relations in the Mindlin theory in terms of distortion $\partial_j u_i$ and microdeformation tensor ψ_{ji}. Accordingly, the stresses are represented as

$$\sigma'_{ij} \equiv \frac{\partial \overline{W}}{\partial(\partial_i u_j)} = \lambda\delta_{ij}\partial_k u_k + \mu(\partial_i u_j + \partial_j u_i) + \\ + g_1\delta_{ij}(\partial_k u_k - \psi_{kk}) + g_2\left(\partial_i u_j - \psi_{ij} + \partial_j u_i - \psi_{ji}\right) + \\ + b_1\delta_{ij}(\partial_k u_k - \psi_{kk}) + b_2(\partial_i u_j - \psi_{ij}) + b_3(\partial_j u_i - \psi_{ji}), \tag{56}$$

$$\tau'_{ij} \equiv \frac{\partial \overline{W}}{\partial \psi_{ij}} = -g_1\delta_{ij}\partial_k u_k - g_2(\partial_i u_j + \partial_j u_i) - \\ - b_1\delta_{ij}(\partial_k u_k - \psi_{kk}) - b_2(\partial_i u_j - \psi_{ij}) - b_3(\partial_j u_i - \psi_{ji}). \tag{57}$$

The double stress remains unchanged. At last, equations of motion take on the form

$$\rho \ddot{u}_j = \partial_i \sigma'_{ij}, \tag{58}$$

$$\frac{1}{3}\rho' d^2_{ij}\ddot{\psi}_{ik} = \partial_i \mu_{ijk} - \tau'_{jk} + \Phi_{jk}. \tag{59}$$

Now we consider the microdeformation tensor ψ_{ij} as an internal variable $\boldsymbol{\alpha}$ and apply the formalism developed in Section 3. The microdeformation gradient $\varkappa_{ijk}$ plays the role of the gradient of the internal variable $\boldsymbol{\alpha}$, and we introduce a dual internal variable $\boldsymbol{\beta}$ in the same way as in Section 3.

In the non-dissipative case, the evolution equation for the internal variable $\boldsymbol{\alpha}$ can be symbolically written as

$$\ddot{\boldsymbol{\alpha}} = (\mathbf{L}^{12} \cdot \mathbf{L}^{12}) \cdot \widetilde{\mathscr{A}} = (\mathbf{L}^{12} \cdot \mathbf{L}^{12}) \cdot \left(-\frac{\partial \overline{W}}{\partial \boldsymbol{\alpha}} + \operatorname{Div} \frac{\partial \overline{W}}{\partial(\nabla \boldsymbol{\alpha})}\right). \tag{60}$$

In terms of components of the microdeformation tensor ψ_{ij} the latter evolution equation obtains the form

$$\left(\mathbf{L}^{12} \cdot \mathbf{L}^{12}\right)^{-1}_{ji} \ddot{\psi}_{ik} = \left(-\frac{\partial \overline{W}}{\partial \psi_{jk}} + \operatorname{Div} \frac{\partial \overline{W}}{\partial(\nabla \psi_{jk})}\right) = \partial_i \mu_{ijk} - \tau'_{jk}. \tag{61}$$

As one can see, the evolution equation for the microdeformation is practically the same as in the Mindlin theory. The volume double force Φ_{jk} can appear if we consider a more general case than the pure nondissipative one.

5 Conclusions

The internal variables theory is extended to cover both internal variables of state and internal degrees of freedom by the generalization of its formal structure exploiting the possible coupling between the dual internal variables. The canonical thermomechanics provides the best framework for this generalization. It should be emphasized, however, that any new balance laws has not been introduced; only the Clausius–Duhem inequality was exploited for the derivation of evolution equations for internal variables.

References

1. Rice, J.R., Inelastic constitutive relations for solids: An internal-variable theory and its application to metal plasticity. *J. Mech. Phys. Solids* **19**, 1971, 433–455.
2. Mandel, J.P., Thermodynamics and plasticity. In: Domingos, J.J., Nina, M.N.R. and Whitelaw, J.H. (Eds.), *Foundations of Continuum Thermodynamics*. MacMillan, London, 1974, pp. 283–304.
3. Kestin, J., Internal variables in the local-equilibrium approximation. *J. Non-Equilibr. Thermodyn.* **18**, 1993, 360–379.
4. Muschik, W., Comment to J. Kestin: Internal variables in the local-equilibrium approximation. *J. Non-Equilibr. Thermodyn.* **18**, 1993, 380–388.
5. Maugin, G.A. and Muschik, W., Thermodynamics with internal variables. *J. Non-Equilib. Thermodyn.* **19**, 1994, 217–249. (See also Part II. Applications, ibid., 250–289.)
6. Maugin, G.A., *The Thermomechanics of Nonlinear Irreversible Behaviors*. World Scientific, Singapore, 1999.
7. Houlsby, G.T. and Puzrin, A.M., A thermomechanical framework for constitutive models for rate-independent dissipative materials. *Int. J. Plasticity* **16**, 2000, 1017–1047.
8. Benallal, A. and Bigoni, D., Effects of temperature and thermo-mechanical couplings on material instabilities and strain localization of inelastic materials. *J. Mech. Phys. Solids* **52**, 2004, 725–753.
9. Xiao, H., Bruhns, O.T. and Meyers, A., Elastoplasticity beyond small deformations. *Acta Mechanica* **182**, 2006, 31–111.
10. Zhao, J. and Sheng, D., Strain gradient plasticity by internal-variable approach with normality structure. *Int. J. Solids Struct.* **43**, 2006, 5836–5850.
11. Magnenet, V., Rahouadj, R., Ganghoffer, J.-F. and Cunat, C., Continuous symmetries and constitutive laws of dissipative materials within a thermodynamic framework of relaxation. Part I: Formal aspects. *Int. J. Plasticity* **23**, 2007, 87–113.
12. Maugin, G.A., On the thermomechanics of continuous media with diffusion and/or weak nonlocality. *Arch. Appl. Mech.* **75**, 2006, 723–738.
13. Ván, P., Berezovski, A. and Engelbrecht, J., Internal variables and dynamic degrees of freedom. *J. Non-Equilib. Thermodyn.* **33**, 2008, 235–254.
14. Maugin, G.A., *Material Inhomogeneities in Elasticity*. Chapman and Hall, London, 1993.
15. Maugin, G.A., On canonical equations of continuum thermomechanics. *Mech. Res. Comm.* **33**, 2006, 705–710.
16. Mindlin, R.D., Micro-structure in linear elasticity. Arch. Rat. Mech. Anal. **16**, 1964, 51–78.

Stratified Energies: Ground States with Cracks

M. Giaquinta, P.M. Mariano, G. Modica and D. Mucci

Abstract Elastic bodies admitting cracks are analyzed. Separated pairs of displacement fields and cracks are found as minimizers of the energy in large strain setting. The crack patterns are constructed in terms of varifolds. The discontinuity set of the displacement field is contained in the cracks and may or may not coincide with them.

1 Introduction

By following Griffith's pioneering suggestions, a variational view on the analysis of cracks in simple bodies has been proposed in [3] (see also [1]). Minimality of the energy at every time among all virtual crack-displacement pairs at that time is required. An energy conservation statement throughout the time evolution is imposed. The difficulty of managing crack geometries in finding minimizers has suggested the convenient simplification of identifying cracks with the jump sets of displacement fields (see results in [2,3]). However, appropriate function spaces contain fields with discontinuity sets with closure of positive Lebesgue measure. Theorems allowing the selection of fields with physically significant discontinuity sets – that are

M. Giaquinta
Scuola Normale Superiore, piazza dei Cavalieri 7, I-56126 Pisa, Italy; e-mail: m.giaquinta@sns.it

P.M. Mariano
DICeA, University of Florence, via Santa Marta 3, I-50139 Firenze, Italy;
e-mail: paolo.mariano@unifi.it

G. Modica
Dipartimento di Matematica Applicata "G. Sansone", University of Florence, via Santa Marta 3, I-50139 Firenze, Italy; e-mail: giuseppe.modica@unifi.it

D. Mucci
Dipartimento di Matematica, University of Parma, v.le G.P. Usberti 53/A, I-43100 Parma, Italy;
e-mail: domenico.mucci@unipr.it

P. Steinmann (ed.), IUTAM Symposium on Progress in the Theory and Numerics of Configurational Mechanics, 159–168.

sets that can be appropriate candidates for describing reasonable crack patterns – seem to be not available at least up to now. A new view has been presented in [5]. Separated pairs of displacement fields and cracks are found as minimizers of the energy in simple bodies undergoing large strain. A way for managing the geometry of crack patterns is constructed in terms of special measures, namely varifolds. The discontinuity set of the displacement field is contained in the cracks and may or may not coincide with them. The description of "closed" cracks is also included in this way. Moreover, an essential point is that no crack is prescribed to exist a priori: its possible existence is eventually obtained by the minimization of the energy of the body. A energetic threshold for the formation of a crack arises naturally. Existence theorems are obtained for a non-standard energy functional including a surface energy which depend on the curvature of the possible crack and on the measure localizing it over the body. The present paper anticipates without proof some of the results collected in [5].

2 Curvature Varifolds with Boundary

Some preliminary notions are necessary to the ensuing developments.

Let $\mathcal{B}$ be an open, bounded subset of $\mathbb{R}^n$, $n \geq 2$, with Lipschitz boundary. For a positive integer k, $1 \leq k \leq n$, the Grassmann manifold of k-planes through the origin in $\mathbb{R}^n$ is indicated by $\mathcal{G}_{k,n}$ and is also identified with the set of projectors $\Pi : \mathbb{R}^n \to \mathbb{R}^n$ onto k-planes, characterized by $\Pi^2 = \Pi$, $\Pi^* = \Pi$, $\operatorname{Rank} \Pi = k$, a set which is a compact subset of $\mathbb{R}^n \otimes \mathbb{R}^n$. Consider also the trivial bundle $\mathcal{G}_k(\mathcal{B}) := \mathcal{B} \times \mathcal{G}_{k,n}$ with natural projection $\pi : \mathcal{G}_k(\mathcal{B}) \to \mathcal{B}$. A *$k$-varifold* on $\mathcal{B}$ is a nonnegative Radon measure V over $\mathcal{G}_k(\mathcal{B})$, namely $V \in \mathcal{M}(\mathcal{G}_k(\mathcal{B}))$. The *weight measure* of V is the Radon measure $\mu_V := \pi_\# V$ where $\pi_\#$ is the natural projection of measures associated with the projection π, and the *mass* of V is $\mathbf{M}(V) := V(\mathcal{G}_k(\mathcal{B})) = \mu_V(\mathcal{B})$.

Denote by $\mathcal{H}^k$ the k-dimensional Hausdorff measure in $\mathbb{R}^n$. If $\mathfrak{b}$ is a $\mathcal{H}^k$-measurable, countably k-rectifiable subset of $\mathcal{B}$ and $\theta \in L^1(\mathfrak{b}, \mathcal{H}^k)$, for $\theta\, \mathcal{H}^k \llcorner \mathfrak{b}$ a.e. $x \in \mathcal{B}$ there exists the approximate tangent k-space $T_x\mathfrak{b}$ to $\mathfrak{b}$ at x. Define

$$V_{\mathfrak{b},\theta}(\varphi) := \int_{\mathcal{G}_k(\mathcal{B})} \varphi(x, \Pi)\, dV_{\mathfrak{b},\theta}(x, \Pi) := \int_{\mathfrak{b}} \theta(x)\varphi(x, \Pi(x))\, d\mathcal{H}^k(x) \quad (1)$$

for any $\varphi \in C_c^0(\mathcal{G}_k(\mathcal{B}))$, where $\Pi(x)$ is the orthogonal projection of $\mathbb{R}^n$ onto $T_x\mathfrak{b}$.

Definition 1. V is called a *curvature k-varifold with boundary* if

1. $V = V_{\mathfrak{b},\theta}$ is the integer rectifiable k-varifold associated with $(\mathfrak{b}, \theta, \mathcal{H}^k)$,
2. there exist a function $A \in L^1(\mathcal{G}_k(\mathcal{B}), \mathbb{R}^{n*} \otimes \mathbb{R}^n \otimes \mathbb{R}^{n*})$, $A = (A_j^{\ell i})$, and a vector Radon measure $\partial V \in \mathcal{M}(\mathcal{G}_k(\mathcal{B}), \mathbb{R}^n)$ such that

$$\int_{\mathcal{G}_k(\mathcal{B})} (\Pi D_x\varphi + A^t D_\Pi \varphi + AI\varphi)\, dV(x, \Pi) = -\int_{\mathcal{G}_k(\mathcal{B})} \varphi\, d\partial V(x, \Pi)$$

for every $\varphi \in C_c^\infty(\mathcal{G}_k(\mathcal{B}))$.

Moreover, for $p \geq 1$ the subclass of curvature k-varifolds with boundary such that $A \in L^p(\mathcal{G}_k(\mathcal{B}))$ is indicated by $CV_k^p(\mathcal{B})$.

The function $x \mapsto A(x, \Pi(x)) \in \mathbb{R}^{n*} \otimes \mathbb{R}^n \otimes \mathbb{R}^{n*}$ is called the *curvature* of the varifold V. The vector measure ∂V is called the *varifold boundary measure*.

The following results proven in [7, 8] collect the geometrical properties of curvature k-varifolds with boundary.

Theorem 1. *Let $V = V_{\mathfrak{b},\theta}$ be a k-varifold with boundary ∂V and curvature A, with $A_j^{\ell i} \in L^1(\mathcal{G}_k(\mathcal{B}))$.*

1. *The following symmetry properties hold:*

$$A_j^{\ell i} = A_\ell^{ji}, \quad A_j^{ji} = 0, \quad A_j^{\ell i} = \Pi_h^\ell A_j^{hi} + \Pi_j^h A_h^{\ell i}, \quad V - a.e.$$

2. *$\Pi_h^i A_j^{\ell h} = A_j^{\ell i}$ V-a.e. in such a way that, by setting $H^i(x) := A_j^{ij}(x, \Pi(x))$, one gets $\Pi_i^h H^h = 0$ V-a.e.; in particular, if $\Pi = \Pi(x)$ is the orthogonal projection over $T_x\mathfrak{b}$, then*

$$H(x, \Pi(x)) \perp T_x\mathfrak{b} \qquad \mu_V - a.e.$$

3. *The projection map $x \to \Pi(x)$ is μ_V-a.e approximately differentiable and*

$$(\nabla^{\mathfrak{b}} \Pi_j^\ell(x))^i = A_j^{\ell i}(x, \Pi(x))$$

for μ_V-a.e. x.

4. *The support of $|\partial V|$ is contained in the support of V and $|\partial V| \perp V$.*
5. *∂V is tangential to $\mathfrak{b}$ in the sense that $(\Pi_j^i)_\# \partial^j V = \partial^i V$ as measures on $\mathcal{G}_k(\mathcal{B})$.*
6. *V is a varifold with locally bounded first variation and generalized mean curvature in the sense of Allard with generalized mean curvature vector $H(x) = H(x, \Pi(x))$ and generalized boundary $\pi_\# \partial V$.*

Theorem 2 (Rectifiability of the boundary). *Let V be a curvature k-varifold with boundary ∂V and $k \geq 1$. There exists a $\mathcal{H}^{k-1}$-countably rectifiable set $\mathfrak{C}$ and a function $\sigma \in L^1(\mathfrak{C}, \mathcal{H}^{k-1})$ such that $\pi_\#|\partial V| = \sigma \mathcal{H}^{k-1} \llcorner \mathfrak{C}$. Moreover, one has*

$$\int \varphi(x, \Pi(x))\, d\partial V(x, \Pi) = \int_{\mathfrak{C}} \Big(\int_{\mathcal{G}_{k,n}} \varphi(x, \Pi)\, d\tau_x(\Pi) \Big)\, d\mathcal{H}^{k-1}(x)$$

for every $\varphi \in C_c^\infty(\mathcal{G}_k(\mathcal{B}))$, where for $\mathcal{H}^{k-1}$-a.e. $x \in \mathfrak{C}$ the vector valued measure τ_x on $\mathcal{G}_{k,n}$ has the structure

$$\tau_x = \sum_{i=1}^{i_x} m_i^x\, \alpha_i^x\, \delta_{p_i^x}, \tag{2}$$

where $i_x \in \mathbb{N}$, $\delta_{p_i^x}$ is the Dirac delta supported by a k-plane p_i^x of the Grassmanian $\mathcal{G}_{k,n}$; moreover, the α_i^x's are positive integers and the m_i^x's are unit vectors in $\mathbb{R}^n$. In addition p_i^x contains the tangent $(k-1)$-space $T_x\mathfrak{C}$ to $\mathfrak{C}$ at x and

$p_i^x = \mathrm{Span}\left\{T_x\mathfrak{C}, m_i^x\right\}$. *In the special case of one-dimensional curvature varifold* V *with boundary, the formula* (2) *reduces to*

$$\tau_x := \sum_{j=1}^{j_x} \alpha_j t_j \delta_{P_j}$$

where δ_{P_j} *is the Dirac delta function supported by a straight line* P_j *in* $\mathcal{G}_{1,n}$, t_j *is a unit vector that orients* P_j *and* α_j *a positive integer. As a consequence, for the boundary of a curvature* 1*-varifold one gets*

$$\partial V(x, P) = \sum_{i=1}^{\infty} \delta_{x_i}(x) \times \tau_{x_i}(P).$$

Theorem 3 (Compactness [8]). *For* $1 < p < \infty$, *let* $\{V^{(r)}\} \subset CV_k^p(\mathcal{B})$ *be a sequence of curvature k-varifolds* $V^{(r)} = V_{\mathfrak{b}_r,\theta_r}$ *with boundary. The corresponding curvatures and boundaries are indicated by* $A^{(r)} = \{A^{(r)}{}_j^{\ell i}\}$ *and* $\partial V^{(r)}$, *respectively. Assume that for every open set* $\Omega \subset\subset \mathcal{B}$ *there exists a constant* $c = c(\Omega) > 0$ *such that for every r*

$$\mu_{V^{(r)}}(\Omega) + |\partial V^{(r)}|(\mathcal{G}_k(\Omega)) + \int_{\mathcal{G}_k(\Omega)} |A^{(r)}|^p \, dV^{(r)} \leq c(\Omega).$$

There exists a subsequence $\{V^{(r_s)}\}$ *of* $\{V^{(r)}\}$ *and a curvature k-varifold* $V = V_{\mathfrak{b},\theta} \in CV_k^p(\mathcal{B})$, *with curvature A and boundary* ∂V, *such that*

$$V^{(r_s)} \rightharpoonup V, \quad A^{(r_s)} \, dV^{(r_s)} \rightharpoonup A \, dV, \qquad \partial V^{(r_s)} \rightharpoonup \partial V,$$

in the sense of measures. Moreover, for any convex and l.s.c. function $f : \mathbb{R}^{n*} \otimes \mathbb{R}^n \otimes \mathbb{R}^{n*} \to [0, +\infty]$, *one gets*

$$\int_{\mathcal{G}_k(\mathcal{B})} f(A) \, dV \leq \liminf_{s \to \infty} \int_{\mathcal{G}_k(\mathcal{B})} f(A^{(r_s)}) \, dV^{(r_s)}.$$

3 Transplacement Fields and Bulk Energy

Only Cauchy bodies are called upon in the analyses presented here. They are bodies for which the morphology of each material element is described only by the place in space occupied by its centre of mass. In other words, a body is identified with a region $\mathcal{B}$ of the Euclidean ambient space $\mathbb{R}^3$ that it occupies in a macroscopic reference configuration, taken as reference place. $\mathcal{B}$ is considered here as an open set with Lipshitz boundary. Other configurations are reached by means of transplacements that are usually taken as orientation preserving differentiable bijections

$u : \mathcal{B} \longrightarrow \mathbb{R}^3$ mapping $\mathcal{B}$ in the current configuration $u(\mathcal{B})$, a set that is presumed to be always open and endowed with Lipshitz boundary.

The body occupying $\mathcal{B}$ is said to be an *hyperelastic simple body* when it is endowed by a bulk energy which is absolutely continuous with respect to the volume measure and depends on the deformation gradient only. Ground states of such a type of body are described by minimizers of the overall energy. Such minimizers can be determined in terms of Cartesian currents [6] that are described briefly below.

3.1 Sobolev Maps and Related Cartesian Currents

Let $I(k,n)$ be the space of multi-indices in $(1,\ldots,n)$ of length k. Denote also by 0 the empty multi-index of length 0. For any α, the *complementary* multi-index to α in $(1,\ldots,n)$ is indicated by $\bar{\alpha}$, $\bar{\alpha} \in I(n-k,n)$, and $\sigma(\alpha,\bar{\alpha})$ is the sign of the permutation from $(1,\ldots,n)$ into $(\alpha_1,\ldots,\alpha_k,\bar{\alpha}_1,\ldots,\bar{\alpha}_{n-k})$. For $(e_1, e_2, \ldots, e_n)$ and $(\epsilon_1, \epsilon_2, \ldots, \epsilon_n)$ bases in $\mathbb{R}^n$ and $\mathbb{R}^N$, respectively, $\Lambda_r(\mathbb{R}^n \times \mathbb{R}^N)$ is the vector space of skew-symmetric tensors over $\mathbb{R}^n \times \mathbb{R}^N$ of the form

$$\xi = \sum_{|\alpha|+|\beta|=r} \xi^{\alpha\beta} e_\alpha \wedge \epsilon_\beta = \sum_{\max(0,r-n)}^{\min(r,N)} \xi_{(k)}, \qquad \xi_{(k)} = \sum_{\substack{|\alpha|+|\beta|=r \\ |\beta|=k}} \xi^{\alpha\beta} e_\alpha \wedge \epsilon_\beta .$$

For any linear map $G : \mathbb{R}^n \to \mathbb{R}^N$, the notation $M(G)$ is used for the simple n-vector in $\Lambda_n(\mathbb{R}^n \times \mathbb{R}^N)$ tangent to the graph of G and defined by

$$M(G) := \Lambda_n(\mathrm{Id} \times G)(e_1 \wedge \cdots \wedge e_n) = (e_1, G(e_1)) \wedge \cdots \wedge (e_n, Ge_n)).$$

For $u : \mathcal{B} \to \mathbb{R}^N$ an a.e. approximately differentiable map, denote by Du its approximate gradient. u has a Lusin representative on the subset $\widetilde{\mathcal{B}}$ of Lebesgue points of both u and Du, and $|\mathcal{B} \setminus \widetilde{\mathcal{B}}| = 0$. Let $\widetilde{u}(x)$ and $D\widetilde{u}(x)$ be the Lebesgue values of u and Du at $x \in \widetilde{\mathcal{B}}$, respectively. Assume that $|M(Du)| \in L^1(\mathcal{B})$. By following [6], the *graph* of u, defined by

$$\mathcal{G}_u := \left\{ (x,y) \in \mathcal{B} \times \mathbb{R}^N \,\middle|\, x \in \widetilde{\mathcal{B}},\ y = \widetilde{u}(x) \right\},$$

is a n-rectifiable subset of $\mathcal{B} \times \mathbb{R}^N$ with approximate tangent vector n-space at $(x, \widetilde{u}(x))$ generated by the vectors $(e_1, D\widetilde{u}(x)e_1), \ldots, (e_n, D\widetilde{u}(x)e_n))$ in $\mathbb{R}^n \times \mathbb{R}^N$. The *$n$-current integration over the graph* of u is defined by the linear functional on smooth n-forms $\omega = \omega(x,y)$ with compact support in $\mathcal{B} \times \mathbb{R}^N$ given by

$$G_u(\omega) = \int < \omega, \xi > \, d\mathcal{H}^n \llcorner \mathcal{G}_u, \tag{3}$$

where $\xi(x) := \frac{M(D\widetilde{u}(x))}{|M(D\widetilde{u}(x))|}$, for $x \in \widetilde{\mathcal{B}}$, is the unit n-vector that orients the approximate tangent n-space to $\mathcal{G}_u$ at $(x, \widetilde{u}(x))$; moreover, G_u has finite mass $\mathbf{M}(G_u) := \sup_{||\omega||_\infty \le 1} G_u(\omega) < \infty$, since

$$\mathbf{M}(G_u) = \int_{\mathcal{B}} |M(Du(x))| \, dx = \mathcal{H}^n(\mathcal{G}_u).$$

In particular, G_u is a vector valued measure on $\mathcal{B} \times \mathbb{R}^N$, actually an *integer rectifiable n*-current with multiplicity 1 on $\mathcal{B} \times \mathbb{R}^N$. The *boundary* of the current G_u can be defined by duality as the $(n-1)$-current acting on compactly supported smooth $(n-1)$-forms ω in $\mathcal{B} \times \mathbb{R}^N$, namely $\partial G_u(\omega) := G_u(d\omega)$, $\omega \in \mathcal{D}^{n-1}(\mathcal{B} \times \mathbb{R}^N)$, where $d\omega$ is the differential of ω.

3.2 The Bulk Energy

By taking apart for a while the description of the possible cracks, it is assumed that the external body forces have conservative nature so that the bulk energy of the body has the usual form

$$\mathcal{E}_{\mathcal{B}}(u) := \int_{\mathcal{B}} e(x, u, Du) \, dx$$

where $e(\cdot)$ is the sum of the elastic energy and the potential of external forces. It is assumed that $e = e(x, u, F)$ satisfies common assumptions listed below:

(H1) $e : \mathcal{B} \times \hat{\mathbb{R}}^n \times \mathbb{M}^+_{n\times n} \to [0, +\infty]$ is continuous, where $\mathbb{M}^+_{n\times n}$ is the class of real $(n \times n)$-matrices F such that $\det F > 0$.

(H2) The map $F \mapsto e(x, u, F)$ is *polyconvex*, i.e. there exists a function

$$Pe(x, u, \xi) : \mathcal{B} \times \hat{\mathbb{R}}^n \times \Lambda_n(\mathbb{R}^n \times \hat{\mathbb{R}}^n) \to [0, +\infty]$$

continuous in (x, u) for every ξ, convex and lower semicontinuous in ξ for every (x, u), such that

$$e(x, u, F) = Pe(x, u, M(F)) \qquad \forall F \in \mathbb{M}^+_{n\times n}, \quad \forall (x, u) \in \mathcal{B} \times \hat{\mathbb{R}}^n.$$

(H3) $e = e(x, u, F)$ satisfies the growth conditions

$$e(x, u, F) \ge c_4 \, |M(F)|^q \qquad \forall F \in \mathbb{M}^+_{n\times n}, \quad \forall (x, u) \in \mathcal{B} \times \hat{\mathbb{R}}^n,$$

for some $c_4 > 0$ and $q > 1$.

(H4) For every $x \in \mathcal{B}$ and $F \in \mathbb{M}^+_{n\times n}$ if for some $u \in \hat{\mathbb{R}}^n$ the inequality $e(x, u, F) < +\infty$ is satisfied, then $\det F > 0$.

The assumptions (H1) and (H4) are essentially suggested by physical plausibility. The hypothesis (H2) is an essence an assumption of material stability while the growth condition (H3) has more technical nature.

4 A Skeletal Model Admitting Formation of Cracks

The aim now is to describe the possible presence of cracks. The setting is selected two-dimensional for the sake of simplicity. The generalization to 3D and extensions are in [5]. Cracks are here represented here by 1-dimensional curvature varifolds with curvature in L^p, $p > 1$, which are quite regular. Basically $V \in CV_1^p(\mathcal{B})$ can be essentially described as (the integration over) a locally finite union of $C^{1,1-1/p}$ curves counted with integer multiplicities. Their boundaries are just Dirac measures concentrated at the endpoints with their tangential directions.

Definition 2. A macroscopic configuration of a body $\mathcal{B} \subset \mathbb{R}^2$ with a crack is a pair composed by the bounded connected open set $\mathcal{B}$ with Lipschitz boundary and a curvature 1-varifold with boundary, namely $V = V_{\mathfrak{b},\theta} \in CV_1^p(\mathcal{B})$ for some $p > 1$.

The gross place occupied by the body and the crack are treated as distinct objects. The crack is not part of the initial boundary: it is selected by a measure over $\mathcal{B}$, namely a curvature varifold, and may or may not be an empty set in the reference place. Since the material bonds across the crack margins are broken, along the deformation, the cracks faces may loose contact. The obvious implication is that the graph of the deformation may have nonzero boundary. An appropriate class of admissible deformations has to be defined.

Weak diffeomorphisms have been found to be natural descriptors of deformations of standard elastic bodies [6]. They are orientation-preserving, allow frictionless contact of parts of the boundary while still prevent self-penetration of the matter. However, they satisfy a condition of zero boundary in the sense of currents, a condition avoiding the formation of 'holes' of various nature. To allow fractures, an extended version of them has to be formulated.

Definition 3. Let $\mathcal{B} \subset \mathbb{R}^2$ be a body with crack $V \in CV_1^p(\mathcal{B})$. A *weak diffeomorphism* on $\mathcal{B}$ *admitting cracks* described by V is an a.e. approximately differentiable map $u : \mathcal{B} \to \hat{\mathbb{R}}^2$ such that

1. $|Du|, \det Du \in L^1(\mathcal{B})$;
2. $\pi_{\#}|\partial G_u| \leq \mu_V$, where $\mu_V := \pi_{\#} V$;
3. $\det Du(x) > 0$ for a.e. $x \in \mathcal{B}$;
4. for every compactly supported smooth function $f : \mathcal{B} \times \hat{\mathbb{R}}^2 \to [0, +\infty)$

$$\int_{\mathcal{B}} f(x, u(x)) \det Du(x)\, dx \leq \int_{\hat{\mathbb{R}}^2} \sup_{x \in \mathcal{B}} f(x, y)\, dy.$$

In this case, one writes $u \in \mathrm{dif}^{1,1}(\mathcal{B}, V, \hat{\mathbb{R}}^2)$. Moreover, for $q > 1$ the class $\mathrm{dif}^{q,1}(\mathcal{B}, V, \hat{\mathbb{R}}^2)$ is defined by

$$\mathrm{dif}^{q,1}(\mathcal{B}, V, \hat{\mathbb{R}}^2) := \left\{ u \in \mathrm{dif}^{1,1}(\mathcal{B}, V, \hat{\mathbb{R}}^2) \,\middle|\, |M(Du)| \in L^q(\mathcal{B}) \right\}.$$

Condition (ii) implies that the Green formulas hold true in $\mathcal{B}$ outside the crack and prescribes that the boundary current has finite mass, namely $\mathbf{M}(\partial G_u) < \infty$.

Theorem 4. *Let* $\{V^{(r)}\} \subset CV_1^p(\mathcal{B})$, *with* $p > 1$, *be a sequence of curvature varifolds describing cracks in the body* $\mathcal{B}$, *with equibounded total variations, i.e.* $\sup_r \mu_{V^{(r)}}(\mathcal{B}) < \infty$. *Moreover, assume* $u_r \in \operatorname{dif}^{1,1}(\mathcal{B}, V^{(r)}, \widetilde{\mathbb{R}}^2)$. *Suppose also that there exist* $u \in L^1(\mathcal{B}, \hat{\mathbb{R}}^2)$, $v \in L^1(\mathcal{B}, \Lambda_n(\mathbb{R}^2 \times \hat{\mathbb{R}}^2))$, *and* $V \in CV_1^p(\mathcal{B})$ *such that* $u_r \rightharpoonup u$, $M(Du_r) \rightharpoonup v$ *weakly in* L^1, *and* $V_r \rightharpoonup V$ *as measures. Then* $v = M(Du)$ *and, moreover, if* $\det Du > 0$ *a.e.,* $u \in \operatorname{dif}^{1,1}(\mathcal{B}, V, \hat{\mathbb{R}}^2)$.

4.1 The Energy Functional

Bulk and crack contributions to the energy are involved as usual. The crack in this skeletal model is one dimensional. The part of the energy associated with the crack is then splitted in two contributions: (i) the energy along the margins, which depend on the curvature of the margins themselves and is represented by the curvature of a varifold and (ii) the energy at the tips, the corners and the junctions of the fracture, that are represented by the boundary of the same varifold.

The energy $\mathcal{E}(u, V)$ reads

$$\begin{aligned} \mathcal{E}(u, V) &:= \mathcal{E}(u, V, \mathcal{B}) \\ &= \int_{\mathcal{B}} e(x, u, Du)\, dx + c_1 \int_{\mathcal{G}_1(\mathcal{B})} |A|^p \, dV + c_2 \mathbf{M}(V) + c_3 \mathbf{M}(\partial V) \end{aligned} \tag{4}$$

where the c_i's are positive constants and the hypotheses (H1) (H2), (H3) and (H4) of Section 3 on the bulk energy density $e = e(x, u, F)$ are satisfied.

With regard to crack energy term, the p-norm $|A|^p$ of the curvature can be replaced by $\phi(|A|)$ where $\phi : \mathbb{R}^+ \to \mathbb{R}^+$ is a convex function satisfying $\phi(t) \geq c_5 t^p$.

The term $c_2\mathbf{M}(V)$ is the Griffith-like part of the surface energy of the crack.

4.2 Ground States: Existence Theorems

It may be convenient to prescribe a *comparison varifold* $\widetilde{V} \in CV_1^p(\mathcal{B})$ such that all competing varifolds V satisfy the bound $\mu_{\widetilde{V}} \leq \mu_V$. The comparison varifold $\widetilde{V}$ can be of course zero when an initial crack is absent. In the opposite case, $\widetilde{V}$ describes a crack from which the competing cracks may extend without excluding that they may have portions unrelated with $\widetilde{V}$.

The space

$$\begin{aligned} \mathcal{A}_{q,p,K,\widetilde{V}}(\mathcal{B}) := \Big\{ (u, V) \,\Big|\, & V \in CV_1^p(\mathcal{B}),\ u \in \operatorname{dif}^{q,1}(\mathcal{B}, V, \hat{\mathbb{R}}^2), \\ & \|u\|_{L^\infty(\mathcal{B})} \leq K,\ \mu_{\widetilde{V}} \leq \mu_V \Big\}, \end{aligned}$$

with $K > 0$, is then the natural functional environment for investigating the existence of minimizers (u, V) for the energy $\mathcal{E}$.

Theorem 5. *Consider $\mathcal{B} \subset \mathbb{R}^2$, $q, p > 1$, $K > 0$, $\widetilde{V} \in CV_1^p(\mathcal{B})$. Assume that there exists an element $(u, V) \in \mathcal{A}_{q,p,K,\widetilde{V}}(\mathcal{B})$ that satisfies the prescribed Dirichlet boundary conditions. Then the energy functional* (4) *attains its minimum in the subclass of $\mathcal{A}_{q,p,K,\widetilde{V}}(\mathcal{B})$ of couples (u, V) where u satisfies the prescribed boundary conditions.*

The constant K is selected at will for purposes of physical plausibility: it is only necessary for establishing the boundedness of the L^∞ norm of u. In contrast, the constants p and q and the comparison varifold $\widetilde{V}$ have constitutive nature. The a-priori L^∞ bound on the transplacement field has been relaxed in [2,4] in a different setting, not dealing with the path followed here.

The simpler description of the boundary measure of the one dimensional curvature varifolds allows one to state another existence theorem with a different growth condition for the bulk energy.

Consider the energy functional (4) where the bulk energy density $e(x, u, F)$ satisfies (H1), (H2), (H4) of Section 3 and impose a different growth condition indicated here by

(H3-1)

$$e(x, u, F) \geq c_4|F|^2 \qquad \forall F \in \mathbb{M}^+_{2\times 2}, \quad \forall (x, u) \in \mathcal{B} \times \hat{\mathbb{R}}^2,$$

for some $c_4 > 0$.

For $K > 0$ and $\widetilde{V} \in CV_1^p(\mathcal{B})$ the class

$$\mathcal{A}_{p,\widetilde{V},K} := \Big\{(u, V) \Big| V \in CV_1^p(\mathcal{B}),\ p > 1, u \in \mathrm{dif}^{1,1}(\mathcal{B}, V, \hat{\mathbb{R}}^2),\ Du \in L^2(\mathcal{B}),$$
$$\|u\|_{L^\infty(\mathcal{B})} \leq K,\ \mu_{\widetilde{V}} \leq \mu_V\Big\}, \tag{5}$$

is then the natural functional setting for another existence result.

Theorem 6. *Assume that the bulk energy density of* (4) *satisfies* (H1), (H2), (H4) *of Section* 3 *and* (H3-1). *Suppose that that there is at least one element (u_0, V_0) in the class* (5) *with u_0 satisfying a given Dirichlet data. Then the functional* (4) *has a minimizer in the subclass of* (5) *of couples (u, V) with u satisfying the prescribed Dirichlet boundary conditions.*

In the previous scheme, a sequence of varifolds accumulating at the boundary of $\mathcal{B}$ vanishes at the limit. It is possible to consider a different situation where the propagation of cracks at the boundary of the body $\mathcal{B}$ is taken into account, and a term involving the crack at the boundary may contribute to the limit energy of minimizing sequences. Such a situation has a clear meaning in terms of transplacements for the Dirichlet problem, where the limit crack may be seen as a rupture of the boundary condition.

A related existence theorem again follows.

Acknowledgement

This work has been developed within the programs of the research group in 'Theoretical Mechanics' of the 'Centro di Ricerca Matematica Ennio De Giorgi' of the Scuola Normale Superiore at Pisa.

References

1. Bourdin, B., Francfort, G.A. and Marigo, J.-J., The variational approach to fracture. *J. Elasticity* **91**, 2008, 5–148.
2. Dal Maso, G., Francfort, G.A. and Toader, R., Quasi-static growth in finite elasticity. *Arch. Rat. Mech. Anal.* **176**, 2005, 165–225.
3. Francfort, G.A. and Marigo, J.-J., Revisiting brittle fracture as an energy minimization problem. *J. Mech. Phys. Solids* **46**(8), 1998, 1319–1342.
4. Francfort, G.A. and Mielke, A., Existence results for a class of rate-independent matrerial models with non convex elastic energies. *J. Reine Angew. Math.* **595**, 2006, 55–91, 2006.
5. Giaquinta, M., Mariano, P.M., Modica, G. and Mucci, D., Ground states of simple bodies that may undergo brittle fractures, preprint, 2008.
6. Giaquinta, M., Modica, G. and Souček, J., *Cartesian Currents in the Calculus of Variations*, Vols. I and II. Springer-Verlag, Berlin, 1998.
7. Hutchinson, J.E., Second fundamental form for varifolds and existence of surfaces minimizing curvature, *Indiana Univ. Math. J.* **35**, 1986, 45–71.
8. Mantegazza, C., Curvature varifolds with boundary. *J. Diff. Geom.* **43**, 1996, 807–843.

Crack Curving Based on Configurational Forces and Their Gradients

Henning Schütte

Abstract A numerical scheme is presented to predict crack trajectories in two dimensional components. First a relation between the curvature in mixed-mode crack propagation and the corresponding configurational forces is derived, based on the principle of maximum dissipation. With the help of this, a numerical scheme is presented which is based on a predictor-corrector method using the configurational forces acting on the crack together with their derivatives along real and test paths. With the help of this scheme it is possible to take bigger than usual propagation steps, represented by splines. Essential for this approach is the correct numerical determination of the configurational forces acting on the crack tip. The methods used by other authors are shortly reviewed and an approach valid for arbitrary non-homogenous and non-linear materials with mixed-mode cracks is presented. Numerical examples show, that the method is a able to predict the crack paths in components with holes, stiffeners, etc., with good accuracy.

1 Introduction

Crack path prediction is based on three parts: a criterion for the onset of crack propagation, a criterion for the direction of propagation, a criterion for the propagation speed or step length, for fatigue or quasi-static propagation respectively. There is a wealth of criteria available, but most of the criteria available in the literature cannot consistently by derived without ad-hoc assumptions. Furthermore one has to distinguish between approaches for kinking and curving of cracks. Most criteria available will give a statement for the immediate directional change of the crack path, thus producing a kink. With small step sizes and kinking angles these criteria are then used to represent a curved crack path, as a kink will physically only be ne-

Henning Schütte
Institute of Mechanics, Ruhr-University of Bochum, Universitätsstr. 150, IA 3/38, 44801 Bochum, Germany; e-mail: henning.schuette@rub.de

P. Steinmann (ed.), IUTAM Symposium on Progress in the Theory and Numerics of Configurational Mechanics, 169–178.

cessary, if there is an abrupt change in the loading or the material properties. Among the few approaches using curving segments is the one of Sumi et al. [1], but he is restricting himself to slightly curving cracks. Most of these criteria are unfortunately restricted to linear elastic fracture mechanics, as they are either based on the near tip stress field solution or the stress intensity factors. So for nonlinear elastic, inhomogeneous or plastic crack propagation methods based on configurational forces have drawn attention [2–4]. Unfortunately the direction of the J-Integral or configurational force vector on the crack tip as the directional criterion is not a correct choice as it does not account for the change of the configurational forces induced by the kink [5]. This can be seen directly from the fact that this criterion predicts always straight crack propagation for pure mode II, which is in contrast with experimental observations. Also the calculation of J2 with the help a domain integral type approach used in most of these methods is inaccurate. So these criteria are only valid for small kinking angles, which results in small steps sizes for an accurate representation of curved cracks. The aim of this paper is to present a derivation of a propagation criterion valid for strongly curved cracks with finite propagation step sizes and formulated with the help of configurational forces, so the numerical approaches presented in [3,4] can be used to end up with a numerical scheme that can be generalized to treat inhomogeneous materials at finite deformations. Additionally a method is presented to calculate valid results for J2 from a direct configurational nodal force approach.

2 Crack Curving in LEFM

The derivation of the criterion for curved crack propagation is done with the help of the results obtained by Amestoy and Leblond [6] in the framework of linear elastic fracture mechanics. Linear elastic fracture mechanics is based on the near tip stress field

$$\sigma_{ij} = K_\alpha\, f_{ij}^\alpha(\theta)\, r^{-1/2} + T_\alpha\, g_{ij}^\alpha(\theta) + b_\alpha\, h_{ij}^\alpha(\theta)\sqrt{r} + O(r), \tag{1}$$

where the K_α are the stress intensity factors (SIFs) for the three modi $\alpha = I, II, III$, T the (non-local) T-stresses and the b_α are the coefficients of square-root stress terms also used by Sumi et al. [1]. The f_{ij}, g_{ij}, h_{ij} matrices of angular functions stem from the Williams series solution [7]. A kinked and curved crack (Figure 1) with the elongation of the crack s is described by $y' = a^\star x'^{3/2} + \frac{1}{2}C^\star x'^2$.

The evolution of the SIFs is given by Amestoy and Leblond [6] as

$$K_\alpha(s) = K_\alpha^\star + K_\alpha^{(1/2)}\sqrt{s} + K_\alpha^{(1)}\, s + O(s^{2/3}) \tag{2}$$

with

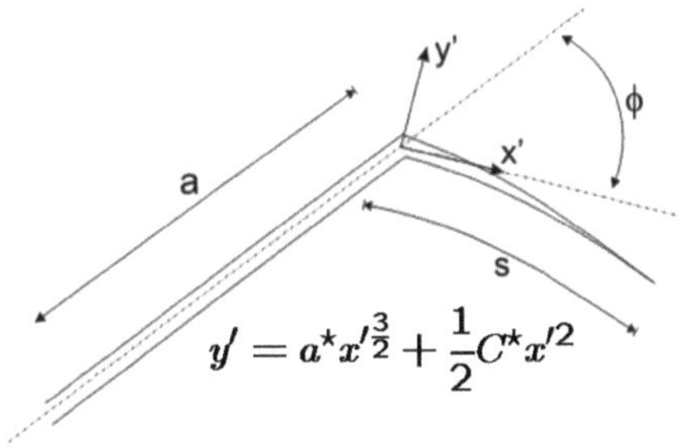

Fig. 1 The kinked and curved crack.

$$K_\alpha^\star = F_{\alpha\beta}(\phi)K_\beta \tag{3}$$

$$K_\alpha^{(1/2)} = G_{\alpha\beta}(\phi)T_\beta + a^\star H_{\alpha\beta}(\phi)K_\beta \tag{4}$$

$$K_\alpha^{(1)} = \left[K_\alpha^{(1)}\right]_{C^\star=0}^{\phi,a^\star} + C^\star M_{\alpha\beta}(\phi)K_\beta \tag{5}$$

where the Greek indices run over I, II, III the three crack modes and the matrices F, G, H, M are universal functions, depending only on the kinking angle and not the special crack problem under consideration. The first term in eq. (3c) involves the b-coefficients of eq. (1), but also some non-universal parts, that means it can only be determined for a special crack problem in a finite body. A special note on the non-universal characteristic of the second-order term for curving cracks seems to be missing in Sumi et al.'s [1] approach, but is pointed out in [6]. Amestoy and Leblond have derived in [6] also the consequences for the crack path of the criterion of local symmetry [8] ($K_{II} = 0$). Here, using the same series approach, the consequences of a maximum dissipation postulate should be derived, motivated by the work of Le et al. [5], where they have shown, that from the variational principle of a body containing a crack the maximum dissipation (or maximum driving force) criterion follows without any ad-hoc assumptions. Furthermore the energetic approach has the advantage that the crack propagation rate and the driving force acting on the crack can accurately be determined for crack kinking and curving and also remain the correct thermodynamic dual quantities for these cases.

3 Maximum Dissipation for Regular Curved Cracks

As the criterion for kinking cracks has already been derived in [5], we will here restrict ourselves to the case of regular crack propagation, i.e. curving without kinking. This implies for all criteria, that mode II has to vanish for the initial crack configuration as it would immediately lead to crack kinking. The starting point is thus the dissipation of a growing crack, based on the driving force acting on the propagating crack tip. Following Le et al. [5] we introduce this driving force with the help of the actual SIFs

$$G(s) = K_\alpha(s)K_\beta(s)\Lambda_{\alpha\beta} = G^\star + G^{(1/2)}\sqrt{s} + G^{(1)}s + O(s^{(3/2)}) \tag{6}$$

with

$$\Lambda_{\alpha\beta} = \frac{1-\nu^2}{E}\begin{pmatrix} 1 & 0 & 0 \\ 0 & 1 & 0 \\ 0 & 0 & \frac{1}{1-\nu} \end{pmatrix} \tag{7}$$

The terms in the series can be given with the help of eq. (2) as

$$G^\star(\phi) = K_\alpha^\star K_\beta^\star \Lambda_{\alpha\beta} = F_{\alpha\gamma}\Lambda_{\alpha\beta}F_{\beta\delta}K_\gamma K_\delta \tag{8}$$

$$G^{(1/2)}(\phi, a^\star) = \left(2K_\alpha^\star K_\beta^{(1/2)} + K_\alpha^{(1/2)}K_\beta^{(1/2)}\sqrt{s}\right)\Lambda_{\alpha\beta} \tag{9}$$

$$G^{(1)}(\phi, a^\star, C^\star) = \left(2K_\alpha^\star K_\beta^{(1)} + 2K_\alpha^{(1/2)}K_\beta^{(1)}\sqrt{s} + K_\alpha^{(1)}K_\beta^{(1)}s\right)\Lambda_{\alpha\beta} \tag{10}$$

The consequences of the postulate of maximum dissipation are here for the sake of simplicity derived from the maximum driving force principle. The consequences of the two are the same, as long as the fracture resistance does not explicitly depend on the direction crack propagation, e.g. through the kinking angle. In the following we restrict ourselves to the two-dimensional case.

3.1 Zeroth Order Approximation

The zeroth order term, depending only on the kinking angle, is the driving force acting on a kinked crack that has already been determined in [5]. Introducing $K_{II} = 0$ into this solution leads to a vanishing kinking angle $\varphi^\star = 0$.

3.2 First Order Approximation

To derive the first curvature parameter $a^\star$ the maximum driving force determined from the series eq. (6) is cut after the square-root term

$$G(s) = G^\star + G^{(1/2)}\sqrt{s} + O(s);\ \phi = \phi^\star. \tag{11}$$

The postulate of maximum driving force

$$\frac{\partial G(s)}{\partial a^\star} = \frac{\partial G^{(1/2)}}{\partial a^\star} = 0 \tag{12}$$

leads to the curvature parameter

$$a^\star = -\frac{1}{s}\frac{\Lambda_{\alpha\beta}\left(K_\alpha^\star H_{\beta\gamma}K_\gamma + \sqrt{s}H_{\beta\gamma}K_\gamma G_\alpha T\right)}{\Lambda_{\alpha\beta}H_{\alpha\gamma}K_\gamma H_{\beta\delta}K_\delta} \tag{13}$$

Inserting $\varphi = 0$ into the matrix H (cf. [6]) and setting $K_{II} = 0$ for regular crack propagation leads to a vanishing first order curvature parameter and thus a vanishing first order driving force

$$a^{\star} = 0 \Rightarrow G^{(1/2)} = 0 \tag{14}$$

3.3 Second Order Approximation

The driving force series eq. (6) is reduced with the help of eq. (refeq:10) and $\varphi = 0$ to

$$G(s) = G^{\star} + G^{(1/2)}\sqrt{s} + G^{(1)}s + O(s^{3/2}) = G + G^{(1)}s + O(s^{3/2}) \tag{15}$$

The second order driving force term can be further simplified with the help of $a^{\star} = 0$ to

$$G^{(1)} = \left(2K_{\alpha}K_{\beta}^{(1)} + K_{\alpha}^{(1)}K_{\beta}^{(1)}s\right)\Lambda_{\alpha\beta} \tag{16}$$

And the second order SIF term appearing here reduces to

$$K_{\alpha}^{(1)} = K_{\alpha}^{(1)}\big|_{\text{straight}} + C^{\star}M_{\alpha\beta}(\phi = 0)K_{\beta} \tag{17}$$

The first non-universal term in eq. (17) is to be understood as the first order term that would appear for a straight (not kinked, not curved) crack propagation. The maximum driving force gives then an equation for the second curvature parameter $C^{\star}$

$$\frac{\partial G(s)}{\partial C^{\star}} = \frac{\partial G^{(1)}}{\partial C^{\star}} = 0 = 2\Lambda_{\alpha\beta}\left(K_{\alpha}M_{\beta\delta}K_{\delta} + K_{\alpha}^{(1)}M_{\beta\delta}K_{\delta}s\right) \tag{18}$$

with the solution (inserting $M_{I,I}(\phi = 0) = M_{II,II}(\phi = 0) = 0$, $M_{II,I} = 1/2$)

$$C^{\star} = \frac{-2K_{II}^{(1)}\big|_{\text{straight}}}{K_I} \Rightarrow C^{\star} = \frac{-2\frac{\mathrm{d}K_{II}}{\mathrm{d}s}\big|_{\text{straight}}}{K_I} \tag{19}$$

where the second interpretation in the above equation is possible because of the vanishing KII for the initial crack. This is a similar result to the one derived in [6] from the principle of local symmetry. Also Sumi presented in [5] a similar result for slightly curved cracks. To be able to use a numerical approach based on configurational nodal forces in the framework of an FEM simulation, this result has to be reformulated in terms of configurational forces.

4 Transition to Configurational Forces

Configurational forces are to be understood as the forces in material space (in opposite to physical space) resulting from the variation in energy due to the change in position of the singularity arising at the crack tip (cf. [9]). For the configurational or Eshelby-stress tensor

$$\mu_{ij} = \psi\delta_{ij} - u_{k,j}\frac{\partial\psi}{\partial u_{k,i}} \tag{20}$$

with the free energy density ψ, u_i the displacement vector and δ_{ij} the Kronecker delta the following balance of material momentum equation is valid

$$\mu_{ij,j} = -\frac{\partial\psi}{\partial X_i}\bigg|_{\text{exp}} \tag{21}$$

where the right hand side term is only non-vanishing, if there exists an explicit dependency of the free energy density with respect to the position X in the material. This is only the case for non-homogeneos materials, e.g. functionally graded materials. For homogenous materials the divergence in eq. (21) is vanishing, giving rise to a path-independent conservation integral, the first component of which is the widely know J-Integral

$$\mathfrak{F}_i = J_i = \lim_{\Gamma\to 0}\oint_{\Gamma}\mu_{ij}n_j ds \tag{22}$$

Please note that eq. (22) shows only an asymptotical path-independency, since the integrand for J_2 is not necessarily vanishing on the crack surfaces. This will be discussed in detail in connection with the numerical approach for the accurate determination of the configurational forces in a finite element framework. In a linear elastic fracture mechanics framework we have the following connection between configurational forces, J-integral vector components and the stress intensity factors

$$\mathfrak{F}_t = J_1 = \frac{1-\nu^2}{E}(K_I^2 + K_{II}^2) \tag{23}$$

$$\mathfrak{F}_n = J_2 = \frac{1-\nu^2}{E}(-2K_I K_{II}) \tag{24}$$

With the help of these relations the main result of the preceeding section, the curvature resulting from maximum dissipation can be rewritten as

$$C^\star = \frac{\nabla_t\cdot\boldsymbol{\mathfrak{F}}\cdot\mathbf{n}}{\boldsymbol{\mathfrak{F}}\cdot\mathbf{t}} = \frac{\partial_t\mathfrak{F}_n}{\|\mathfrak{F}\|} \tag{25}$$

where ∇_t is the tangential part of the gradient of the material force with respect to material space (compare Figure 4). This tangential derivative is to be understood as the derivative of the normal component of the configurational force along a straight crack elongation. The second interpretation of eq. (25) is valid because for the real crack the normal component will always vanish. Equation (25) means the local

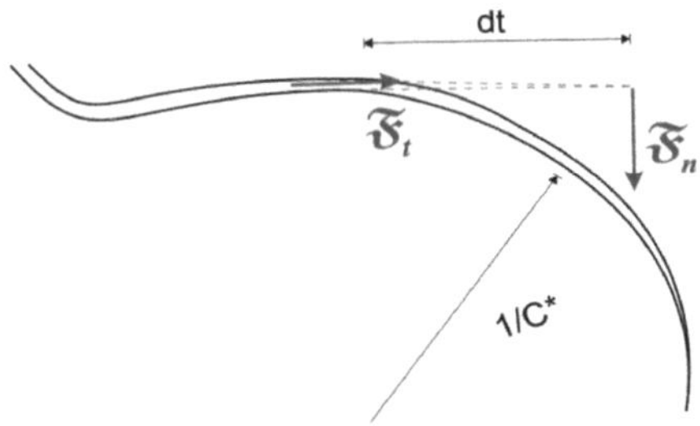

Fig. 2 Curved crack with normal and tangential configurational forces.

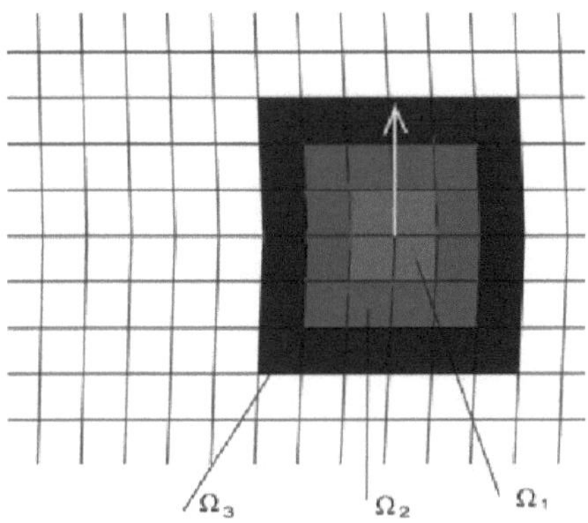

Fig. 3 Integration domains of increasing size.

curvature of the crack trajectory is the same as the local curvature of the material force field. This result seems to be such a natural and straightforward result, that the author believes, it will hold in general without the underlying assumptions made earlier in the linear elastic fracture mechanics framework.

5 Finite Element Framework

A finite element framework making use of nodal configurational forces is used, similar to the ones described in [3,4]. The essential part is, that this approach gives in a simple post-processing step the configuration forces as the thermodyamical dual quantitiy to a variational change of the position of the corresponding node with respect to the material

$$\mathfrak{F}^h = \mathop{\mathbf{A}}_{e=1}^{E} \sum_{n=1}^{n_{en}} \int_{\mathcal{B}_0^e} \boldsymbol{\mu} \cdot \nabla_{\mathbf{X}} N^e \, \mathrm{d}A \tag{26}$$

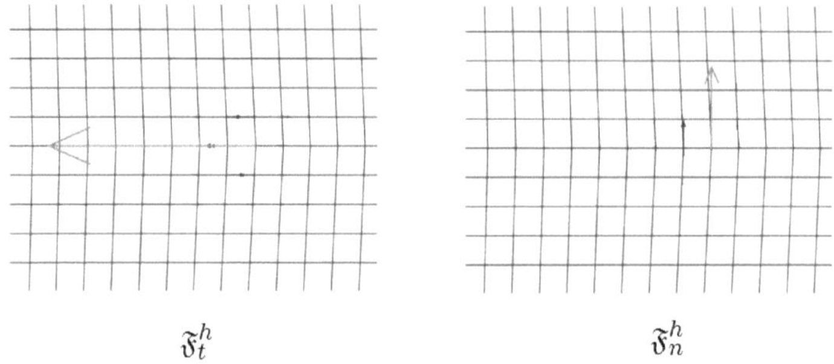

Fig. 4 Normal and tangential components of the configurational forces at the crack tip

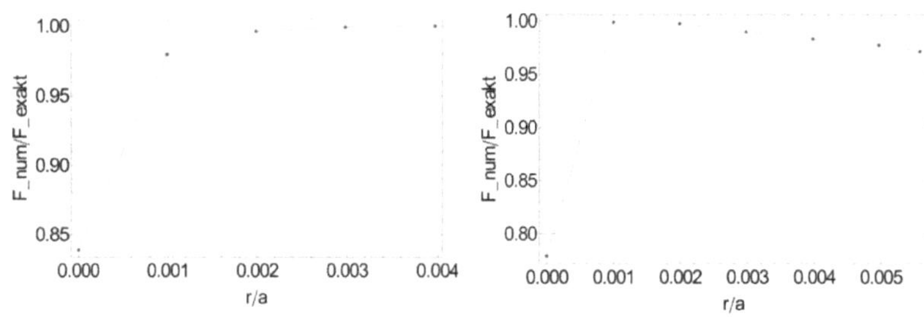

Fig. 5 J_1 and J_2 values for increasing domain size.

Without discretization errors the finite element results for a body with a crack would produce only configurational forcees acting on the nodes representing the crack front (or one force acting on the crack tip). Due to the failure of the shape-functions normally used in an FEM based approach to accurately represent the two singularities involved at the crack tip (namely the stress singularity and the singularity of the Eshelbian-stress, which are of different order) also spurious configurational nodal forces are produced in the vincinity of the crack front (or tip). Figure 4 shows the results from a simulation separated into the normal and tangential parts.

The accuracy of the forces acting directly on the tip usually is low. Thus many authors [3–5] have adopted some method similar to the domain integral method, which in this framework consists simply in adding up the contributions of the nodes contained in a certain area surrounding the crack tip (Figure 5).

$$\mathfrak{F} = \sum_{n_i \in \Omega} \mathfrak{F}_i^h \tag{27}$$

Figure 5 shows clearly, that the value for J_1, but not the value for J_2 is converging, when the size of the domain is increased. Because of this an extrapolation back to a zero area domain is necessary, as suggested by the limit value appearing in eq. (22) hinting to the asymptotic path independence of the J-integral vector.

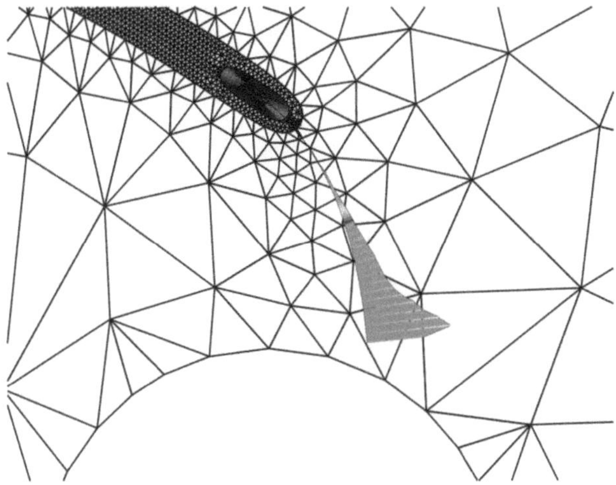

Fig. 6 Forward sensing alaong the predicted crack path.

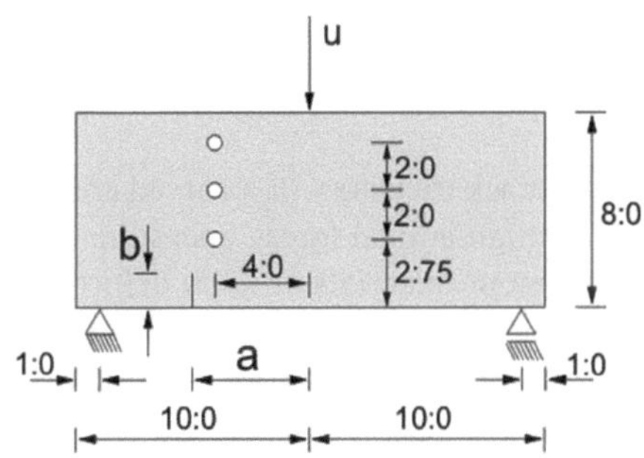

Fig. 7 Specimen by Bittencourt et al.

6 Step by Step Propagation Scheme

A step by step numerical scheme has been implemented in the commercial FEM-code ANSYS. After each step the geometry has been created newly and a new mesh has been created. After that the following scheme has been adopted for each propagation step:

- small test step to determine the curvature 25
- "forward sensing" the ratio σ_{nt}/σ_{nn} to determine the maximum length (Figure 6)
- constant curvature propagation
- small change of the tail slope of the spline by $\Delta\varphi = J_2/J_1$ to get vanishing J_2
- or cut back, if J_2/J_1 is too big.

With the help of this scheme the experiments from Bittencourt et al. [10] (Figure 7) have been simulated. Figure 6 illustrates, that highly accurate results can be attained with a small number of propagation steps.

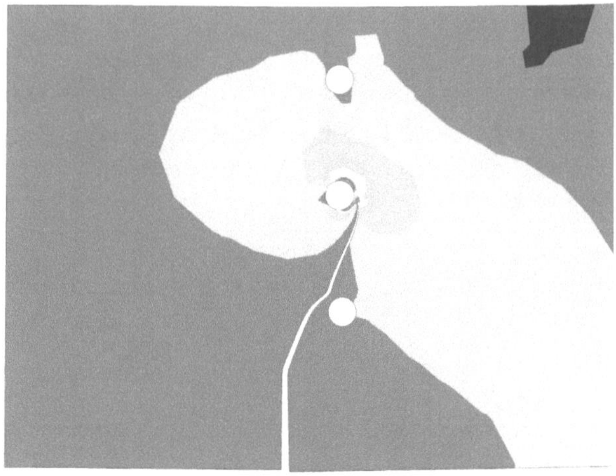

Fig. 8 Computational result from 20 propagation steps.

7 Conclusions

A criterion to describe the crack trajectory of a curved crack has been derived. The formulation in terms of configurational forces opens the door to a applicability to a wider range of inhomogenous materials or finite deformations. The necessity of an extrapolation technique to determine accurate J_2 values has been shown. The numerical scheme based on splines and a predictor-corrector method enables to take large steps in an FEM simulation and thus save computational effort, while keeping or increasing the accuracy of the predicted path.

References

1. Sumi, Y., Yang, C. and Hayashi, S., *Int. J. Fract.* **82**, 1996, 205–220.
2. Steinmann, P., Ackermann, D. and Barth, F.J., *Int. J. Solids Struct.* **38**, 2001, 5509–5526.
3. Miehe, C., Gürses, E. and Birkle, M., *Int. J. Fract.* **145**, 2007, 245–259.
4. Denzer, R., Barth, F.J. and Steinmann, P., *Int. J. Num. Meth. Engng.* **58**, 2003, 1817–1835.
5. Le, K.C., Schütte, H. and Stumpf, H., *Arch. Appl. Mech.* **69**, 1999, 337–344.
6. Amestoy, M. and Leblond, J.B., *Int. J. Solids Struct.* **29**(4), 1992, 465–501.
7. Williams, M., *J. Appl. Mech.* **24**, 1957, 109–114.
8. Goldstein, R.V. and Salganik, R.L., *Int. J. Fract.* **10**, 1974, 507–523.
9. Maugin, G.A., *Material Inhomogeneities in Elasticity*, Chapman and Hall, London, 1993.
10. Bittencourt, T.N., Wawrzynek, P.A., Ingraffea, A.R. and Sousa, J.L., *Engng. Fract. Mech.* **55**, 1996, 321–334.

Anisotropic Elasticity of Grade Three: Conservation and Balance Laws

Eleni Agiasofitou and Markus Lazar

Abstract In this work we derive conservation and balance laws in the context of linear, anisotropic elasticity of grade three including cohesive forces. More particularly, for a homogeneous medium without external forces we derive the conservation laws of translation and addition of solutions as well as the balance laws that stem from the rotation and scaling transformations. The Eshelby stress tensor of such a gradient theory of higher order is determined. On the other hand, we calculate all the corresponding balance laws for an inhomogeneous medium in the presence of external forces. The dynamical reciprocal theorem for anisotropic elasticity of grade three is derived and its relationship to the balance law of addition of solutions is examined.

1 Introduction

Elasticity of grade three is an extension of elasticity in which there is a microstructure associated with material points. The theory of gradient elasticity enriches the classical elasticity with additional material-characteristic length scales in order to describe the size effects resulting from the underlying microstructure. Such a consideration is based on the constitutive relations in which the strain energy density W is a function not only of the displacement gradient but also of the gradients of the displacement up to the third order $W = W(\nabla \mathbf{u}, \nabla^2 \mathbf{u}, \nabla^3 \mathbf{u})$ including surface energy as originally proposed by Mindlin [10]. For crystals, anisotropic elasticity of grade three is a proper candidate to describe their microstructure. The aim of this

Eleni Agiasofitou
Emmy Noether Research Group, Department of Physics, Darmstadt University of Technology, Hochschulstr. 6, D-64289 Darmstadt, Germany; e-mail: agiasofitou@mechanik.tu-darmstadt.de

Markus Lazar
Emmy Noether Research Group, Department of Physics, Darmstadt University of Technology, Hochschulstr. 6, D-64289 Darmstadt, Germany; e-mail: lazar@fkp.tu-darmstadt.de

P. Steinmann (ed.), IUTAM Symposium on Progress in the Theory and Numerics of Configurational Mechanics, 179–191.

paper is to derive the conservation as well as the balance laws for linear, anisotropic elasticity of grade three including the so-called cohesive forces [10, 11, 15]. The present work is a generalization of the paper [2] to anisotropic materials.

2 Anisotropic Elasticity of Grade Three

In the framework of linear elasticity of grade-3 for a homogeneous, anisotropic material in the absence of external forces, the *strain energy* density is considered as a function of the first, second and third gradient of the displacement field $u_\alpha = u_\alpha(\mathbf{x}),\ \alpha = 1, 2, 3$ that is

$$W = W(u_{\alpha,i}, u_{\alpha,ij}, u_{\alpha,ijk}), \tag{1}$$

where $u_{\alpha,i} = \partial u_\alpha / \partial x_i$. We denote by $\mathbf{x} = (x_1, x_2, x_3, t)$ that is the first three components of $\mathbf{x}$ are the spatial coordinates and the fourth one corresponds to the time variable t. Considering the dynamical case, we represent the *kinetic energy* density by

$$T = \frac{1}{2}\, p_\alpha \dot{u}_\alpha, \tag{2}$$

where p_α is the *linear momentum vector* and a superposed dot will denote the time derivative. Then, the *Lagrange* density can be expressed as

$$\mathcal{L} = T - W. \tag{3}$$

Therefore, for the problem under study we have

$$\mathcal{L} = \mathcal{L}(\dot{u}_\alpha, u_{\alpha,i}, u_{\alpha,ij}, u_{\alpha,ijk}). \tag{4}$$

In general, a Lagrangian of third order for an inhomogeneous body in the presence of body forces can be of the form

$$\mathcal{L} = \mathcal{L}(x_i, u_\alpha, \dot{u}_\alpha, u_{\alpha,i}, u_{\alpha,ij}, u_{\alpha,ijk}). \tag{5}$$

A third order variational problem consists of finding the extrema of the so-called *action integral*

$$J[\mathbf{u}] = \int_\Omega \mathcal{L}(\mathbf{x}, \mathbf{u}^{(3)})\mathrm{d}\mathbf{x}, \tag{6}$$

where the integrand $\mathcal{L}(\mathbf{x}, \mathbf{u}^{(3)})$ is a smooth function of $\mathbf{x}$, $\mathbf{u}$ and the derivatives of $\mathbf{u}$ up to third order, $\mathbf{u}^{(3)}$, and Ω is an open, connected subset of $\mathbb{R}^4$ with smooth boundary $\partial\Omega$ and $\mathrm{d}\mathbf{x} = \mathrm{d}x_1\mathrm{d}x_2\mathrm{d}x_3\mathrm{d}t$.

For a smooth function $\mathbf{u}(\mathbf{x})$ to be an extremum of the action integral $J[\mathbf{u}]$ it is necessary that it satisfies the Euler–Lagrange equations

$$E_\alpha(\mathcal{L}) = 0, \quad \alpha = 1, 2, 3. \tag{7}$$

Throughout the paper, we assume the usual convention of summation over a repeated index. The *Euler–Lagrange equations* for a third order variational problem are given by the formula

$$\begin{aligned} E_\alpha(\mathcal{L}) &= \frac{\partial \mathcal{L}}{\partial u_\alpha} - \mathrm{D}_i \frac{\partial \mathcal{L}}{\partial u_{\alpha,i}} + \mathrm{D}_i \mathrm{D}_j \frac{\partial \mathcal{L}}{\partial u_{\alpha,ij}} - \mathrm{D}_i \mathrm{D}_j \mathrm{D}_k \frac{\partial \mathcal{L}}{\partial u_{\alpha,ijk}} \\ &= 0, \quad i, j, k = 1, \ldots, 4, \end{aligned} \tag{8}$$

where D_i is the total derivative.

Namely, the strain energy density for a linear, anisotropic material can have the form

$$W = \frac{1}{2}\tau_{\alpha j} u_{\alpha,j} + \frac{1}{2}\tau_{\alpha jk} u_{\alpha,jk} + \frac{1}{2}\tau_{\alpha jkl} u_{\alpha,jkl} + \frac{1}{2}A^0_{\alpha jk} u_{\alpha,jk} + \frac{1}{2}B^0_{\alpha jkl} u_{\alpha,jkl}, \tag{9}$$

where $\tau_{\alpha j}$ is the *force stress tensor*, $\tau_{\alpha jk}$ and $\tau_{\alpha jkl}$ are the *double* and the *triple stress tensors*, respectively and can be defined as follows

$$\tau_{\alpha j} = \frac{\partial W}{\partial u_{\alpha,j}}, \qquad \tau_{\alpha jk} = \frac{\partial W}{\partial u_{\alpha,jk}}, \qquad \tau_{\alpha jkl} = \frac{\partial W}{\partial u_{\alpha,jkl}}, \tag{10}$$

or more specifically [7, 8, 11]

$$\begin{aligned} \tau_{\alpha j} &= C_{\alpha j\beta k} u_{\beta,k} + F_{\alpha j\beta kl} u_{\beta,kl} + E_{\alpha j\beta klm} u_{\beta,klm}, \\ \tau_{\alpha jk} &= F_{\beta l\alpha jk} u_{\beta,l} + A_{\alpha jk\beta lm} u_{\beta,lm} + G_{\alpha jk\beta mnp} u_{\beta,mnp} + A^0_{\alpha jk}, \\ \tau_{\alpha jkl} &= E_{\beta m\alpha jkl} u_{\beta,m} + G_{\beta mn\alpha jkl} u_{\beta,mn} + B_{\alpha jkl\beta mnp} u_{\beta,mnp} + B^0_{\alpha jkl}, \end{aligned} \tag{11}$$

where $C_{\alpha j\beta k}$, $F_{\alpha j\beta kl}$, $E_{\alpha j\beta klm}$, $A_{\alpha jk\beta lm}$, $G_{\alpha jk\beta mnp}$, $B_{\alpha jkl\beta mnp}$, $A^0_{\alpha jk}$ and $B^0_{\alpha jkl}$ are the tensors of the material constants and especially $A^0_{\alpha jk}$ and $B^0_{\alpha jkl}$ are the *cohesive forces*. To complete the set of the constitutive relations (10) for the considered Lagrangian we give the following relation for the linear momentum vector

$$p_\alpha = \frac{\partial T}{\partial \dot{u}_\alpha} = \rho \dot{u}_\alpha, \tag{12}$$

where ρ is the mass density.

For the considered model, the Euler–Lagrange equations (8) via the relations (10) and (12) take the form

$$\mathrm{D}_t\, p_\alpha - \mathrm{D}_i\big(\tau_{\alpha i} - \mathrm{D}_j \tau_{\alpha ij} + \mathrm{D}_j \mathrm{D}_k \tau_{\alpha ijk}\big) = 0, \qquad i, j, k = 1, 2, 3. \tag{13}$$

3 Conservation – Balance Laws in Elasticity of Grade Three

In this section, we start giving a brief mathematical setting for the derivation of a conservation law for a third order variational problem. The next subsections deal with the derivation of conservation or balance laws for specific group of transformations for a homogeneous body without external forces.

3.1 Structure of Conservation Laws

This subsection gives the absolutely necessary material concerning the construction of the currents which enter into a conservation law. The calculation is based on the Noether theorem [13] and especially in the prolongation method as it is described in the books by Olver [14] and Bluman and Kumei [3]. For the detailed calculation of the currents for a third order variational problem, the reader is referred to [2].

Let us consider an n-parameter group of transformations G acting on both independent $\mathbf{x}$ and dependent $\mathbf{u}$ variables

$$(\mathbf{x}^*, \mathbf{u}^*) = (\Xi_{\mathbf{w}}(\mathbf{x}, \mathbf{u}), \Phi_{\mathbf{w}}(\mathbf{x}, \mathbf{u}))\,, \tag{14}$$

where $\mathbf{x} \in X \equiv \mathbb{R}^4$, $\mathbf{u} \in U \equiv \mathbb{R}^3$ and the parameter $\mathbf{w} = (w_1, w_2, \dots, w_n)$. Each parameter of an n-parameter Lie group of transformations leads to an infinitesimal generator. The *infinitesimal generator* $\mathbf{v}_r$, corresponding to the parameter w_r, $r = 1, \dots, n$ of the group G is a vector field defined on an open subset $M \subset X \times U$

$$\mathbf{v}_r = \xi_i^r(\mathbf{x}, \mathbf{u})\,\frac{\partial}{\partial x_i} + \phi_\alpha^r(\mathbf{x}, \mathbf{u})\,\frac{\partial}{\partial u_\alpha}, \qquad i = 1 \dots, 4, \quad \alpha = 1, 2, 3 \tag{15}$$

where

$$\xi_i^r(\mathbf{x}, \mathbf{u}) = \left.\frac{\partial\,\Xi^i_{\mathbf{w}}(\mathbf{x}, \mathbf{u})}{\partial w_r}\right|_{\mathbf{w}=\mathbf{0}}, \qquad \phi_\alpha^r(\mathbf{x}, \mathbf{u}) = \left.\frac{\partial\,\Phi^\alpha_{\mathbf{w}}(\mathbf{x}, \mathbf{u})}{\partial w_r}\right|_{\mathbf{w}=\mathbf{0}}, \tag{16}$$

where $\Xi^i_{\mathbf{w}}$, $\Phi^\alpha_{\mathbf{w}}$ are the components of $\Xi_{\mathbf{w}}$, $\Phi_{\mathbf{w}}$ and

$$Q_\alpha^r(\mathbf{x}, \mathbf{u}^{(1)}) = \phi_\alpha^r(\mathbf{x}, \mathbf{u}) - \xi_i^r(\mathbf{x}, \mathbf{u}) u_{\alpha,i} \tag{17}$$

are the *characteristics* of the vector fields $\mathbf{v}_r$.

Noether's theorem states that if G is a symmetry group of the variational problem (6), $\mathbf{v}_r, r = 1, \dots, n$ are the infinitesimal generators of G and Q_α^r are the characteristics of $\mathbf{v}_r$, then Q_α^r are also the characteristics of conservation laws for the Euler–Lagrange equations $E_\alpha(\mathcal{L}) = 0$; in other words there are $A_{ri}, i = 1, \dots, 4$ such that

$$\mathrm{D}_i A_{ri} = -Q_\alpha^r E_\alpha(\mathcal{L}) \tag{18}$$

are conservation laws in characteristic form for the Euler–Lagrange equations.

In our case, that is for a third order variational problem, the general form of the currents are given by the following formulas (see [2])

$$A_{ri} = Q^r_\alpha \left[\frac{\partial \mathcal{L}}{\partial u_{\alpha,i}} - \mathrm{D}_j \frac{\partial \mathcal{L}}{\partial u_{\alpha,ij}} + \mathrm{D}_j \mathrm{D}_k \frac{\partial \mathcal{L}}{\partial u_{\alpha,ijk}} \right] + \mathrm{D}_j Q^r_\alpha \left[\frac{\partial \mathcal{L}}{\partial u_{\alpha,ij}} - \mathrm{D}_k \frac{\partial \mathcal{L}}{\partial u_{\alpha,ijk}} \right]$$
$$+ \mathrm{D}_j \mathrm{D}_k Q^r_\alpha \frac{\partial \mathcal{L}}{\partial u_{\alpha,ijk}} + \mathcal{L} \xi^r_i, \quad i, j, k = 1, 2, 3 \tag{19}$$

and

$$A_{r4} = Q^r_\alpha \frac{\partial \mathcal{L}}{\partial \dot{u}_\alpha} + \mathcal{L} \xi^r_4. \tag{20}$$

The statement of Noether's theorem holds also in the case that the vector fields $\mathbf{v}_r$ generate a divergence symmetry instead of a variational one for the functional J. In that case, Noether's theorem gives the following currents

$$A_{ri} = Q^r_\alpha \left[\frac{\partial \mathcal{L}}{\partial u_{\alpha,i}} - \mathrm{D}_j \frac{\partial \mathcal{L}}{\partial u_{\alpha,ij}} + \mathrm{D}_j \mathrm{D}_k \frac{\partial \mathcal{L}}{\partial u_{\alpha,ijk}} \right] + \mathrm{D}_j Q^r_\alpha \left[\frac{\partial \mathcal{L}}{\partial u_{\alpha,ij}} - \mathrm{D}_k \frac{\partial \mathcal{L}}{\partial u_{\alpha,ijk}} \right]$$
$$+ \mathrm{D}_j \mathrm{D}_k Q^r_\alpha \frac{\partial \mathcal{L}}{\partial u_{\alpha,ijk}} + \mathcal{L} \xi^r_i - B_{ri}, \quad i, j, k = 1, \ldots, 3 \tag{21}$$

and

$$A_{r4} = Q^r_\alpha \frac{\partial \mathcal{L}}{\partial \dot{u}_\alpha} + \mathcal{L} \xi^r_4 - B_{r4}, \tag{22}$$

where $B_{ri}(\mathbf{x}, \mathbf{u}^{(m)})$ are functions of $\mathbf{x}$, $\mathbf{u}$ and derivatives of $\mathbf{u}$.

In what follows, we find the conserved quantities that enter every time into the specific conservation law calculating firstly the components of the infinitesimal generators and the corresponding characteristics and substituting these quantities into Eqs. (19) and (20) if the group of transformations is a variational symmetry or to Eqs. (21) and (22) if the group of transformations is a divergence symmetry.

3.2 Translations in Space and Time

Let us consider the following transformation corresponding to translations of the independent variables in space and time, namely

$$\begin{aligned} x^*_i &= x_i + w_r \delta_{ri}, \quad & r, i &= 1, \ldots, 4 \\ u^*_\alpha &= u_\alpha, & \alpha &= 1, 2, 3 \end{aligned} \tag{23}$$

where δ_{ri} is the usual Kronecker delta.

Due to the fact that the group (23) is a 4-parameter one, Eq. (18) can be read as follows

$$\mathrm{D}_i A_{ri} = 0, \quad i = 1, \dots, 4, \quad r = 1, 2, 3 \tag{24}$$

and

$$\mathrm{D}_i A_{4i} = 0, \quad i = 1, \dots, 4. \tag{25}$$

Eq. (24), using the relations (19), (20), (10) and (12) concludes to the *translational conservation law* in the linear, anisotropic elasticity of grade-3

$$\mathrm{D}_t \mathcal{P}_r - \mathrm{D}_l P_{rl} = 0, \tag{26}$$

where

$$\mathcal{P}_r := A_{r4} = -p_\alpha u_{\alpha,r} \tag{27}$$

and

$$\begin{aligned} P_{rl} &:= -A_{rl} \\ &= -\mathcal{L}\delta_{rl} - u_{\alpha,r}\left(\tau_{\alpha l} - \tau_{\alpha lj,j} + \tau_{\alpha ljk,jk}\right) - u_{\alpha,rj}\left(\tau_{\alpha lj} - \tau_{\alpha ljk,k}\right) - u_{\alpha,rjk}\tau_{\alpha ljk} \end{aligned} \tag{28}$$

are the *pseudomomentum vector* and the *Eshelby stress tensor* [4] for the linear elasticity of grade-3, respectively.

On the other hand, the analysis of Eq. (25) leads to the *conservation law of energy* for the linear, anisotropic elasticity of grade-3

$$\mathrm{D}_t \mathcal{H} - \mathrm{D}_l S_l = 0, \tag{29}$$

where

$$\mathcal{H} := A_{44} = p_\alpha \dot{u}_\alpha - \mathcal{L} = T + W \tag{30}$$

and

$$S_l := -A_{4l} = \dot{u}_\alpha\left(\tau_{\alpha l} - \tau_{\alpha lj,j} + \tau_{\alpha ljk,jk}\right) + \dot{u}_{\alpha,j}\left(\tau_{\alpha lj} - \tau_{\alpha ljk,k}\right) + \dot{u}_{\alpha,jk}\tau_{\alpha ljk} \tag{31}$$

are the *Hamiltonian density* and the *material flux vector* or *field intensity* [12] for the linear, anisotropic elasticity of grade-3, correspondingly.

Furthermore, we can take the integral form of the equations (26) and (29) integrating over a volume V of a regular bounded body with surface S using the divergence theorem

$$\int_S P_{rl} n_l \mathrm{d}S - \int_V \mathrm{D}_t \mathcal{P}_r \mathrm{d}V = 0, \tag{32}$$

$$\int_S S_l n_l \mathrm{d}S - \int_V \mathrm{D}_t \mathcal{H} \mathrm{d}V = 0, \tag{33}$$

where $\mathbf{n}$ is the unit outward normal vector to S.

3.3 Rotations in Space

We continue to obtain actually the balance law that comes from the rotation of the dependent and the independent variables in the space

$$\begin{aligned} x_i^* &= x_i + e_{irk} w_r x_k, \quad i,k,r = 1,2,3 \\ t^* &= t, \\ u_\alpha^* &= u_\alpha + e_{\alpha rk} w_r u_k, \quad \alpha,k,r = 1,2,3 \end{aligned} \tag{34}$$

where e_{irk} is the permutation tensor or Levi-Civita tensor.

The expressions of A_{rl} and A_{r4} (Eqs. (19) and (20)) for the group of rotations are

$$\begin{aligned} M_{rl} := -A_{rl} = e_{rm\alpha} \Big(& x_m P_{\alpha l} + u_m (\tau_{\alpha l} - \tau_{\alpha lj,j} + \tau_{\alpha ljk,jk}) + u_{m,j} (\tau_{\alpha lj} - \tau_{\alpha lkj,k}) \\ & + u_{m,jk} \tau_{\alpha ljk} + u_{j,m} (\tau_{jl\alpha} - \tau_{jlk\alpha,k}) + 2 u_{j,mk} \tau_{jlk\alpha} \Big) \end{aligned} \tag{35}$$

and

$$\mathcal{M}_r := A_{r4} = e_{rm\alpha} (x_m \mathcal{P}_\alpha + u_m p_\alpha) \tag{36}$$

and construct the following *rotational balance law*

$$\begin{aligned} & \mathrm{D}_t \mathcal{M}_r - \mathrm{D}_l M_{rl} = \\ & \quad e_{r\alpha m} \big(u_{j,m} \tau_{j\alpha} + u_{m,j} \tau_{\alpha j} + u_{j,ml} \tau_{j\alpha l} + u_{j,ml} \tau_{jl\alpha} + u_{m,jl} \tau_{\alpha lj} \\ & \qquad + u_{j,mlk} \tau_{j\alpha lk} + u_{m,jkl} \tau_{\alpha ljk} + u_{j,mkl} \tau_{jl\alpha k} + u_{k,mjl} \tau_{klj\alpha} \big), \end{aligned} \tag{37}$$

where M_{rl} is the *total angular pseudomomentum vector* and $\mathcal{M}_r$ is the *total angular momentum tensor* for the linear, anisotropic elasticity of grade-3.

It is clear that in the above calculation of Eq. (37) we have not taken into account the symmetries of $\tau_{\alpha jl}$ and $\tau_{\alpha jlk}$ with respect to the last two and three indices, correspondingly. If these symmetries are considered, Eq. (37) is simplified to

$$\begin{aligned} \mathrm{D}_t \mathcal{M}_r - \mathrm{D}_l M_{rl} = e_{r\alpha m} \big(& u_{j,m} \tau_{j\alpha} + u_{m,j} \tau_{\alpha j} + 2\, u_{j,ml} \tau_{j\alpha l} \\ & + u_{m,jl} \tau_{\alpha lj} + 3\, u_{j,mlk} \tau_{j\alpha lk} + u_{m,jkl} \tau_{\alpha ljk} \big). \end{aligned} \tag{38}$$

For an isotropic material the rotational balance law (37) changes to a conservation law since the isotropy condition in this case is fulfilled (see also [2]).

3.4 Scaling

We investigate here the case of the one-parameter group of scaling of independent and depended variables

$$
\begin{aligned}
x_i^* &= (1+w)x_i, \qquad i = 1,2,3 \\
t^* &= (1+w)t, \\
u_\alpha^* &= (1+wd_u)u_\alpha, \quad \alpha = 1,2,3
\end{aligned}
\tag{39}
$$

where d_u denotes the scaling dimension of $\mathbf{u}$.

In this case, the currents (19) and (20) have the specific form

$$
\begin{aligned}
Y_l := -A_l = x_i P_{il} - tS_l + d_u u_\alpha(\tau_{\alpha l} - \tau_{\alpha lj,j} + \tau_{\alpha ljk,jk}) \\
+ (d_u - 1)u_{\alpha,j}(\tau_{\alpha lj} - \tau_{\alpha lkj,k}) + (d_u - 2)u_{\alpha,jk}\tau_{\alpha ljk}
\end{aligned}
\tag{40}
$$

and

$$
\mathcal{Y} := A_4 = x_i \mathcal{P}_i - t\mathcal{H} + d_u u_i p_i . \tag{41}
$$

Having in mind that the scaling group leads to a conservation law in classical linear elasticity we obtain that

$$
d_u = \frac{1-n}{2} = \frac{2-d}{2}, \tag{42}
$$

where n and $d = n+1$ are the space and space-time dimensions, respectively.

Finally, we conclude to the following *equation for the scaling group*

$$
\mathrm{D}_t\mathcal{Y} - \mathrm{D}_l Y_l = \Big(\tau_{\alpha lj} - \frac{d}{2}A^0_{\alpha lj}\Big)u_{\alpha,lj} + \Big(2\tau_{\alpha ljk} - \frac{d}{2}B^0_{\alpha ljk}\Big)u_{\alpha,ljk}. \tag{43}
$$

It is also interesting to see that the above equation in the dynamical case, that is $d = 4$, using also the constitutive relations (11), becomes

$$
\begin{aligned}
\mathrm{D}_t\mathcal{Y} - \mathrm{D}_l Y_l = (F_{\beta k\alpha lj}u_{\beta,k} + A_{\alpha lj\beta km}u_{\beta,km} + G_{\alpha lj\beta mnp}u_{\beta,mnp} - A^0_{\alpha lj})u_{\alpha,lj} \\
+ 2\left(E_{\beta m\alpha ljk}u_{\beta,m} + G_{\beta mn\alpha ljk}u_{\beta,mn} + B_{\alpha ljk\beta mnp}u_{\beta,mnp}\right)u_{\alpha,ljk},
\end{aligned}
\tag{44}
$$

where it is obvious that $B^0_{\alpha ljk}$ does not influence the breaking of symmetry instead of $A^0_{\alpha lj}$.

3.5 Addition of Solutions

Next, we examine the case of a divergence symmetry as it is the group of addition of solutions

$$
\begin{aligned}
x_i^* &= x_i, \qquad i = 1,2,3 \\
t^* &= t, \\
u_\alpha^* &= u_\alpha + w\,v_\alpha, \quad \alpha = 1,2,3
\end{aligned}
\tag{45}
$$

where $v_\alpha = v_\alpha(\mathbf{x})$ is an arbitrary solution of the field equation (13).

Using the relations (21) and (22) we find the expressions for the quantities

$$B_i = -u_\alpha\big(\tau_{\alpha i}(\mathbf{v}) - \tau_{\alpha ik,k}(\mathbf{v}) + \tau_{\alpha ikl,kl}(\mathbf{v})\big) - u_{\alpha,k}\big(\tau_{\alpha ik}(\mathbf{v}) - \tau_{\alpha ikl,l}(\mathbf{v})\big) \\ - u_{\alpha,kl}\tau_{\alpha ikl}(\mathbf{v}) + A^0_{\alpha ik}(u_{\alpha,k} - v_{\alpha,k}) + B^0_{\alpha ikl}(u_{\alpha,kl} - v_{\alpha,kl}), \tag{46}$$

$$B_4 = u_\alpha p_\alpha(\mathbf{v}) \tag{47}$$

and consequently for the conserved currents

$$A_i = -v_\alpha\big(\tau_{\alpha i}(\mathbf{u}) - \tau_{\alpha ik,k}(\mathbf{u}) + \tau_{\alpha ikl,kl}(\mathbf{u})\big) - v_{\alpha,k}\big(\tau_{\alpha ik}(\mathbf{u}) - \tau_{\alpha ikl,l}(\mathbf{u}) - A^0_{\alpha ik}\big) \\ - v_{\alpha,kl}\big(\tau_{\alpha ikl}(\mathbf{u}) - B^0_{\alpha ikl}\big) \\ + u_\alpha\big(\tau_{\alpha i}(\mathbf{v}) - \tau_{\alpha ik,k}(\mathbf{v}) + \tau_{\alpha ikl,kl}(\mathbf{v})\big) + u_{\alpha,k}\big(\tau_{\alpha ik}(\mathbf{v}) - \tau_{\alpha ikl,l}(\mathbf{v}) - A^0_{\alpha ik}\big) \\ + u_{\alpha,kl}\big(\tau_{\alpha ikl}(\mathbf{v}) - B^0_{\alpha ikl}\big) \tag{48}$$

and

$$A_4 = v_\alpha p_\alpha(\mathbf{u}) - u_\alpha p_\alpha(\mathbf{v}). \tag{49}$$

The *conservation law* that comes from the transformation of *addition of solutions* has the form

$$\mathrm{D}_t \mathcal{I} - \mathrm{D}_i I_i = 0, \tag{50}$$

where $\mathcal{I} := A_4$ and $I_i := -A_i$.

Furthermore, Eq. (50) can be written in the following integral form

$$\int_S I_i n_i \mathrm{d}S - \int_V \mathrm{D}_t \mathcal{I} \mathrm{d}V = 0. \tag{51}$$

4 Balance Laws in Elasticity of Grade Three

This section is devoted to the derivation of the balance laws that correspond to the former groups of transformations in the presence of material inhomogeneities and external body forces. We conclude giving the dynamical reciprocal theorem in the anisotropic elasticity of grade-3. It is also discussed its connection with the balance law of addition of solutions and the restrictions under which it is valid.

4.1 Balance Laws and the J, L *and M-Integrals*

The Lagrangian for an inhomogeneous medium with external body forces is given as

$$\mathcal{L} = T - W - V, \tag{52}$$

where V is the potential of external forces, therefore $\mathcal{L}$ will be of the general form (5). The *external body forces* and the *material* or *inhomogeneity forces* [9] are respectively defined by

$$F_\alpha := -\frac{\partial V}{\partial u_\alpha}, \qquad f_i^{\text{inh}} := \frac{\partial \mathcal{L}}{\partial x_i}. \tag{53}$$

In this case, the tensors $C_{\alpha j\beta k}$, $F_{\alpha j\beta kl}$, $E_{\alpha j\beta klm}$, $A_{\alpha jk\beta lm}$, $G_{\alpha jk\beta mnp}$, $B_{\alpha jkl\beta mnp}$, $A^0_{\alpha jk}$ and $B^0_{\alpha jkl}$ can be considered as continuously differentiable functions of the position. The associated Euler–Lagrange equations are given by the formula (8) or in the terms of the stress tensors as follows

$$\mathrm{D}_t\, p_\alpha - \mathrm{D}_i(\tau_{\alpha i} - \mathrm{D}_j\tau_{\alpha ij} + \mathrm{D}_j\mathrm{D}_k\tau_{\alpha ijk}) = F_\alpha, \qquad \alpha, i, j, k = 1, 2, 3. \tag{54}$$

Starting from the translational conservation law (26) with the aid of the Euler–Lagrange equations (54), we can compute the *translational balance law* for the linear, anisotropic elasticity of grade-3

$$\mathrm{D}_t\mathcal{P}_r - \mathrm{D}_l P_{rl} = f_r^{\text{inh}}, \tag{55}$$

where it is evident the existence of a source term breaking the translational symmetry.

Repeating the same procedure for the case of the rotation group, Eq. (37) via Eqs. (54) and (55) leads to the following *rotational balance law*

$$\begin{aligned}\mathrm{D}_t\mathcal{M}_r - \mathrm{D}_l M_{rl} = e_{rm\alpha}\big(&x_m f_\alpha^{\text{inh}} + u_m F_\alpha - u_{l,m}\tau_{l\alpha} - u_{m,l}\tau_{\alpha l}\\ &-u_{l,mj}\tau_{l\alpha j} - u_{m,jl}\tau_{\alpha lj} - u_{j,ml}\tau_{jl\alpha}\\ &-u_{l,mjk}\tau_{l\alpha jk} - u_{m,jkl}\tau_{\alpha ljk} - u_{j,mkl}\tau_{jl\alpha k} - u_{k,mjl}\tau_{klj\alpha}\big).\end{aligned} \tag{56}$$

Moreover, Eq. (43) through Eqs. (26), (29) and (54) gives us the following *scaling balance law*

$$\begin{aligned}\mathrm{D}_t\mathcal{Y} - \mathrm{D}_l Y_l = \Big(\tau_{\alpha lj} - \frac{d}{2}A^0_{\alpha lj}\Big)u_{\alpha,lj} + \Big(2\tau_{\alpha ljk} - \frac{d}{2}B^0_{\alpha ljk}\Big)u_{\alpha,ljk}\\ + x_l f_l^{\text{inh}} + \frac{2+d}{2}u_\alpha F_\alpha.\end{aligned} \tag{57}$$

Finally, the *balance law* that corresponds to the *addition of solutions* is given as follows

$$\mathrm{D}_t\mathcal{I} - \mathrm{D}_i I_i = v_\alpha F_\alpha(\mathbf{u}) - u_\alpha F_\alpha(\mathbf{v}) + A^0_{\alpha ik,i}(v_{\alpha,k} - u_{\alpha,k}) + B^0_{\alpha ikl,i}(v_{\alpha,kl} - u_{\alpha,kl}), \tag{58}$$

where it is easy to see that the gradients of the cohesive forces give additional source terms due to the inhomogeneity of the material.

The integral forms of the balance laws (55), (56) and (57) lead respectively to the generalizations of *the dynamical* **J**, **L** *and* M*-integrals* in the linear, compatible, anisotropic elasticity of grade-3

$$J_r := \int_S P_{rl} n_l \mathrm{d}S - \int_V \mathrm{D}_t \mathcal{P}_r \mathrm{d}V = -\int_V f_r^{\mathrm{inh}} \mathrm{d}V, \tag{59}$$

$$L_r := \int_S M_{rl} n_l \mathrm{d}S - \int_V \mathrm{D}_t \mathcal{M}_r \mathrm{d}V \tag{60}$$

$$= -\int_V e_{rm\alpha}\big(x_m f_\alpha^{\mathrm{inh}} + u_m F_\alpha - u_{l,m}\tau_{l\alpha} - u_{m,l}\tau_{\alpha l} - u_{l,mj}\tau_{l\alpha j} - u_{m,jl}\tau_{\alpha lj}$$
$$- u_{j,ml}\tau_{jl\alpha} - u_{l,mjk}\tau_{l\alpha jk} - u_{m,jkl}\tau_{\alpha ljk} - u_{j,mkl}\tau_{jl\alpha k} - u_{k,mjl}\tau_{klj\alpha}\big)\mathrm{d}V,$$

$$M := \int_S Y_{rl} n_l \mathrm{d}S - \int_V \mathrm{D}_t \mathcal{Y}_r \mathrm{d}V \tag{61}$$

$$= -\int_V \Big[\Big(\tau_{\alpha lj} - \frac{d}{2}A^0_{\alpha lj}\Big)u_{\alpha,lj} + \Big(2\tau_{\alpha ljk} - \frac{d}{2}B^0_{\alpha ljk}\Big)u_{\alpha,ljk}$$
$$+ x_l f_l^{\mathrm{inh}} + \frac{2+d}{2}u_\alpha F_\alpha\Big)\Big]\mathrm{d}V.$$

4.2 Dynamical Reciprocal Theorem

We proceed to show that in the case of vanishing cohesive forces, the balance law of addition of solutions (58) is actually the *dynamical reciprocal theorem* in linear, anisotropic elasticity of grade-3.

Indeed, following Achenbach [1], one can show (see also [2]) that if we have the displacements fields $\mathbf{u}(\mathbf{x})$ and $\mathbf{v}(\mathbf{x})$ produced by body forces $F_\alpha(\mathbf{u})$ and $F_\alpha(\mathbf{v})$, then the following relation holds

$$u_\alpha\big(F_\alpha(\mathbf{v}) - \mathrm{D}_t p_\alpha(\mathbf{v})\big) - v_\alpha\big(F_\alpha(\mathbf{u}) - \mathrm{D}_t p_\alpha(\mathbf{u})\big) =$$
$$\mathrm{D}_i\big[v_\alpha\big(\tau_{\alpha i}(\mathbf{u}) - \tau_{\alpha ik,k}(\mathbf{u}) + \tau_{\alpha ikl,kl}(\mathbf{u})\big) + v_{\alpha,k}\big(\tau_{\alpha ik}(\mathbf{u}) - \tau_{\alpha ikl,l}(\mathbf{u})\big) + v_{\alpha,kl}\tau_{\alpha ikl}(\mathbf{u})$$
$$- u_\alpha\big(\tau_{\alpha i}(\mathbf{v}) - \tau_{\alpha ik,k}(\mathbf{v}) + \tau_{\alpha ikl,kl}(\mathbf{v})\big) - u_{\alpha,k}\big(\tau_{\alpha ik}(\mathbf{v}) - \tau_{\alpha ikl,l}(\mathbf{v})\big) - u_{\alpha,kl}\tau_{\alpha ikl}(\mathbf{v})\big]$$
$$+ A^0_{\alpha ik}(u_{\alpha,ik} - v_{\alpha,ik}) + B^0_{\alpha ikl}(u_{\alpha,ikl} - v_{\alpha,ikl}). \tag{62}$$

We mention here that for the derivation of the above formula the symmetries $C_{\alpha j\beta k} = C_{\beta k\alpha j}$, $A_{\alpha jk\beta lm} = A_{\beta lm\alpha jk}$ and $B_{\alpha jkl\beta mnp} = B_{\beta mnp\alpha jkl}$ are essential. It is easy to check that Eq. (62) coincides with the balance law (58).

If $A^0_{\alpha ik} = 0$ and $B^0_{\alpha ikl} = 0$, then Eq. (62) is nothing but *the reciprocal theorem for the linear, anisotropic elastodynamics of grade-3* in local form. The integration of Eq. (62) for vanishing cohesive forces concludes to the following relation

$$
\begin{aligned}
&\int_V u_\alpha\big(F_\alpha(\mathbf{v}) - \mathrm{D}_t p_\alpha(\mathbf{v})\big)\mathrm{d}V + \int_S \Big(u_\alpha\big(\tau_{\alpha i}(\mathbf{v}) - \tau_{\alpha ik,k}(\mathbf{v}) + \tau_{\alpha ikl,kl}(\mathbf{v})\big)\\
&\quad + u_{\alpha,k}\big(\tau_{\alpha ik}(\mathbf{v}) - \tau_{\alpha ikl,l}(\mathbf{v})\big) + u_{\alpha,kl}\tau_{\alpha ikl}(\mathbf{v})\Big)\mathrm{d}S_i\\
&= \int_V v_\alpha\big(F_\alpha(\mathbf{u}) - \mathrm{D}_t p_\alpha(\mathbf{u})\big)\mathrm{d}V + \int_S \Big(v_\alpha\big(\tau_{\alpha i}(\mathbf{u}) - \tau_{\alpha ik,k}(\mathbf{u}) + \tau_{\alpha ikl,kl}(\mathbf{u})\big)\\
&\quad + v_{\alpha,k}\big(\tau_{\alpha ik}(\mathbf{u}) - \tau_{\alpha ikl,l}(\mathbf{u})\big) + v_{\alpha,kl}\tau_{\alpha ikl}(\mathbf{u})\Big)\mathrm{d}S_i, \qquad (63)
\end{aligned}
$$

which is the generalization of the reciprocal theorem in global form, akin to the theorems of Betti and Rayleigh (see e.g. [1]), to the linear, anisotropic elastodynamics of grade-3.

From the above analysis it is clear that in the case of anisotropic elasticity of grade-2 or dipolar gradient elasticity, that is $E_{\alpha j\beta klm} = 0$, $G_{\alpha ij\beta klm} = 0$ and $B_{\alpha jkl\beta mnp} = 0$ in the constitutive relations (11), the reciprocal theorem holds without to demand the vanishing of the tensor $F_{\alpha j\beta k}$ as erroneously claimed in [5, 6].

5 Conclusions

To sum up, for a homogeneous material in the absence of external forces only the groups of translations and addition of solutions lead to conservation laws in contrast to the groups of rotations and scaling that conclude just to balance laws. Moreover, the cohesive forces appear explicitly in the balance law of scaling even in the case of a homogeneous material. On the other hand, for an inhomogeneous material, the cohesive forces influence explicitly the balances of scaling and addition of solutions giving additional source terms. Finally, the balance law of addition of solutions is actually the dynamical reciprocal theorem under the restriction of vanishing cohesive forces.

Acknowledgement

The authors have been supported by an Emmy-Noether grant of the Deutsche Forschungsgemeinschaft (Grant No. La1974/1-2).

References

1. Achenbach, J.D., *Reciprocity in Elastodynamics*. Cambridge University Press, 2003.
2. Agiasofitou, E.K. and Lazar, M., Conservation and balance laws in linear elasticity of grade three. *J. Elast.* **94**, 2009, 69–85.

3. Bluman, G.W. and Kumei, S., *Symmetries and Differential Equations*. Springer-Verlag, New York, 1989.
4. Eshelby, J.D., The elastic energy-momentum tensor. *J. Elast.* **5**, 1975, 321–335.
5. Georgiadis, H.G. and Grentzelou, C.G., Energy theorems and the J-integral in dipolar gradient elasticity. *Int. J. Solids Struct.* **43**, 2006, 5690–5712.
6. Giannakopoulos, A.E., Amanatidou, E. and Aravas, N., A reciprocity theorem in linear gradient elasticity and the corresponding Saint-Venant principle. *Int. J. Solids Struct.* **43**, 2006, 3875–3894.
7. Kröner, E. and Datta, B.K., Nichtlokale Elastostatik: Ableitung aus der Gittertheorie. *Z. Phys.* **196**, 1966, 203–211.
8. Lazar, M., Maugin, G.A. and Aifantis, E.C., Dislocations in second strain gradient elasticity. *Int. J. Solids Struct.* **43**, 2006, 1787–1817.
9. Maugin, G.A., *Material Inhomogeneities in Elasticity*. Chapman and Hall, London, 1993.
10. Mindlin, R.D., Second gradient of strain and surface-tension in linear elasticity. *Int. J. Solids Struct.* **1**, 1965, 417–438.
11. Mindlin, R.D., Elasticity, piezoelectricity and crystal lattice dynamics. *J. Elast.* **2**, 1972, 217–282.
12. Morse, P.M. and Feshbach, H., *Methods of Theoretical Physics I*. McGraw-Hill, New York, 1953.
13. Noether, E., Invariante Variationsprobleme. *Nachr. König. Gesell. Wissen. Göttingen, Math.-Phys. Kl.*, 1918, 235–257. (English translation: Invariant variation problems. *Transport Theory Stat. Phys.* **1**, 1971, 186–207.)
14. Olver, P.J., *Applications of Lie Groups to Differential Equations*. Springer, New York, 1986.
15. Wu, C.H., Cohesive elasticity and surface phenomena. *Quart. Appl. Math. L* **1**, 1992, 73–103.

Evaluation of Crack-Driving Forces at Finite Viscoelasticity: Theory and Experiment

Michael Kaliske, Christiane Netzker and Bastian Näser

Abstract This contribution presents an approach to determine the fracture mechanical parameters energy release rate and crack-driving force efficiently with the material force method expanded for dissipative materials. A reliable test procedure for the evaluation of these parameters is presented. Another focus lies on the investigation of time-dependent effects in this context induced by viscoelasticity.

1 Introduction

During the last two decades, numerical simulation has been growing to one of the most important tools in the development and optimisation of new construction elements, structural parts and their assemblies. The Finite Element Method represents the most common numerical procedure for evaluating the structural behaviour of a component. It allows to determine both the displacements as well as the stress and strain conditions in a part due to certain load conditions. The reliability of these investigations depends on the underlying constitutive model of the observed material. In order to investigate the complex constitutive behaviour of elastomers, which show a combination of nonlinear elasticity at finite strains, dissipative phenomena like history dependent reduction of stiffness as well as rate-dependent and rate-independent inelasticity, it is therefore necessary to adopt physically based material

Michael Kaliske
Institute for Structural Analysis, Technische Universität Dresden, Nürnberger Str. 31a, 01062 Dresden, Germany; e-mail: michael.kaliske@tu-dresden.de

Christiane Netzker
Institute for Structural Analysis, Technische Universität Dresden, Nürnberger Str. 31a, 01062 Dresden, Germany; e-mail: christiane.netzker@tu-dresden.de

Bastian Näser
Institute of Structural Mechanics, University of Leipzig, Marschnerstr. 31, 04109 Leipzig, Germany

P. Steinmann (ed.), IUTAM Symposium on Progress in the Theory and Numerics of Configurational Mechanics, 193–202.

models since purely phenomenologically inspired constitutive equations can only account for single characteristics.

Even though the Finite Element Method offers the possibility to perform both static and dynamic simulations of highly materially and geometrically nonlinear problems, it yields no direct information on the fracture mechanical behaviour of the investigated components. However, the consideration of fracture mechanical aspects is of special importance since the safety and serviceability of most products can be mainly attributed to the initiation and growth of cracks. Therefore, further fracture mechanical investigations are inevitable in order to evaluate the sensitivity of these components with respect to crack growth.

Conventional fracture mechanical concepts were derived for brittle materials as well as metals and hold true only for purely elastic material characteristics. Considering the complex properties of elastomeric material, fracture mechanical investigations are very challenging and cannot be conducted with these standard approaches. According to this, only few publications regarding the fracture behaviour of elastomers exist, most of them assuming pure elasticity. Another approach regarding the determination of the crack-sensitivity of a component is offered by the concept of material forces. The fracture mechanical investigations presented in this contribution base on the extension of the material force method to inelastic material behaviour.

2 Viscoelasticity of Elastomers

A realistic material formulation of elastomers has to take into account stress softening and damage, respectively, as well as phenomena like rate-dependent and rate-independent dissipation of energy.

Viscoelastic material models basing on linear evolution laws proved to be insufficient for reproducing the rate-dependent behaviour of elastomers. A nonlinear formulation was proposed by Bergström and Boyce [1]. On basis of this finite viscoelasticity model, an algorithmic approach to model the rate-dependent behaviour of elastomeric material in the context of the Finite Element Method was derived [2, 3]. This micromechanically inspired model describes the material by a Helmholtz free energy function consisting of an equilibrium part representing the pure elastic material behaviour and a non-equilibrium part modelling the viscous effects

$$W = W^e + W^v. \tag{1}$$

Assuming nearly incompressibility, the elastic response is split into a volumetric and an isochoric part, while the viscous deformations are assumed to be purely isochoric. For viscoelastic, incompressible material behaviour, the evolution of the dissipated energy is expressed by the reduced dissipation inequality

$$\mathcal{D}^v = \overline{\underline{\tilde{\mathbf{M}}}}^v : \underline{\dot{\mathbf{F}}}^v \tag{2}$$

$$= \underline{\overline{\tau}}^v : \underline{\tilde{\mathbf{d}}}^v \tag{3}$$

$$\geq 0 \tag{4}$$

with the isochoric part of the Mandel tensor in the intermediate configuration $\overline{\underline{\tilde{\mathbf{M}}}}^v$ and the rate of the viscous deformation gradient $\underline{\dot{\mathbf{F}}}^v$.

The inelastic rate of the deformation tensor in the current configuration $\underline{\dot{\mathbf{F}}}^v$

$$\underline{\tilde{\mathbf{d}}}^v = -\frac{1}{2}\mathscr{L}_v\left(\underline{\mathbf{b}}^e\right)\,\underline{\mathbf{b}}^{e\text{-}1} \tag{5}$$

can be obtained by the Lie time derivative $\mathscr{L}_v\left(\underline{\mathbf{b}}^e\right)$ of the elastic finger tensor $\underline{\mathbf{b}}^e$. The direction of the evolution is represented by

$$\underline{\mathbf{n}} = \frac{\underline{\tau}^v_{\text{iso}}}{\left\|\underline{\tau}^v_{\text{iso}}\right\|} \quad \text{with} \quad \left\|\underline{\tau}^v_{\text{iso}}\right\| = \sqrt{\underline{\tau}^v_{\text{iso}} : \underline{\tau}^v_{\text{iso}}} \text{ and } \underline{\tau}^v_{\text{iso}} = \underline{\underline{\mathbb{P}}} : \underline{\overline{\tau}}^v, \tag{6}$$

where $\underline{\overline{\tau}}^v$ denotes the viscoelastic, incompressible part of the Kirchhoff stress and

$$\underline{\underline{\mathbb{P}}} = \underline{\underline{\mathbf{1}}} - \frac{1}{3}\,\underline{\mathbf{1}} \otimes \underline{\mathbf{1}} \tag{7}$$

the fourth order deviatoric projection tensor. Introducing furthermore the effective creep rate $\dot{\gamma}$ leads to the following evolution law

$$\underline{\tilde{\mathbf{d}}}^v = \dot{\gamma}\ \underline{\mathbf{n}}. \tag{8}$$

Based on the tube-model of Doi and Edwards [4], which restricts the parameter c to $c < 0$, and assuming an energy activated creep process, Bergström and Boyce derive the effective creep rate

$$\dot{\gamma} := \dot{\gamma}_0 \left[\lambda^v_c - 1\right]^c \left(\frac{\tau^v}{\hat{\tau}}\right)^m \quad \text{with } \tau^v = \frac{\left\|\underline{\tau}^v_{\text{iso}}\right\|}{\sqrt{2}} \tag{9}$$

depending on the viscous network stretch $\lambda^v_c = \sqrt{I^v_1/3}$, where $I^v_1 = tr\underline{\mathbf{C}}^v$ with the inelastic metric in the reference configuration $\overline{\underline{\mathbf{C}}}^v := \underline{\mathbf{F}}^{v\mathrm{T}}\ \underline{\mathbf{F}}^v$. With regard to the dissipation inequality, the parameters $\dot{\gamma}_0/\hat{\tau}^m$ and m must be positive.

The integration of the evolution law depends on an operator split of the material time derivative into an elastic predictor and an inelastic corrector step

$$\underline{\dot{\mathbf{b}}}^e := \underbrace{\underline{\mathbf{l}}_{\text{iso}}\underline{\mathbf{b}}^e + \underline{\mathbf{b}}^e\underline{\mathbf{l}}^{\mathrm{T}}_{\text{iso}}}_{\text{elastic}} + \underbrace{\mathscr{L}_v\left(\underline{\mathbf{b}}^e\right)}_{\text{inelastic}} \tag{10}$$

where

$$\mathscr{L}_v\left(\underline{\mathbf{b}}^e\right) = \overline{\underline{\mathbf{F}}}\ \underline{\dot{\mathbf{C}}}^{v\text{-}1}\ \overline{\underline{\mathbf{F}}}^{\mathrm{T}} \tag{11}$$

with the incompressible part of the deformation gradient $\underline{\overline{\mathbf{F}}} = J^{-1/3}\ \underline{\mathbf{F}}$. With $\dot{\underline{\mathbf{C}}}^{v\text{-}1} = 0$ during the elastic trial step leading to

$$\underline{\mathbf{b}}^{e,tr} = \underline{\overline{\mathbf{F}}}\ \left(\underline{\mathbf{C}}^{v\text{-}1}\right)_{t_n}\ \underline{\overline{\mathbf{F}}}^{\mathrm{T}} \tag{12}$$

and with $\mathscr{L}_v\left(\underline{\mathbf{b}}^e\right) = \dot{\underline{\mathbf{b}}}^e$ during the elastic corrector step, the evolution law is finally derived as

$$\dot{\underline{\mathbf{b}}}^e = \left[-2\dot{\gamma}\ \underline{\mathbf{n}}\right]\ \underline{\mathbf{b}}^{e,\mathrm{tr}}. \tag{13}$$

The elastic Finger tensor $\underline{\mathbf{b}}^e$ is derived on basis of an exponential mapping algorithm by an iterative solution of the nonlinear set of equations. The consistent tangent moduli are computed from a virtual intermediate stress expression by employing a pull-back operation of the Kirchhoff stress to the intermediate configuration by the trial elastic deformation gradient [2, 3].

3 Fracture Mechanical Parameters

Every infinitesimal crack growth in a structural component involves a release of energy. This change of potential energy is represented by the energy release rate. According to the Griffith fracture criterion [6], the crack will propagate if the crack growth releases more energy than required for the development of new surfaces.

The energy release rate is one fracture mechanical parameter. In case of elastic, non-dissipative material, it can be entirely attributed to the tearing of the material.

Inelastic, dissipative materials on the other hand show a distinctive process zone around the crack tip leading to a change of potential energy due to the movement of the dissipative zone around the crack tip in addition to the change of potential energy due to the displacement of the crack tip itself. The total amount of released energy is represented by the energy release rate which can be understood as the energy flow into the dissipative zone. This energy release rate offers no information on the part of energy released solely by the creation of new surfaces. Therefore, a second fracture mechanical parameter has to be distinguished. The energy flow to the crack tip is represented by the so-called crack-driving force which is related to the change of the elastic potential with respect to a movement of the crack tip within a fixed dissipative zone. The decision which of the two parameters has to be applied depends on the purpose of investigation.

The fracture process on the micro-scale can be described by the crack-driving force because it yields only the energy flow to the crack tip contributing entirely to the tearing of the material. Due to the extent of the dissipative zone, the value of the energy release rate depends on the evaluated area around the crack tip resulting in a path-dependency of the corresponding contour integrals. This is of special importance regarding the fact that the dissipative zone cannot always be completely

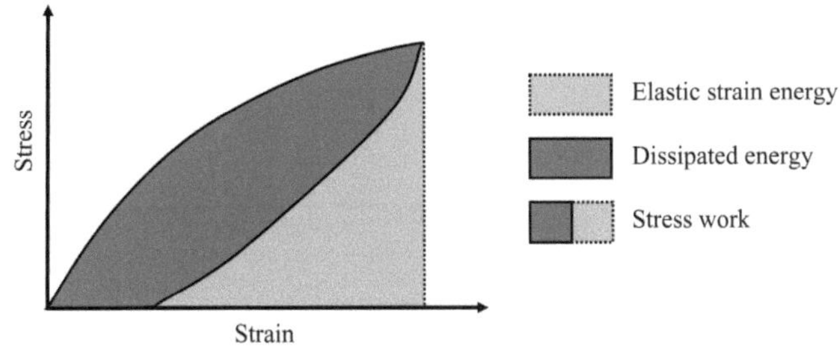

Fig. 1 Internal mechanical work and elastic strain energy.

included into the contour. In contrast, the crack tip is entirely included into every contour so that the crack-driving force is path-independent.

3.1 Classical Fracture Mechanical Approaches

Both fracture mechanical parameters can be derived with variations of the J-integral formulation. This line integral can be evaluated either in terms of the internal mechanical work as originally defined by Rice [10] yielding the crack-driving force or in terms of the strain energy as implemented in most commercial Finite Element programs (cf. Figure 1). The latter definition yields only the energy release rate in case of dissipative material while the definition according to Rice is restricted to monotonic loading due to the incremental evaluation of the internal mechanical work. Therefore, both approaches are not suited to identify the crack-driving force of inelastic, dissipative materials under realistic loading conditions.

Both, the crack-driving force and the energy release rate can be determined by an energy balance of two model states with different crack lengths. Since this method is again restricted to monotonic loading, requires extensive numerical efforts and allows only to investigate one crack at a time, it can only be applied under certain restrictions.

3.2 Material Force Method

Within the material force method, the described fracture mechanical parameters can be computed efficiently in the context of the Finite Element Method. This approach can be adapted to linear and non-linear elastic as well as inelastic material behaviour at small and finite strains and is not restricted to monotonic loading.

The material forces due to the Eshelby-stress tensor $\underline{\Sigma}$, which was introduced by Eshelby [5], captures only the change of the elastic strain energy. Evaluated inside

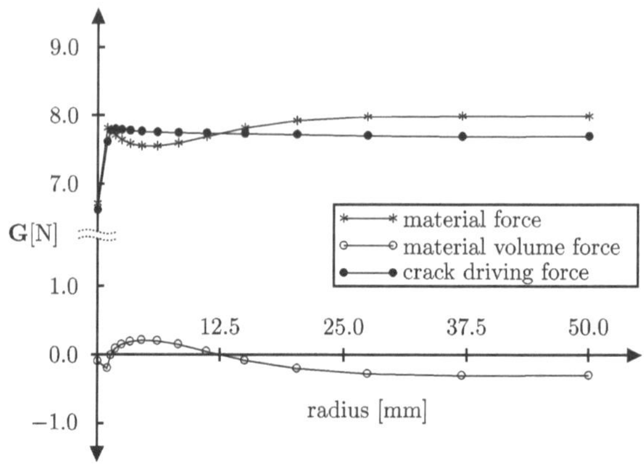

Fig. 2 Path-dependency of material force, material volume force and crack-driving force.

a domain around the crack tip, these material forces correspond therefore to the energy release rate.

In order to ensure the local balance of the material motion problem

$$\mathrm{Div}\,\underline{\Sigma} + \mathbf{G} = \mathbf{0} \quad \text{with} \quad \mathbf{G} = \mathbf{G}_0 + \mathbf{G}^v, \tag{14}$$

material volume forces due to inelastic effects $\mathbf{G}^v$ arise in addition to the material body forces resulting from the elastic problem (cf. Section 4). These additional material volume forces can be related to the negative driving forces on the corresponding dissipative field. Thus, it is possible to separate the pure crack-driving force from the material forces without any restrictions regarding the loading or the number of cracks.

The importance of taking into account the material volume forces which arise from inelastic material behaviour is emphasized by Figure 2. The diagram shows the sum of material forces due to the Eshelby-stress tensor, the material volume forces and the crack-driving forces, respectively, as a function of the distance from the crack tip. Obviously, the crack-driving force converges even in close vicinity of the crack tip. The crack-driving force can be stated as path-independent. On the contrary, the material force and the material volume force are a function of the integration path. Further examination showed also a small discretization-dependency of both the material force and material volume force. Depending on the integration path, disregard of the material volume forces will lead to an over- or underestimation of the crack-driving force as the fracture mechanical parameter.

4 Material Forces of Viscoelastic Material

According to the definition of the material forces, the material gradient of the strain energy has to be computed. For purely elastic material, the strain energy density depends only on the deformation gradient $\underline{\mathbf{F}}$ and – in case of inhomogeneities – on the spatial place $\mathbf{X}$.

The strain energy is only dependent on the elastic deformations. Applying a multiplicative split of the deformation gradient $\underline{\mathbf{F}} = \underline{\mathbf{F}}^e\ \underline{\mathbf{F}}^v$ into an elastic and a viscoelastic part, it can be therefore expressed in terms of the total deformation gradient $\underline{\mathbf{F}}$ and the viscoelastic deformations $\underline{\mathbf{F}}^v$ in case of viscoelastic material behaviour

$$W^v = W^v\left(\underline{\mathbf{F}}^e, \mathbf{X}\right) = W^v\left(\underline{\mathbf{F}}\ \underline{\mathbf{F}}^{v\text{-}1}, \mathbf{X}\right) = W^v\left(\underline{\mathbf{F}}, \underline{\mathbf{F}}^v, \mathbf{X}\right). \tag{15}$$

The material gradient of the strain energy density

$$\nabla_{\mathbf{X}} W^v = \nabla_{\mathbf{X}}(J^v\ \tilde{W}^v) \tag{16}$$

$$= \frac{\partial J^v}{\partial \mathbf{X}}\ \tilde{W}^v + J^v\ \underbrace{\frac{\partial \tilde{W}^v}{\partial \underline{\mathbf{F}}^e}}_{\underline{\tilde{\mathbf{P}}}^v} : \frac{\partial \underline{\mathbf{F}}^e}{\partial \mathbf{X}} + \left.\frac{\partial \psi}{\partial \mathbf{X}}\right|_{\text{expl}} \tag{17}$$

contains with

$$W^v = J^v\ \tilde{W}^v \quad \text{and} \quad \underline{\mathbf{P}}^v = J^v\ \underline{\tilde{\mathbf{P}}}^v\ \underline{\mathbf{F}}^{v\text{-T}} \tag{18}$$

the strain energy density and the first Piola–Kirchhoff stress tensor in the intermediate configuration.

Assuming nearly incompressibility $J^v = \det \underline{\mathbf{F}}^v \equiv 1$, as shown by all elastomeric materials of pratical relevance, and considering $\nabla_{\mathbf{X}} J^v = \mathbf{0}$, Eq. (17) reduces to

$$\nabla_{\mathbf{X}} W^v = \underbrace{\frac{\partial \tilde{W}^v}{\partial \underline{\mathbf{F}}^e}}_{\underline{\tilde{\mathbf{P}}}^v} : \frac{\partial \underline{\mathbf{F}}^e}{\partial \mathbf{X}} + \left.\frac{\partial \psi}{\partial \mathbf{X}}\right|_{\text{expl}}. \tag{19}$$

Rearranging Eq. (19) leads to

$$\nabla_{\mathbf{X}} W^v = \underline{\tilde{\mathbf{P}}} : \frac{\partial\left(\underline{\mathbf{F}}\ \underline{\mathbf{F}}^{v\text{-}1}\right)}{\partial \mathbf{X}} + \left.\frac{\partial \psi}{\partial \mathbf{X}}\right|_{\text{expl}} \tag{20}$$

$$= \underline{\tilde{\mathbf{P}}} : \left(\frac{\partial \underline{\mathbf{F}}}{\partial \mathbf{X}}\ \underline{\mathbf{F}}^{v\text{-}1}\right) + \underline{\tilde{\mathbf{P}}} : \left(\underline{\mathbf{F}}\ \frac{\partial \underline{\mathbf{F}}^{v\text{-}1}}{\partial \mathbf{X}}\right) + \left.\frac{\partial \psi}{\partial \mathbf{X}}\right|_{\text{expl}} \tag{21}$$

$$= \underline{\tilde{\mathbf{P}}} : (\nabla_{\mathbf{X}} \underline{\mathbf{F}}\ \underline{\mathbf{F}}^{v\text{-}1}) + \underline{\tilde{\mathbf{P}}} : (\underline{\mathbf{F}}\ \nabla_{\mathbf{X}} \underline{\mathbf{F}}^{v\text{-}1}) + \left.\frac{\partial \psi}{\partial \mathbf{X}}\right|_{\text{expl}}. \tag{22}$$

Since the method of material forces is based on the material motion balance, which can be derived from the spatial motion problem

$$\mathrm{Div}\,\underline{\mathbf{P}} + \mathbf{B} = \mathbf{0} \tag{23}$$

with the physical volume forces $\mathbf{B}$ by a pull-back operation into the reference configuration [9]

$$\underline{\mathbf{F}}^{\mathrm{T}}\,\mathrm{Div}\,\underline{\mathbf{P}} + \underline{\mathbf{F}}^{\mathrm{T}}\,\mathbf{B} = 0, \tag{24}$$

a pull-back operation is employed to the first term of Eq. (22)

$$\tilde{\underline{\mathbf{P}}} : \left(\nabla_{\mathrm{X}}\underline{\mathbf{F}}\;\underline{\mathbf{F}}^{v\text{-}1}\right) = \underline{\mathbf{P}} : \nabla_{\mathrm{X}}\underline{\mathbf{F}} = \mathrm{Div}\left(\underline{\mathbf{F}}^{\mathrm{T}}\underline{\mathbf{P}}\right) - \underline{\mathbf{F}}^{\mathrm{T}}\;\underbrace{\mathrm{Div}\,\underline{\mathbf{P}}}_{-\mathbf{B}}. \tag{25}$$

This term corresponds to the derivation of material forces for elastic material.

The second term of Eq. (22), depending on the inverse viscoelastic deformation gradient, can be expressed as

$$\underline{\mathbf{F}}\;\nabla_{\mathrm{X}}\underline{\mathbf{F}}^{v\text{-}1} = \underline{\mathbf{F}}\;\frac{\partial\underline{\mathbf{F}}^{v\text{-}1}}{\partial\underline{\mathbf{F}}^{v}} : \frac{\partial\underline{\mathbf{F}}^{v}}{\partial\mathbf{X}} = \underline{\mathbf{F}}\left[-\,\underline{\mathbf{F}}^{v\text{-}1}\otimes\underline{\mathbf{F}}^{v\text{-}\mathrm{T}}\right]^{\mathrm{T}_{2,3}} : \nabla_{\mathrm{X}}\underline{\mathbf{F}}^{v} \tag{26}$$

where $[\bullet]^{\mathrm{T}_{2,3}}$ denotes an exchange of the components attributed to the second and third index, yielding finally

$$\tilde{\underline{\mathbf{P}}} : \left(\underline{\mathbf{F}}\;\nabla_{\mathrm{X}}\underline{\mathbf{F}}^{v\text{-}1}\right) = -\underline{\mathbf{F}}^{v\text{-}\mathrm{T}}\left(\underline{\mathbf{F}}^{\mathrm{T}}\;\underline{\mathbf{P}}\right) : \nabla_{\mathrm{X}}\underline{\mathbf{F}}^{v}. \tag{27}$$

Considering Eq. (14) and taking into account Eqs. (22) as well as (25), the material volume forces from the elastic derivation include a term resulting from physical volume forces, like gravitational or centrifugal forces, and a second term accounting for elastic inhomogeneities

$$\mathbf{G}_0 = \underline{\mathbf{F}}^{\mathrm{T}}\;\mathbf{B} + \left.\frac{\partial\psi}{\partial\mathbf{X}}\right|_{\mathrm{expl}}. \tag{28}$$

The material volume forces due to viscoelastic effects result in

$$\begin{aligned}
\mathbf{G}^{v} &= \underline{\mathbf{F}}^{v\text{-}\mathrm{T}}\;\underbrace{\left(\underline{\mathbf{F}}^{\mathrm{T}}\;\underline{\mathbf{P}}\right)}_{\underline{\mathbf{M}}^{v}} : \nabla_{\mathrm{X}}\underline{\mathbf{F}}^{v} &(29)\\
&= \underline{\mathbf{F}}^{v\text{-}\mathrm{T}}\;\underline{\mathbf{M}}^{v} : \nabla_{\mathrm{X}}\underline{\mathbf{F}}^{v} &(30)\\
&= \tilde{\underline{\mathbf{M}}}^{v} : \nabla_{\mathrm{X}}\underline{\mathbf{F}}^{v} &(31)
\end{aligned}$$

with the Mandel-stress tensor defined in the intermediate configuration, $\tilde{\underline{\mathbf{M}}}^{v}$, as the work conjugated quantity of the history variable $\underline{\mathbf{F}}^{v}$.

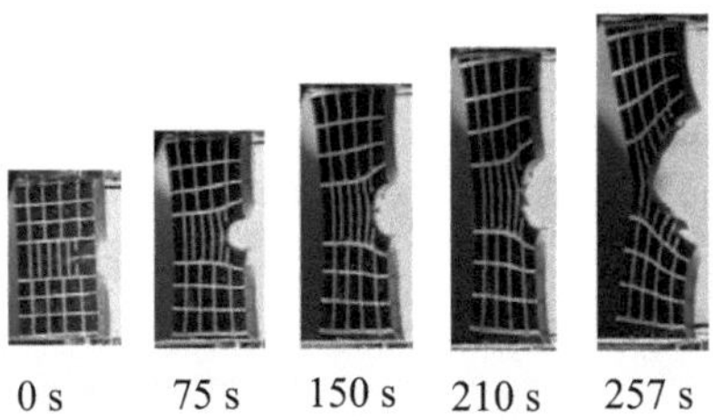

Fig. 3 Multiple-specimen experiment at increasing load levels and time.

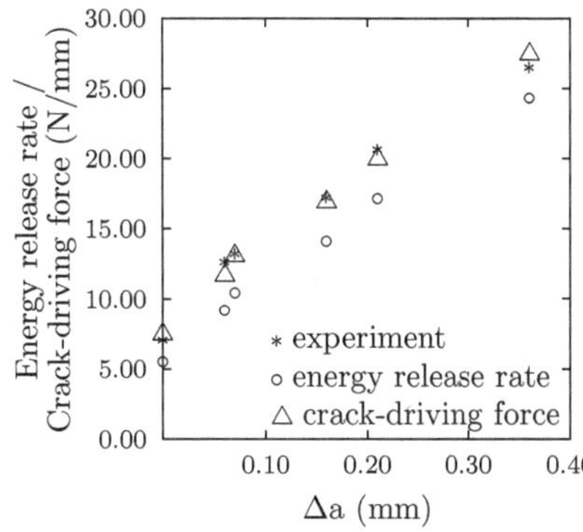

Fig. 4 Experimentally and numerically determined crack resistance curves.

5 Example

In cooperation with partners within the research group "Fracture Mechanics and Statistical Mechanics of Reinforced Elastomeric Blends", conventional test configurations were examined with regard to their applicability in terms of the interpretation of the two fracture mechanical parameters. Eventually, a quasi-static fracture mechanical test with the multiple-specimen method was applied. In this test, several identical single edge notched tension specimens were subjected to different load levels causing different lengths of stable crack growth Δa (Figure 3). The dimension of Δa was measured. The corresponding J-value was determined with the help of energy values calculated from the recorded load-displacement dependencies considering the internal mechanical work as well as the elastic strain energy (cf. Figure 1). The according J-Δa data pairs were plotted as crack resistance curves.

For the numerical simulation of this test, the viscoelastic material model presented in Section 2 was applied. For each crack length, $a + \Delta a$, the energy release rate (without consideration of material body forces) and the crack-driving force (with consideration of material body forces) were determined by the material force method. The comparison of experimental and numerical results (Figure 4) proved for the first time that the material forces with consideration of material body forces can be related to the measured J-integral.

6 Conclusions

Although the context of material forces is widely used in today's fracture mechanics (e.g. [7–9, 11]), the publications concentrate mainly on the examination of elastic material features. The extension and adaption of the material force method to inelastic effects and stress softening of rubber-like materials under finite strains is a new development, which allows to model the fracture mechanical behaviour of filled elastomers more realistically. By taking into account the material volume forces, different fracture mechanical effects of viscoelastic material, for example the driving force on a crack tip and the driving forces on the viscoelastic zone, can be evaluated.

It was shown that a realistic numerical investigation of the fracture mechanical behaviour of elastomeric material requires the consideration of the time-dependent inelastic properties.

Acknowledgements

The support of this work by the research group "Fracture Mechanics and Statistical Mechanics of Reinforced Elastomeric Blends" FOR597 and by the Deutsche Forschungsgemeinschaft under contract KA1163/3-1 is gratefully acknowledged.

References

1. Bergström, J.S. and Boyce, M.C., Constitutive modeling of the large strain time-dependent behavior of elastomers. *J. Mech. Phys. Solids* **46**, 1998, 931–954.
2. Dal, H. and Kaliske, M., Bergström–Boyce model for nonlinear finite rubber viscoelasticity: Theoretical aspects and algorithmic treatment for FE method. *Comp. Mech.*, 2008, submitted.
3. Dal, H., Kaliske, M. and Nasdala, L., Computational aspects of Bergström–Boyce finite viscoelasticity model. In: Boukamel, A., Laiarinandrasana, L., Meo, S. and Verron, E. (Eds.), *Proceedings of Fifth European Conference on Constitutive Models for Rubber*, Paris, 2007.
4. Doi, M. and Edwards, S.F., *The Theory of Polymer Dynamics*. Clarendon Press, Oxford, 1986.
5. Eshelby, J.D., The elastic energy-momentum tensor. *J. Elast.* **5**, 1975, 321–335.
6. Griffith, A.A., The phenomenon of rupture and flow in solids. *Philos. Trans. Roy. Soc. London* **221**, 1921, 163–198.
7. Gross, G., Müller, R. and Kolling, S., Configurational forces – morphology evolution and finite elements. *Mech. Res. Comm.* **29**, 2002, 529–536.
8. Kienzler, R. and Herrmann, G., *Mechanics in Material Space*. Springer, Berlin, 2000.
9. Maugin, G.A., *Material Inhomogeneities in Elasticity*. Chapman & Hall, London, 1993.
10. Rice, J.R., A path independent integral and the approximate analysis of strain concentration by notches and cracks. *J. Appl. Mech.* **35**, 1968, 379–386.
11. Steinmann, P., Application of material force to hyperelastostatic fracture mechanics. I. Continuum mechanical setting. *Int. J. Solids Struct.* **37**, 2000, 7371–7391.

On Configurational Forces within Green–Naghdi Thermo-Hyperelasticity

Swantje Bargmann, Ralf Denzer and Paul Steinmann

Abstract The contribution's main objective is the examination of configurational forces in non-classical nonlinear thermoelasticity based on the approach of Green and Naghdi. In the early 1990s, Green and Naghdi introduced a theory attracting interest as heat propagates as thermal waves at finite speed, does not necessarily involve energy dissipation and fully integrates the classical theory. A wide range of heat flow problems can be modeled and the classical theory is fully embedded. As configurational forces have proven to be well suited for the examination of defect mechanics, a numerical example from that research area is discussed. The numerical realization is based on Galerkin finite elements in space as well as in time.

1 Kinematics of Spatial and Material Motion Problem

A body B is considered to be a collection of continuum particles. A finite, isotropic and homogeneous body is considered. Let $\boldsymbol{X}$ denote the fixed position of a particular particle in the undeformed and stress-free configuration $\mathcal{B}_0$, the region the body occupies at the initial time t_0. $\boldsymbol{X}$ and $\mathcal{B}_0$ are called the particle's material position and the material configuration, respectively. Then the deformation of this configuration is described by the time-dependent vector field of the nonlinear spatial deformation map $\boldsymbol{\varphi}$, whereas the material deformation map $\boldsymbol{\Phi}$, with which continuum particles

Swantje Bargmann
Department of Applied Mechanics, Chalmers University of Technology, 412 96 Göteburg, Sweden; e-mail: swantje.bargmann@tu-dortmund.de

Ralf Denzer
Department of Mechanical and Structural Engineering, University of Trento, 38050 Trento, Italy; e-mail: ralf.denzer@ing.unitn.it

Paul Steinmann
Chair of Applied Mechanics, University of Erlangen-Nuremberg, 91058 Erlangen, Germany; e-mail: paul.steinmann@ltm.uni-erlangen.de

P. Steinmann (ed.), IUTAM Symposium on Progress in the Theory and Numerics of Configurational Mechanics, 203–214.

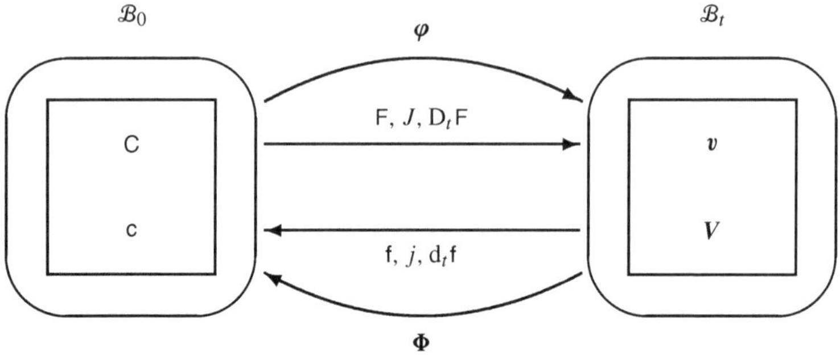

Fig. 1 Spatial motion problem: kinematics.

are followed through the material at fixed spatial coordinates $\boldsymbol{x}$, describes the inverse motion problem

$$\boldsymbol{x} = \boldsymbol{\varphi}\,(\,\boldsymbol{X}, t\,) \qquad \text{and} \qquad \boldsymbol{X} = \boldsymbol{\Phi}\,(\,\boldsymbol{x}, t\,). \tag{1}$$

$\boldsymbol{x}$ denotes the spatial position of the particle $\boldsymbol{X}$ in the deformed (spatial) configuration $\mathcal{B}_t$ at time t.

The deformation gradient F of the spatial motion map and its counterpart f of the material motion problem, are defined by

$$\mathsf{F} = \nabla_X \boldsymbol{\varphi}\,(\boldsymbol{X}, t) \qquad \text{and} \qquad \mathsf{f} = \nabla_x \boldsymbol{\Phi}\,(\boldsymbol{x}, t). \tag{2}$$

The positive Jacobians will be denoted as $J = \det \mathsf{F} > 0$ and $j = \det \mathsf{f} > 0$. Furthermore, we introduce the right Cauchy–Green strain tensor $\mathsf{C} = \mathsf{F}^t \cdot \mathsf{F}$ as a typical strain measure of the spatial motion problem. And in complete analogy, the material motion right Cauchy–Green strain tensor is given by $\mathsf{c} = \mathsf{f}^t \cdot \mathsf{f}$.

In what follows, the material time derivative of an arbitrary quantity $\{\bullet\}$ at fixed material placement $\boldsymbol{X}$ will be denoted as $\mathrm{D}_t\{\bullet\} = \partial_t\{\bullet\}|_X$. Accordingly, the spatial velocity $\boldsymbol{v}$ can be introduced as the material time derivative of the spatial motion map as $\boldsymbol{v} = \mathrm{D}_t\boldsymbol{\varphi}\,(\boldsymbol{X}, t)$. The gradient and the divergence of an arbitrary quantity $\{\bullet\}$ with respect to the material placement will be denoted as $\nabla_X\{\bullet\}$ and $\mathrm{Div}\,\{\bullet\}$, respectively. Whereas, the spatial time derivative of a quantity $\{\bullet\}$ at fixed spatial placements $\boldsymbol{x}$ is denoted as $\mathrm{d}_t\{\bullet\} = \partial_t\{\bullet\}|_x$. Thus, it defines the material velocity $\boldsymbol{V}$: $\boldsymbol{V} = \mathrm{d}_t\,\boldsymbol{\Phi}\,(\boldsymbol{x}, t)$. Furthermore, $\nabla_x\{\bullet\}$ and $\mathrm{div}\,\{\bullet\}$ denote the gradient and divergence of an arbitrary quantity $\{\bullet\}$ with respect to the spatial placement.

Remark: Spatial vs. material motion kinematics
The spatial and the material motion problem are related through the identity maps in $\mathcal{B}_0$ and $\mathcal{B}_t$: $id_{\mathcal{B}_0} = \boldsymbol{\Phi}\,(\,\boldsymbol{\varphi}(\boldsymbol{X}, t), t\,)$ and $id_{\mathcal{B}_t} = \boldsymbol{\varphi}\,(\,\boldsymbol{\Phi}(\boldsymbol{x}, t), t\,)$. Their deformation gradients are simply related via the inverses as $\mathsf{F}^{-1} = \mathsf{f}\,(\,\boldsymbol{\varphi}(\boldsymbol{X}, t), t\,)$ and $\mathsf{f}^{-1} = \mathsf{F}\,(\,\boldsymbol{\Phi}(\boldsymbol{x}, t), t\,)$ and e.g. the velocities via $\boldsymbol{v} = -\boldsymbol{V} \cdot \mathsf{F}$. The Euler theorem relates

the material and spatial time derivative of any scalar– or vector–valued function $\{\bullet\}$: $\mathrm{D}_t\{\bullet\} = \mathrm{d}_t\{\bullet\} + \nabla_x\{\bullet\} \cdot \boldsymbol{v}$ and $\mathrm{d}_t\{\bullet\} = \mathrm{D}_t\{\bullet\} + \nabla_X\{\bullet\} \cdot \boldsymbol{V}$.

2 Thermoelastic Model

The non-classical theory of thermoelasticity of Green and Naghdi [1] is based on balance laws, i.e. the balance of energy and the balance of momentum, like in the classical case. The novelty of the Green–Naghdi approach is the introduction of a new thermal variable, i.e. the thermal displacement α:

$$\alpha\,(\boldsymbol{X}, t) := \int_{t_0}^{t} T\,(\boldsymbol{X}, \tau)\;\mathrm{d}\tau + \alpha_0 \quad \text{or} \quad \mathrm{D}_t\alpha = T. \tag{3}$$

T represents the empirical temperature and α_0 denotes the initial value of the thermal displacement α at reference time t_0. For the sake of simplicity and with out loss of generality, the absolute temperature θ equals the empirical temperature: $\theta = T$.

The existence of a scalar-valued, positive, polyconvex, objective, material specific and path independent free energy density ψ is postulated. As in the classical case, Green–Naghdi thermo-hyperelasticity has a conservative structure implying that ψ plays the role of a potential for some of the constitutive relations, e.g. for the stress. For the derivation of the energetically conjugated quantities in case of Green–Naghdi thermo-hyperelasticity the reader is referred to the publication of Bargmann and Steinmann [11].

The promising approach of Green and Naghdi [1] is based on the introduction of three varying state spaces leading to three types of heat conduction, labeled type I, II and III. Thereof, type I corresponds to Fourier's approach. Consequently, the classical theory is fully included. In the following, we introduce and compare the classical Fourier theory (Green–Naghdi type I) to the non-classical Green–Naghdi theory of type III. Type III is the most general case: it contains type II as a limiting case and allows thermal wave propagation. In the following, the thermal theories of type I as well as type III are coupled with a classical Neo-Hookean hyperelastic theory.

2.1 Type I – Classical Theory

Usually, heat conduction in solids is based on Fourier's law which describes a diffusive process and leads to the well-known parabolic heat equation. This classical theory hypothesizes that the heat flux is proportional to the temperature gradient. Most engineering applications are described accurately by this parabolic equation, but at cryogenic temperatures material behavior can completely differ from that at room temperature. Predictions of the classical theory may diverge conspicuously from experimental data. One of these situations occurs at temperatures near absolute zero, where the phenomenon of second sound was discovered in the 20th century. Moreover, Fourier's theory predicts unnatural infinite transmission speed for parts of local heat pulses.

In the following, the basic equations are revised. The constitutive equations may depend on the right Cauchy–Green strain tensor C, the temperature $\mathrm{D}_t\alpha = T$ and the temperature gradient $\nabla_X \mathrm{D}_t\alpha = \nabla_X T$, thus

$$\psi = \psi\left(\mathrm{D}_t\alpha, \nabla_X \mathrm{D}_t\alpha, \mathsf{C}\right) \quad \text{and} \quad \boldsymbol{H} = \boldsymbol{H}\left(\mathrm{D}_t\alpha, \nabla_X \mathrm{D}_t\alpha, \mathsf{C}\right). \tag{4}$$

The constitutive equations for the free energy density ψ (acting as a potential for the entropy density η and the mechanical stresses, i.e. the Piola stress P) reads:

$$\begin{aligned} \rho_0\psi := {} & \frac{\mu}{2}\left[\mathsf{C}-\mathsf{I}\right]:\mathsf{I}+\frac{\lambda}{2}\ln^2 J-\mu\ln J+\rho_0 c\left[T-T_0-T\ln\frac{T}{T_0}\right] \\ & -3wK\left[T-T_0\right]\frac{\ln J}{J}-\left[T-T_0\right]S_0, \end{aligned} \tag{5}$$

with λ and μ denoting the Lamé constants and w, K and T_0 are the thermal expansion coefficient, the bulk modulus and the reference temperature. ρ_0 denotes the body's density in the reference configuration $\mathcal{B}_0$ and c the constant specific heat. The absolute entropy density S_0,

$$S_0 = S_0(T_0) = \int_0^{T_0} c \,\mathrm{d}T. \tag{6}$$

is added to the free energy in order to scale the entropy. Otherwise the change of entropy and not the entropy itself is denoted by η which would lead to reference temperature depending internal configurational forces. The entropy flux vector $\boldsymbol{H} := -\frac{\kappa_1}{T}\nabla_X T$ is assumed to be isotropic in the reference configuration $\mathcal{B}_0$ and leads to an isotropic heat flux vector $\boldsymbol{Q}$ following Fourier's law $\boldsymbol{Q} = T\boldsymbol{H} = -\kappa_1\nabla_X T$. Consequently, the spatial heat flux vector represents an anisotropic behavior, as $\boldsymbol{q} = -j\kappa_1 \mathsf{F}\cdot\mathsf{F}^t\cdot\nabla_x T$.

2.2 Type III – Non-Classical Theory

Fourier's law is not suitable for modeling second sound as has already been shown in the papers by Bargmann and Steinmann [2, 3]. The numerical results diverge tremendously from the experimental data. The theory of type III (being a gradient theory in time and in space) overcomes this drawback. It is a combination of the classical Fourier theory (type I) and the Green–Naghdi theory without energy dissipation (type II). The free energy density ψ and the heat flux $\boldsymbol{Q}$ are assumed to be functions of the thermal displacement gradient $\nabla_X\alpha$, the temperature $\mathrm{D}_t\alpha = T$, the temperature gradient $\nabla_X \mathrm{D}_t \alpha = \nabla_X T$ and the right Cauchy–Green strain tensor C.[1]

$$\psi = \psi\left(\mathrm{D}_t\alpha, \nabla_X\alpha, \nabla_X\mathrm{D}_t\alpha, \mathsf{C}\right) \quad \text{and} \quad \boldsymbol{H} = \boldsymbol{H}\left(\mathrm{D}_t\alpha, \nabla_X\alpha, \nabla_X\mathrm{D}_t\alpha, \mathsf{C}\right) \tag{7}$$

We conclude the following relations for ψ and $\boldsymbol{Q}$, based on those stated in [1]:

$$\begin{aligned}
\rho_0\psi &:= \frac{\mu}{2}\left[\mathsf{C} - \mathsf{I}\right] : \mathsf{I} + \frac{\lambda}{2}\ln^2 J - \mu \ln J + \rho_0 c\left[T - T_0 - T\ln\frac{T}{T_0}\right] \\
&\quad - 3wK\left[T - T_0\right]\frac{\ln J}{J} - \left[T - T_0\right]S_0 + \frac{\kappa_3}{2bT_0}\nabla_X\alpha \cdot \nabla_X\alpha, \\
\boldsymbol{H} &:= -\frac{1}{b}\left[\frac{\kappa_3}{T_0}\nabla_X\alpha + \frac{\kappa_4}{T}\nabla_X T\right],
\end{aligned} \tag{8}$$

where b denotes a constant material parameter.

In the following, the governing equations of thermo-hyperelasticity are introduced for the spatial as well as for the material motion problem. The mechanical problem is primarily described by the balance of momentum, whereas the thermal problem is characterized by the balance of entropy.

2.3 Spatial Motion Problem

The balance of momentum of the spatial motion problem reads

$$\mathrm{D}_t\left(\rho_0\boldsymbol{v}\right) = \mathrm{Div}\mathsf{P} + \rho_0\boldsymbol{b} \tag{9}$$

P represents the Piola stress tensor and $\boldsymbol{b}$ is the momentum source.

The thermal problem is governed by the balance of entropy

$$\mathrm{D}_t\left(\rho_0\eta\right) = -\mathrm{Div}\boldsymbol{H} + \rho_0[s + \xi], \tag{10}$$

where η, $\boldsymbol{H}$, s and $\xi \geq 0$ are the entropy density, the material entropy flux vector, the entropy source and the non-negative entropy production, respectively. Multiplying the balance of entropy (10) by the temperature T results in

$$T\mathrm{D}_t\left(\rho_0\eta\right) = \boldsymbol{H} \cdot \nabla_X T - \mathrm{Div}\,\boldsymbol{Q} + \rho_0 T\left[s + \xi\right]. \tag{11}$$

Here, the material heat flux vector is related to the material entropy flux vector by $\boldsymbol{Q} = T\boldsymbol{H}$, see [11]. The entropy density η is thermodynamically conjugated to the empirical temperature T. Moreover, the classical relation between the Piola stress tensor P and the free energy density ψ holds:

[1] Actually, as introduced by Green and Naghdi [1], the constitutive equations may also depend on the thermal displacement α. However, we follow the usual procedure and define them independent of α, leading to Eq. (7).

$$\frac{\partial T}{\partial(\mathrm{D}_t\alpha)}\eta := -\frac{\partial\psi}{\partial T} \quad \text{and} \quad \mathsf{P} := \rho_0\frac{\partial\psi}{\partial \mathsf{F}}, \tag{12}$$

see [11]. Thus, the material time derivative of the entropy density renders $\mathrm{D}_t\,(\rho_0\eta) = \rho_0\partial_T\eta\mathrm{D}_t T - \partial_T\mathsf{P} : \mathrm{D}_t\mathsf{F}$. Since the specific heat capacity c can be defined by $c = T\partial_T\eta$ we finally obtain

$$\rho_0 c\mathrm{D}_t T = \boldsymbol{H}\cdot\nabla_X T - \mathrm{Div}\,\boldsymbol{Q} + \rho_0 T\,[s+\xi] + \mathcal{Q}_0^{\mathrm{mech}}. \tag{13}$$

The thermodynamic coupling effect, i.e. the Gough–Joule effect, is caused by the thermomechanical coupling term $\mathcal{Q}_0^{\mathrm{mech}} = T\partial_T\mathsf{P} : \mathrm{D}_t\mathsf{F}$.

2.4 Material Motion Problem

The so-called configurational forces are established by the balance of linear momentum of the material motion problem. The latter is defined as a projection of the spatial balance of linear momentum onto the material manifold, i.e. by a premultiplication of $-j\mathsf{F}^t$:

$$j\mathrm{D}_t\boldsymbol{P}_0 = \mathrm{div}\mathsf{p} + \rho_t\boldsymbol{B}, \tag{14}$$

i.e. it is postulated analogously to (9) [15–17]. It balances the time rate of change of the material motion momentum $\boldsymbol{P}_0$ with the material motion momentum flux p and the momentum source $\rho_t\boldsymbol{B}$. The mass specific kinetic energy density k reads $k = \boldsymbol{V}\cdot\mathsf{C}\cdot\boldsymbol{V}/2$ and the material volume force is denoted by $\rho_t\boldsymbol{B}$. Moreover, the two-point material momentum flux p is related to the classical material Eshelby stress tensor via $\Sigma = J\mathsf{p}\cdot\mathsf{f}^t$. We leave out the scalar-valued balance of entropy of the material motion problem for the sake of brevity as it will not be used in the computations.

Remark: Spatial vs. material quantities

At this point, relations between the material and the spatial quantities are stated more detailed. Due to the vector-valued nature of the balance of linear momentum, its material formulation differs more from the spatial one in contrast to scalar-valued balance equations. The thermoelastic balance of pseudomomentum is derived following the way of Dascalu and Maugin [8] who present a rational and well documented derivation in case of type II. Thus, in case of type III, the entropy equation (10) is multiplied by the thermal displacement gradient $\nabla_X\alpha$ and then inserted into the balance of pseudomomentum.

Then, the Eshelby stress tensor Σ reads $\Sigma = \rho_0[\psi - k]\mathsf{I} - \mathsf{F}^t\cdot\mathsf{P}$ for the classical type I and $\Sigma = \rho_0[\psi - k]\mathsf{I} - \mathsf{F}^t\cdot\mathsf{P} + \nabla_X\alpha\otimes\boldsymbol{H}$ for the non-classical type III.

The total thermoelastic pseudomomentum of type I is $\boldsymbol{P}_0 = \rho_0\mathsf{C}\cdot\boldsymbol{V}$ and $\boldsymbol{P}_0 = \rho_0\mathsf{C}\cdot\boldsymbol{V} - \rho_0\eta\nabla_X\alpha$ in case of type III. At first, the appearance of the entropy density η in the pseudomomentum might be seen a bit awkward. But as α is introduced

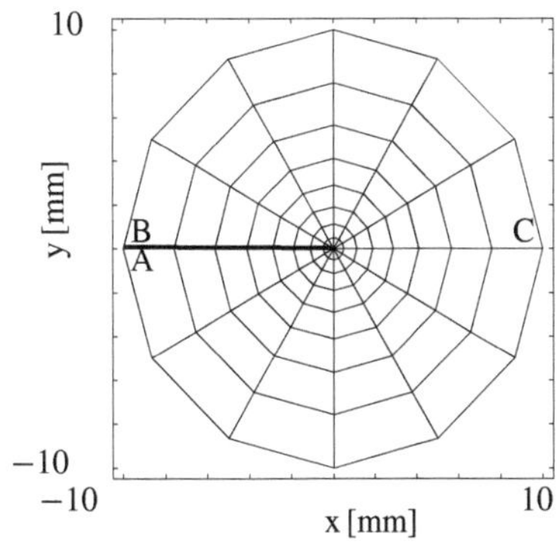

density ρ	10	$[\mathrm{kg/m^3}]$
specific heat c	100	$[\mathrm{W/kgK}]$
thermal conductivity κ_1	100	$[\mathrm{W/mK}]$
thermal conductivity κ_3	100	$[\mathrm{W/s^2mK}]$
thermal conductivity κ_4	0.001	$[\mathrm{W/mK}]$
thermal exp. coeff. w	10^{-5}	$[1/\mathrm{K}]$
Young's modulus E	210000	$[\mathrm{N/m^2}]$
Poisson's ratio ν	0.4	
reference temperature T_0	10	[K]
absolute entropy S_0	1	$[\mathrm{N/m^2K}]$
material parameter b	1	

Fig. 2 Discretization and material parameters of two dimensional disc. The disc is fixed at nodes A, B and C in both directions. The bold line indicates the crack of the specimen.

in analogy with the mechanical displacement $\boldsymbol{u}$, why not regard η as some kind of thermal momentum in analogy with the mechanical momentum? See also [9, 10].

The classical spatial volume force $\boldsymbol{b}$ does not contain any contribution from the kinetic energy. Contrarily, the material volume force $\rho_0 \boldsymbol{B}$ can be expressed as

$$\rho_0 \boldsymbol{B} = -\rho_0 \mathsf{F}^t \cdot \boldsymbol{b} + \rho_0 \eta \nabla_X T - \frac{\partial \rho_0}{\partial \boldsymbol{X}} [\psi - k] , \tag{15}$$

in case of type I, with $\rho_0 \boldsymbol{B}_{\text{ext}} = -\rho_0 \mathsf{F}^t \cdot \boldsymbol{b}$ being the external force. The internal force $\rho_0 \boldsymbol{B}_{\text{int}} = \rho_0 \eta \nabla_X T - \rho_0 \partial_{\boldsymbol{X}} \psi$ can be interpreted as a measure of the material's inhomogeneity in the material motion context [15]. For type III, the material volume force $\rho_0 \boldsymbol{B}$ reads

$$\rho_0 \boldsymbol{B} = -\rho_0 \mathsf{F}^t \cdot \boldsymbol{b} - \rho_0 s \nabla_X \alpha - \nabla_X \left(\frac{\nabla_X \alpha}{T} \right) \cdot \boldsymbol{Q}_I - \frac{\partial \rho_0}{\partial \boldsymbol{X}} [\psi - k] , \tag{16}$$

where $\boldsymbol{Q}_I$ refers to part of the heat flux vector with is proportional to the temperature gradient analogous to the theory of type I.

3 Discretization and Numerical Example

The numerical discretization is done with Galerkin finite element methods. First the equations are discretized in space with a Bubnov–Galerkin finite element approach and subsequently the temporal discretization, applying a Galerkin time finite element method, is carried out. The complete derivation of the discretization procedure is given in [13]. The result of the discretization of Eq. (14) is the discrete balance of pseudomomentum of the material motion problem, which formally reads

$$\mathsf{F}^{\Phi h}_{\text{dyn}} + \mathsf{F}^{\Phi h}_{\text{int}} - \mathsf{F}^{\Phi h}_{\text{sur}} - \mathsf{F}^{\Phi h}_{\text{vol}} = 0. \tag{17}$$

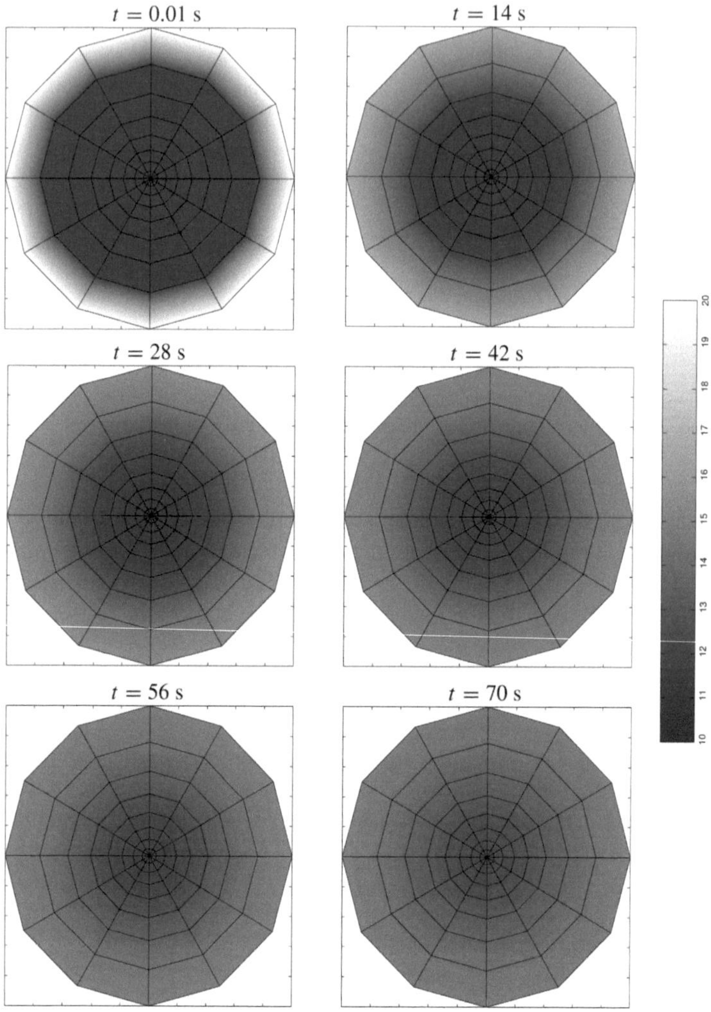

Fig. 3 Two dimensional disc with crack. Heat conduction of type I. The temperature distribution is plotted at times $t = 0.01$, 14, 28, 42, 56, 70 s.

The configurational node point inertia, internal, surface and volume forces take the form

$$\mathsf{F}_{\mathrm{dyn}}^{\Phi h} = \mathop{\mathbf{A}}_{e=1}^{n_{\mathrm{el}}} \sum_{i=1}^{n_{\mathrm{en}}} \int_{\mathcal{B}_t^e} N_{\Phi}^i j \mathrm{D}_t \boldsymbol{P}_0 \, \mathrm{d}v, \qquad \mathsf{F}_{\mathrm{int}}^{\Phi h} = \mathop{\mathbf{A}}_{e=1}^{n_{\mathrm{el}}} \sum_{i=1}^{n_{\mathrm{en}}} \int_{\mathcal{B}_t^e} \nabla_x N_{\Phi}^i \cdot \mathsf{p}^t \, \mathrm{d}v,$$

$$\mathsf{F}_{\mathrm{sur}}^{\Phi h} = \mathop{\mathbf{A}}_{e=1}^{n_{\mathrm{el}}} \sum_{i=1}^{n_{\mathrm{en}}} \int_{\partial \mathcal{B}_t^e} N_{\Phi}^i \mathsf{p} \cdot \boldsymbol{n} \, \mathrm{d}a, \qquad \mathsf{F}_{\mathrm{vol}}^{\Phi h} = \mathop{\mathbf{A}}_{e=1}^{n_{\mathrm{el}}} \sum_{i=1}^{n_{\mathrm{en}}} \int_{\mathcal{B}_t^e} N_{\Phi}^i \rho_t \boldsymbol{B} \, \mathrm{d}v \tag{18}$$

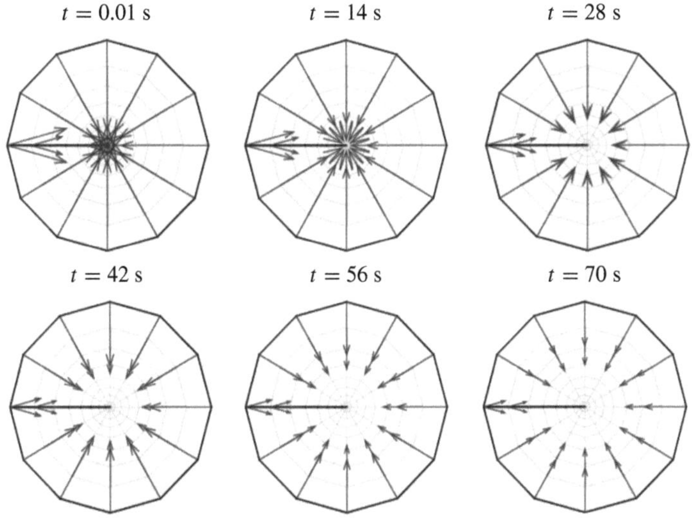

Fig. 4 Two dimensional disc with crack. Heat conduction of type I. The discrete configurational node point forces $\mathsf{F}^{\Phi h}$ are plotted at different times.

with N_{Φ}^{i} the shape function of the spatial discretization and A the assembly operator.

As a numerical example a cracked disc consisting of coupled geometrically non-linear thermo-hyperelastic material is discussed. Figure 2 illustrates the spatial discretization of the specimen with a radius of 10 mm and the material parameters stated in the table aside. In space, it is discretized with 12 constant strain triangular and 84 bilinear quadrilateral finite elements. Moreover, the spatial mesh is refined towards the crack tip. For the temporal discretization 10000 equidistant finite elements are applied. The specimen is fixed at nodes A, B and C (as indicated in Figure 2), meaning the movement of the nodes mentioned is limited in x as well as in y-direction. The disc is observed for 10 seconds, thus the time step equals $h_n = 0.01$ s.

The disc has a temperature distribution of 10 K with a Dirichlet boundary of 20 K. The displacements are assumed to be zero in the beginning. In case of type III, the thermal displacement α is initialized to be zero at every node. Figure 3 depicts the temperature development in the disc in case of thermoelasticity type I at times $t = 0.01$ s, $t = 14$ s, $t = 28$ s, $t = 42$ s, $t = 56$ s and $t = 70$ s. The heat propagates diffusively from the boundary to the interior.

In Figure 4 the corresponding discrete configurational node point forces $\mathsf{F}^{\Phi h} = \mathsf{F}_{\text{dyn}}^{\Phi h} + \mathsf{F}_{\text{int}}^{\Phi h} + \mathsf{F}_{\text{vol}}^{\Phi h}$ are shown. The dynamic configurational force does not vanish in case of coupled thermoelastodynamics as considered in this example, i.e. $\mathsf{F}_{\text{dyn}}^{\Phi h} \neq \mathbf{0}$. As the heat and the temperature gradient $\nabla_X T$ decrease at the boundary, the boundary discrete node point material forces become smaller, too. Moreover, as the center heatens up, configurational forces develop in this region.

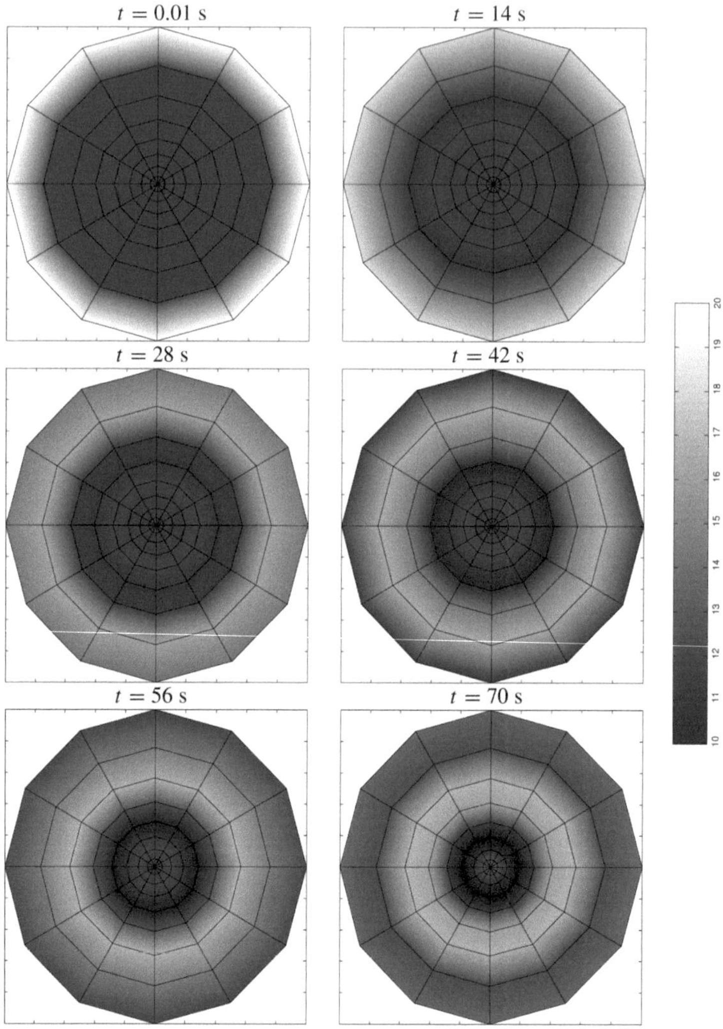

Fig. 5 Two dimensional disc with crack. Heat conduction of type III. The temperature distribution is plotted at times $t = 0.01, 14, 28, 42, 56, 70$ s. Heat propagates as a thermal wave from the warmer boundary to the cooler middle of the disc.

The temperature distribution according to type III is illustrated as contour plots in Figure 5. After the boundary is heated, the heat propagates as a thermal wave towards the disc's interior. However, due to the existing diffusive part of the heat flux, the wave's amplitude declines in the beginning. Because of the smaller diameter, it starts to increase again as it travels towards the disc's center point. Furthermore, the computed discrete configurational node point forces $\mathsf{F}^{\Phi h}$ are shown in Figure 6. The configurational node point forces decrease over time due to the decreasing heat flux

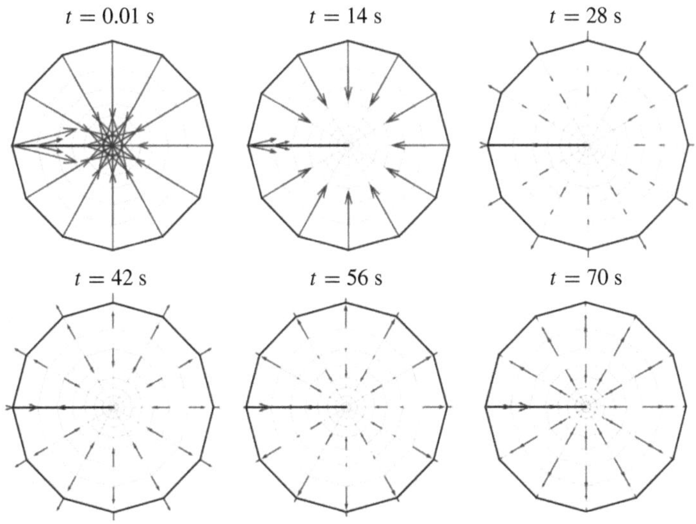

Fig. 6 Two dimensional disc with crack. Heat conduction of type III. The discrete configurational node point forces $\mathsf{F}^{\Phi h}$ are plotted at different times.

Q and propagate along with the wave. Moreover, while pointing to the middle in the beginning, their direction changes when the wave detaches from the boundary.

4 Conclusions

The objective of this contribution was the investigation of non-classical Green–Naghdi thermo-hyperelasticity in context of the material force method. The dynamic, geometrically nonlinear, fully coupled problem of the classical theory of type I and the non-classical theory of type III were considered and compared. Due to the two different ways of heat propagation, the behaviors of the configurational forces differs considerably when examining type I or III.

In general, configurational forces appear at places where thermal effects can be understood as a source of continuously distributed defects. The fully coupled problem was studied in a cracked specimen. The configurational forces indicated the tendency of the material inhomogeneity to move relative to their surrounding material and, therefore, present a powerful computational tool when modeling thermoelastic defect mechanics.

Acknowledgement

This research was mainly carried out while all authors were researchers at the University of Kaiserslautern, Germany. The financial support of the German Science Foundation (Deutsche Forschungsgemeinschaft, DFG), grant STE 544/23, is gratefully acknowledged.

References

1. Green, A.E. and Naghdi, P.M., A re-examination of the basic postulates of thermomechanics. *Proc. R. Soc. Lond.* **432**, 1991, 171–194.
2. Bargmann, S. and Steinmann, P., Finite element approaches to non-classical heat conduction in solids. *Comp. Meth. Eng. Sci.* **9**(2), 2005, 133–150.
3. Bargmann, S. and Steinmann, P., Modeling and simulation of first and second sound in solids. *Int. J. Solid Struct.* **45**, 2008, 6067–6073.
4. Eshelby, J.D., The force on an elastic singularity. *Phil. Trans. Royal Soc. Lond. A* **244**, 1951, 87–112.
5. Eshelby, J.D., The elastic energy-momentum tensor. *J. Elasticity* **5**, 1975, 321–335.
6. Gurtin, M.E., *Configurational Forces as Basic Concepts of Continuum Physics*, Springer, Berlin/Heidelberg, 1999.
7. Bargmann, S. and Steinmann, P., An incremental variational formulation of dissipative and non-dissipative coupled thermoelasticity for solids. *Heat Mass Trans.* **45**, 2008, 107–116.
8. Dascalu, C. and Maugin, G.A., The thermoelastic material-momentum equation. *J. Elasticity* **39**, 1995, 201–212.
9. Maugin, G.A. and Kalpadikes, V.K., The slow march towards an analytical mechanics of dissipative materials. *Techn. Mechanik* **22**, 2002, 98–103.
10. Maugin, G.A. and Kalpadikes, V.K., A Hamiltonian formulation for elasticity and thermoelasticity. *J. Phys. A: Math. Gen.* **35**, 2002, 10775–10788.
11. Bargmann, S. and Steinmann, P., A classical result for a non-classical theory: Remarks on Green–Naghdi thermo-hyperelasticity. *Cont. Mech. Thermodyn.* **19**(1-2), 2007, 59–66.
12. Kalpadikes, V.K. and Maugin, G.A., Canonical formulation and conservation laws of thermoelasticity without energy dissipation. *Rep. Math. Physics* **53**, 2004, 371–391.
13. Bargmann, S., Denzer, R. and Steinmann, P., Material forces in non-classical thermo-hyperelasticity. *J. Therm. Stress.* **32**(4), 2009, 361–393.
14. Maugin, G.A., *Material Inhomogeneities in Elasticity*, Chapman & Hall, London, 1993.
15. Kuhl, E., Denzer, R., Barth, F.J. and Steinmann, P., Application of the material force method to thermo-hyperelasticity. *Comp. Meth. Appl. Mech. Engng.* **193**, 2004, 3303–3325.
16. Steinmann, P., On spatial and material settings of thermo-hyper-elastodynamics. *J. Elasticity* **66**, 2002, 109–157.
17. Steinmann, P., On spatial and material settings of hyperelastodynamics. *Acta Mechanica* **156**, 2002, 193–218.
18. Podio-Guidugli, P., A virtual power format for thermomechanics. *Continuum Mechanics and Thermodynamics*, DOI 10.1007/s00161-009-0093-5, 2009.

Translational Conservation and Balance Laws in the Gauge Theory of Dislocations

Markus Lazar and Charalampos Anastassiadis

Abstract We present the translation gauge theory of moving dislocations which is mathematically analogue to the Maxwell theory of electromagnetic fields. Such a theory describes dislocations in an incompatible elastic continuum. The physical state quantities in the dislocation gauge theory are the physical velocity vector, the elastic distortion tensor, the dislocation density tensor and the dislocation current tensor. We use gauge-principles like minimal replacement known from field theories. We derive the equations of motion of dislocations. We give the isotropic as well as the anisotropic constitutive relations for the present theory. Translational conservation and balance laws are derived. We give the canonical as well as the gauge-invariant currents. We show that the dynamical Peach–Koehler force is mathematically analogous to the Lorentz force in Maxwell's theory of electromagnetic fields.

1 Introduction

Dislocations are important crystal defects. After bending and twisting crystals contain lots of dislocations. The dislocation is a defect breaking the translational symmetry. Furthermore, dislocations cause plasticity.

Why do we need generalized elasticity for dislocations? Elasticity breaks down at small distances far from crystal defects and it leads to singularities in strain and stress fields. Elasticity is a scale-free theory without characteristic length scales. Moreover, 'classical' dislocation theory misses constitutive relations for the char-

Markus Lazar
Emmy Noether Research Group, Department of Physics, Darmstadt University of Technology, Hochschulstr. 6, D-64289 Darmstadt, Germany, e-mail: lazar@fkp.tu-darmstadt.de

Charalampos Anastassiadis
Emmy Noether Research Group, Department of Physics, Darmstadt University of Technology, Hochschulstr. 6, D-64289 Darmstadt, Germany, e-mail: anastassiadis@fkp.tu-darmstadt.de

P. Steinmann (ed.), IUTAM Symposium on Progress in the Theory and Numerics of Configurational Mechanics, 215–227.

acteristic fields of dislocations. A promising candidate of an improved dynamical theory of dislocations is the so-called translational gauge theory of dislocations [1, 2, 8, 9]. The translational gauge theory of dislocations was introduced by Edelen et al. [1, 2] and later improved by Lazar [5, 6] and Lazar and Anastassiadis [8,9]. Such a theory uses the field theoretical framework which is well accepted in theoretical physics. In physics fundamental theories are gauge theories. Therefore, we consider a gauge theory of dislocations which is analogue to the Maxwell theory of electromagnetic fields. We want to investigate the (dynamical) Peach–Koehler force in the framework of dislocation gauge theory and from the mathematical point of view if it is a Lorentz-type force. The analysis of the Lie-point symmetries (translation, rotation, scaling), the symmetry of addition of solutions and the gauge symmetry can be found in [8].

2 Dislocation Gauge Theory

In this section we want to explain how we can extend the theory of elasticity to a gauge theory of dislocations. We consider the three-dimensional translation group $T(3)$ as the gauge group. In elasticity we deal with a displacement vector, u_i, transforming from the undeformed to the deformed states. The distortion tensor and the velocity vector are defined by

$$\beta_{ij} = u_{i,j}, \qquad v_i = \dot{u}_i \tag{1}$$

which are obviously invariant under a 'rigid' translation transformation:

$$u_i^* = u_i + f_i, \qquad f_i - \text{constant translations.} \tag{2}$$

As a consequence the elastic Lagrangian is also invariant under a constant translation: $\mathcal{L}(v_i^*, \beta_{ij}^*) = \mathcal{L}(v_i, \beta_{ij})$ due to $v_i^* = v_i$, $\beta_{ij}^* = \beta_{ij}$. Thus, elasticity has a *rigid* translational symmetry. On the other hand, one question arises: What about *local* translational symmetry?

If we postulate a local translation transformation for u_i:

$$u_i^* = u_i + f_i(t,x), \quad f_i(t,x) - \text{local translations,} \tag{3}$$

the invariance of $\dot{u}_i$, $u_{i,j}$ and $\mathcal{L}$ is lost. The extension from the rigid to soft translation symmetry affects the transformation behavior of the Lagrangian $\mathcal{L}$ which contains derivatives $\dot{u}_i$ and $u_{i,j}$ of the field u_i. The soft symmetry transformations on $\dot{u}_i$ and $u_{i,j}$ generate terms containing derivatives $\dot{f}_i(t,x)$ and $f_{i,j}(t,x)$ of the spacetime-dependent group parameters which spoil the former rigid invariance. In order to compensate these terms, one is forced to introduce so-called gauge fields. The translational gauge fields φ_i and ϕ_{ij} have the following inhomogeneous transformation laws:

$$\varphi_i^* = \varphi_i - \dot{f}_i(t,x), \qquad \phi_{ij}^* = \phi_{ij} - f_{i,j}(t,x), \tag{4}$$

where φ_i is a vector gauge field and ϕ_{ij} is a tensor gauge field. The gauge fields couple to the field u_i by $T(3)$-*gauge-covariant derivatives*:

$$v_i := \nabla_t u_i \equiv \dot{u}_i + \varphi_i, \qquad \beta_{ij} := \nabla_j u_i \equiv u_{i,j} + \phi_{ij}. \tag{5}$$

Thus, we have redefined the physical momentum v_i and the elastic distortion β_{ij} by means of the gauge-covariant derivatives. Now v_i and β_{ij} are *gauge invariant* under local $T(3)$:

$$v_i^* = v_i, \qquad \beta_{ij}^* = \beta_{ij}. \tag{6}$$

Therefore, they are physical state quantities. In addition they are now incompatible fields. Also we replace in the Lagrangian the partial derivatives of the field u_i by gauge-covariant derivatives

$$\mathcal{L}(\dot{u}_i, u_{i,j}) \longrightarrow \mathcal{L}(\dot{u}_i, u_{i,j}, \varphi_i, \phi_{ij}) = \mathcal{L}(v_i, \beta_{ij}) = \mathcal{L}(v_i^*, \beta_{ij}^*). \tag{7}$$

This is called *minimal replacement* (see, e.g., [1,2]). The translational gauge fields φ_i and ϕ_{ij} give rise to additional physical state quantities. These physical state quantities are the following translational field strengths

$$T_{ijk} = \phi_{ik,j} - \phi_{ij,k}, \qquad I_{ij} = -\varphi_{i,j} + \dot{\phi}_{ij} \tag{8}$$

or in terms of β_{ij} and v_i

$$T_{ijk} = \beta_{ik,j} - \beta_{ij,k}, \qquad I_{ij} = -v_{i,j} + \dot{\beta}_{ij} \tag{9}$$

called the dislocation density tensor (torsion tensor) and the dislocation current tensor, respectively. Since they are state quantities, they have to be gauge-invariant:

$$T_{ijk}^* = T_{ijk}, \qquad I_{ij}^* = I_{ij}. \tag{10}$$

In addition, they have to fulfill the translational Bianchi identities (square brackets indicate skewsymmetrization)

$$\epsilon_{jkl} T_{ijk,l} = 0, \qquad \dot{T}_{ijk} + 2\, I_{i[j,k]} = 0. \tag{11}$$

The first equation means that dislocations do not have sources and the second one represents that the circulation of the dislocation current is proportional to the time-derivative of the dislocation density.

Finally, we have four physical state quantities in the translation gauge theory of dislocations, namely the physical velocity v_i, the elastic distortion β_{ij}, the dislocation density tensor T_{ijk} and the dislocation current tensor I_{ij}. Thus, the Lagrangian is of the form: $\mathcal{L} = \mathcal{L}(v_i, \beta_{ij}, I_{ij}, T_{ijk})$. The Lagrangian of dislocation gauge theory is given by

$$\mathcal{L} = T - W = \frac{1}{2}\, p_i v_i - \frac{1}{2}\, \sigma_{ij}\beta_{ij} + \frac{1}{2}\, D_{ij} I_{ij} - \frac{1}{4}\, H_{ijk} T_{ijk}. \tag{12}$$

The canonical conjugate quantities read

$$p_i := \frac{\partial \mathcal{L}}{\partial v_i}, \qquad \sigma_{ij} := -\frac{\partial \mathcal{L}}{\partial \beta_{ij}}, \qquad D_{ij} := \frac{\partial \mathcal{L}}{\partial I_{ij}}, \qquad H_{ijk} := -2\frac{\partial \mathcal{L}}{\partial T_{ijk}}, \tag{13}$$

where p_i, σ_{ij}, I_{ij}, and H_{ijk} are the physical momentum vector, the force stress tensor, the dislocation momentum flux tensor, and the pseudomoment stress tensor, respectively. The Euler–Lagrange equations derived from the Lagrangian $\mathcal{L} = \mathcal{L}(v_i, \beta_{ij}, I_{ij}, T_{ijk})$ are given by

$$E_i^u(\mathcal{L}) = \mathrm{D}_t \frac{\partial \mathcal{L}}{\partial \dot{u}_i} + \mathrm{D}_j \frac{\partial \mathcal{L}}{\partial u_{i,j}} - \frac{\partial \mathcal{L}}{\partial u_i} = 0, \tag{14}$$

$$E_i^\varphi(\mathcal{L}) = \mathrm{D}_t \frac{\partial \mathcal{L}}{\partial \dot{\varphi}_i} + \mathrm{D}_j \frac{\partial \mathcal{L}}{\partial \varphi_{i,j}} - \frac{\partial \mathcal{L}}{\partial \varphi_i} = 0, \tag{15}$$

$$E_{ij}^\phi(\mathcal{L}) = \mathrm{D}_t \frac{\partial \mathcal{L}}{\partial \dot{\phi}_{ij}} + \mathrm{D}_k \frac{\partial \mathcal{L}}{\partial \phi_{ij,k}} - \frac{\partial \mathcal{L}}{\partial \phi_{ij}} = 0, \tag{16}$$

where D_t and D_i are the so-called total derivatives. Written in terms of the canonical conjugate quantities (13), Eqs. (14)–(16) take the form

$$\mathrm{D}_t p_i - \mathrm{D}_j \sigma_{ij} = 0 \qquad \text{(force balance)}, \tag{17}$$

$$\mathrm{D}_j D_{ij} + p_i = 0 \qquad \text{(momentum balance of dislocations)}, \tag{18}$$

$$\mathrm{D}_t D_{ij} + \mathrm{D}_k H_{ijk} + \sigma_{ij} = 0 \qquad \text{(stress balance of dislocations)}. \tag{19}$$

Equation (17) represents the force balance. The time-derivative of the physical momentum is the source of the force stress. It is a kind of continuity equation. Equation (18) is the momentum balance law of dislocations, where the physical momentum is the source of the dislocation momentum flux. Equation (19) represents the stress balance of dislocations. Thus, the force stress and the time derivative of the dislocation momentum flux are the sources of the pseudomoment stress. The coupled system (17)–(19) together with the Bianchi identities (11) are analogous to the inhomogeneous Maxwell equations of electromagnetic fields.

If we neglect the effects that the stresses may depend on the velocity v_i and the dislocation flux I_{ij} and that the momenta may depend on the elastic distortion β_{ij} and the dislocation density T_{ijk}, the anisotropic constitutive relations for a homogeneous material are given by

$$p_i = M_{ij} v_j + F_{ijk} I_{jk}, \tag{20}$$

$$D_{ij} = F_{kij} v_k + E_{ijkl} I_{kl}, \tag{21}$$

$$\sigma_{ij} = C_{ijkl} \beta_{kl} + B_{ijklm} T_{klm}, \tag{22}$$

$$H_{ijk} = 2B_{lmijk} \beta_{lm} + A_{ijklmn} T_{lmn}, \tag{23}$$

with the material tensors: M_{ij}, F_{ijk}, E_{ijkl}, C_{ijkl}, B_{ijklm} and A_{ijklmn}. They possess the symmetries:

$$M_{ij} = M_{ji}, \tag{24}$$
$$E_{ijkl} = E_{klij}, \tag{25}$$
$$C_{ijkl} = C_{klij}, \tag{26}$$
$$B_{ijklm} = -B_{ijkml}, \tag{27}$$
$$A_{ijklmn} = A_{lmnijk} = -A_{ikjlmn} = -A_{ijklnm}. \tag{28}$$

Substituting the constitutive relations (20)–(23) into the field equations (17)–(19), we obtain

$$M_{ij}\dot{v}_j + F_{ijk}\dot{I}_{jk} - C_{ijkl}\beta_{kl,j} - B_{ijklm}T_{lkm,j} = 0, \tag{29}$$
$$F_{kij}v_{k,j} + E_{ijkl}I_{kl,j} + M_{ij}v_j + F_{ijk}I_{jk} = 0, \tag{30}$$
$$\begin{aligned} F_{kij}\dot{v}_k + E_{ijkl}\dot{I}_{kl} + A_{ijklmn}T_{lmn,k} + 2B_{lmijk}\beta_{lm,k} \\ + B_{ijklm}T_{klm} + C_{ijkl}\beta_{kl} = 0, \end{aligned} \tag{31}$$

or just in terms of v_i and β_{ij}

$$M_{ij}\dot{v}_j + F_{ijk}(\ddot{\beta}_{jk} - \dot{v}_{j,k}) - C_{ijkl}\beta_{kl,j} + B_{ijklm}(\beta_{lm,kj} - \beta_{lk,mj}) = 0, \tag{32}$$
$$F_{kij}v_{k,j} + E_{ijkl}(\dot{\beta}_{kl,j} - v_{k,lj}) + M_{ij}v_j + F_{ijk}(\dot{\beta}_{jk} - v_{j,k}) = 0, \tag{33}$$
$$\begin{aligned} F_{kij}\dot{v}_k + E_{ijkl}(\ddot{\beta}_{kl} - \dot{v}_{k,l}) + A_{ijklmn}(\beta_{ln,km} - \beta_{lm,kn}) + 2B_{lmijk}\beta_{lm,k} \\ + B_{ijklm}(\beta_{km,l} - \beta_{kl,m}) + C_{ijkl}\beta_{kl} = 0. \end{aligned} \tag{34}$$

Equations (32)–(34) is a system of coupled partial differential equations for v_i and β_{ij}.

For an isotropic and centrosymmetric medium the constitutive relations (20)–(23) reduce to

$$p_i = \rho v_i, \tag{35}$$
$$D_{ij} = d_1\delta_{ij}I_{kk} + (d_2 + d_3)I_{ij} + (d_2 - d_3)I_{ji}, \tag{36}$$
$$\sigma_{ij} = \lambda\delta_{ij}\beta_{kk} + (\mu + \gamma)\beta_{ij} + (\mu - \gamma)\beta_{ji}, \tag{37}$$
$$H_{ijk} = c_1T_{ijk} + c_2(T_{jki} + T_{kij}) + c_3(\delta_{ij}T_{llk} + \delta_{ik}T_{ljl}), \tag{38}$$

where ρ is the mass density and with 9 material constants μ, λ, γ, $c_1, \ldots, c_3$ and $d_1, \ldots, d_3$. Then the field equations simplify to

Table 1 This table lists the material parameters c_I, d_I, μ, λ and γ for the translational gauge theory of dislocations.

	Edelen et al. [1,2]	Lazar and Anastassiadis [8]
μ	μ	μ
λ	λ	λ
γ	0	γ
c_1	$2s_1$	c_1
c_2	0	c_2
c_3	0	c_3
d_1	0	d_1
d_2	$s_2/2$	d_2
d_3	$s_2/2$	d_3

$$\rho \dot{v}_i - \lambda \beta_{j,ji} - (\mu+\gamma)\beta_{ij,j} - (\mu-\gamma)\beta_{ji,j} = 0, \tag{39}$$

$$\begin{aligned} & d_1(\dot{\beta}_{jj,i} - v_{jj,i}) + (d_2+d_3)(\dot{\beta}_{ij,j} - v_{i,jj}) \\ & \quad + (d_2-d_3)(\dot{\beta}_{ji,j} - v_{j,ji}) + \rho v_i = 0, \end{aligned} \tag{40}$$

$$\begin{aligned} & d_1\delta_{ij}(\ddot{\beta}_{kk} - \dot{v}_{k,k}) + (d_2+d_3)(\ddot{\beta}_{ij} - \dot{v}_{i,j}) + (d_2-d_3)(\ddot{\beta}_{ji} - \dot{v}_{j,i}) \\ & \quad + c_1(\beta_{ik,jk} - \beta_{ij,kk}) + c_2(\beta_{ji,kk} - \beta_{jk,ik} + \beta_{kj,ik} - \beta_{ki,jk}) \\ & \quad + c_3\big[\delta_{ij}(\beta_{lk,lk} - \beta_{ll,kk}) + (\beta_{kk,ji} - \beta_{kj,ki})\big] \\ & \quad + \lambda\delta_{ij}\beta_{kk} + (\mu+\gamma)\beta_{ij} + (\mu-\gamma)\beta_{ji} = 0. \end{aligned} \tag{41}$$

The differences between the gauge theories of dislocations are mainly the choice of the constitutive relations. In Table 1 we have listed the choice of the constitutive relations given by Edelen et al. [1,2]. It can be seen that Edelen's constitutive relations are very special.

3 Conservation Laws

The Lagrangian $\mathcal{L}$ depends on the first derivatives of the dependent fields v_i, φ_i and ϕ_{ij}. In this case the infinitesimal criterion of invariance [11] says that a Lie group G is a variational symmetry if and only if

$$\mathrm{pr}^{(1)}\mathbf{v}(\mathcal{L}) + \mathcal{L}\,(\mathrm{D}_i X_i + \mathrm{D}_t \tau) = 0, \tag{42}$$

where $\mathrm{pr}^{(1)}$ denotes the first prolongation (see [8]). After some standard calculations (see, e.g., [7, 8]) we find the conservation law in characteristic form

$$\mathrm{D}_i A_i + \mathrm{D}_t A_4 + Q^u_\alpha E^u_\alpha(\mathcal{L}) + Q^\varphi_\alpha E^\varphi_\alpha(\mathcal{L}) + Q^\phi_{\alpha\beta} E^\phi_{\alpha\beta}(\mathcal{L}) = 0, \tag{43}$$

where the characteristics are defined by

$$Q_\alpha^u = U_\alpha - X_j u_{\alpha,j} - \tau \dot{u}_\alpha, \tag{44}$$

$$Q_\alpha^\varphi = \Psi_\alpha - X_j \varphi_{\alpha,j} - \tau \dot{\varphi}_\alpha, \tag{45}$$

$$Q_{\alpha\beta}^\phi = \Phi_{\alpha\beta} - X_j \phi_{\alpha\beta,j} - \tau \dot{\phi}_{\alpha\beta}. \tag{46}$$

Here the infinitesimal generators are defined by

$$X_i(\boldsymbol{x}, t, \boldsymbol{u}, \boldsymbol{\varphi}, \boldsymbol{\phi}) := \left.\frac{\partial x_i^*}{\partial \boldsymbol{\varepsilon}}\right|_{\boldsymbol{\varepsilon}=0}, \qquad \tau(\boldsymbol{x}, t, \boldsymbol{u}, \boldsymbol{\varphi}, \boldsymbol{\phi}) := \left.\frac{\partial t^*}{\partial \boldsymbol{\varepsilon}}\right|_{\boldsymbol{\varepsilon}=0}, \tag{47}$$

$$U_\alpha(\boldsymbol{x}, t, \boldsymbol{u}, \boldsymbol{\varphi}, \boldsymbol{\phi}) := \left.\frac{\partial u_\alpha^*}{\partial \boldsymbol{\varepsilon}}\right|_{\boldsymbol{\varepsilon}=0}, \qquad \Psi_\alpha(\boldsymbol{x}, t, \boldsymbol{u}, \boldsymbol{\varphi}, \boldsymbol{\phi}) := \left.\frac{\partial \varphi_\alpha^*}{\partial \boldsymbol{\varepsilon}}\right|_{\boldsymbol{\varepsilon}=0}, \tag{48}$$

$$\Phi_{\alpha\beta}(\boldsymbol{x}, t, \boldsymbol{u}, \boldsymbol{\varphi}, \boldsymbol{\phi}) = \left.\frac{\partial \phi_{\alpha\beta}^*}{\partial \boldsymbol{\varepsilon}}\right|_{\boldsymbol{\varepsilon}=0}.$$

Therefore, if the Euler-Lagrange equations (14)–(16) are fulfilled, then Eq. (43) is a conservation law

$$\mathrm{D}_t A_4 + \mathrm{D}_i A_i = 0, \tag{49}$$

where A_i is the associated flux and A_4 is the conserved density. The corresponding components A_4 and A_i of the conservation law (49) are given by

$$A_4 = \mathcal{L}\tau + Q_\alpha^u \frac{\partial \mathcal{L}}{\partial \dot{u}_\alpha} + Q_\alpha^\varphi \frac{\partial \mathcal{L}}{\partial \dot{\varphi}_\alpha} + Q_{\alpha\beta}^\phi \frac{\partial \mathcal{L}}{\partial \dot{\phi}_{\alpha\beta}}, \tag{50}$$

$$A_i = \mathcal{L}X_i + Q_\alpha^u \frac{\partial \mathcal{L}}{\partial u_{\alpha,i}} + Q_\alpha^\varphi \frac{\partial \mathcal{L}}{\partial \varphi_{\alpha,i}} + Q_{\alpha\beta}^\phi \frac{\partial \mathcal{L}}{\partial \phi_{\alpha\beta,i}}. \tag{51}$$

4 Canonical Currents of Translations in Space and Time

The translation acts on the independent variables. The Lie-point group transformation of the translation in space and time is given by the formulas

$$x_i^* = x_i + \varepsilon_k \delta_{ki}, \qquad t^* = t + \varepsilon_4 \delta_{44}, \tag{52}$$

leaving the field variables unchanged

$$u_\alpha^* = u_\alpha, \qquad \varphi_\alpha^* = \varphi_\alpha, \qquad \phi_{\alpha\beta}^* = \phi_{\alpha\beta}. \tag{53}$$

The components of the generators (47) corresponding to the infinitesimal transformations (52) and (53) take the form

$$X_{ki} = \delta_{ki}, \qquad \tau = \delta_{44}, \qquad U_\alpha = 0, \qquad \Psi_\alpha = 0, \qquad \Phi_{\alpha\beta} = 0. \tag{54}$$

Using Eqs. (50), (51) and (54), we obtain for the translational density and flux quantities

$$A_{ki} = \mathcal{L}\,\delta_{ki} - u_{\alpha,k}\frac{\partial\mathcal{L}}{\partial u_{\alpha,i}} - \varphi_{\alpha,k}\frac{\partial\mathcal{L}}{\partial\varphi_{\alpha,i}} - \phi_{\alpha\beta,k}\frac{\partial\mathcal{L}}{\partial\phi_{\alpha\beta,i}}, \tag{55}$$

$$A_{k4} = -u_{\alpha,k}\frac{\partial\mathcal{L}}{\partial\dot{u}_{\alpha}} - \varphi_{\alpha,k}\frac{\partial\mathcal{L}}{\partial\dot{\varphi}_{\alpha}} - \phi_{\alpha\beta,k}\frac{\partial\mathcal{L}}{\partial\dot{\phi}_{\alpha\beta}}, \tag{56}$$

$$A_{4i} = -\dot{u}_{\alpha}\frac{\partial\mathcal{L}}{\partial u_{\alpha,i}} - \dot{\varphi}_{\alpha}\frac{\partial\mathcal{L}}{\partial\varphi_{\alpha,i}} - \dot{\phi}_{\alpha\beta}\frac{\partial\mathcal{L}}{\partial\phi_{\alpha\beta,i}}, \tag{57}$$

$$A_{44} = \mathcal{L} - \dot{u}_{\alpha}\frac{\partial\mathcal{L}}{\partial\dot{u}_{\alpha}} - \dot{\varphi}_{\alpha}\frac{\partial\mathcal{L}}{\partial\dot{\varphi}_{\alpha}} - \dot{\phi}_{\alpha\beta}\frac{\partial\mathcal{L}}{\partial\dot{\phi}_{\alpha\beta}}. \tag{58}$$

In terms of the physical momentum p_i, the dislocation momentum flux D_{ij}, the force stress σ_{ij} and pseudomoment stress H_{ijk} they read

$$P_{ki} := -A_{ki} = -\mathcal{L}\,\delta_{ki} - \sigma_{\alpha i}\,u_{\alpha,k} - D_{\alpha i}\,\varphi_{\alpha,k} + H_{\alpha\beta i}\,\phi_{\alpha\beta,k}, \tag{59}$$

$$\mathcal{P}_k := A_{k4} = -p_{\alpha}\,u_{\alpha,k} - D_{\alpha\beta}\,\phi_{\alpha\beta,k}, \tag{60}$$

$$S_i := A_{4i} = \sigma_{\alpha i}\,\dot{u}_{\alpha} + D_{\alpha i}\,\dot{\varphi}_{\alpha} - H_{\alpha\beta i}\,\dot{\phi}_{\alpha\beta}, \tag{61}$$

$$\mathcal{H} := -A_{44} = -\mathcal{L} + p_{\alpha}\,\dot{u}_{\alpha} + D_{\alpha\beta}\,\dot{\phi}_{\alpha\beta}. \tag{62}$$

Here P_{ki} is the canonical Eshelby stress tensor of dislocation gauge theory. The vector $\mathcal{P}_k$ is the canonical pseudomomentum vector, S_i denotes the canonical field intensity or Poynting vector. The scalar $\mathcal{H}$ is the canonical Hamiltonian. The conservation laws of the canonical pseudomomentum and the canonical energy are

$$\mathrm{D}_t\,\mathcal{P}_k - \mathrm{D}_i\,P_{ki} = 0, \tag{63}$$

$$\mathrm{D}_t\,\mathcal{H} - \mathrm{D}_i\,S_i = 0. \tag{64}$$

First we observe that indeed they are conservation laws because no sources like configurational forces appear. The total energy and momentum of the system are conserved. However, they are not conserved separately. There is an exchange of energy and momentum between the two subsystems (elastic continuum, dislocations). Therefore, in order to obtain configurational forces we have to investigate the two subsystems.

4.1 Elastic Subsystem

First we deal with the subsystem of the elastic continuum. This system is usually used in continuum mechanics. For the elastic subsystem we use the Lagrangian of the elastic continuum

$$\mathcal{L}_{\rm e} = \frac{1}{2}\, p_i v_i - \frac{1}{2}\, \sigma_{ij} \beta_{ij}. \tag{65}$$

So we obtain from (59)–(62) the following canonical quantities of the elastic subsystem

$$P^{(\rm e)}_{ki} := -\mathcal{L}_{\rm e}\, \delta_{ki} - \sigma_{\alpha i}\, u_{\alpha,k}, \tag{66}$$

$$\mathcal{P}^{(\rm e)}_{k} := -p_\alpha\, u_{\alpha,k}, \tag{67}$$

$$S^{(\rm e)}_{i} := \sigma_{\alpha i}\, \dot{u}_\alpha, \tag{68}$$

$$\mathcal{H}^{(\rm e)} := -\mathcal{L}_{\rm e} + p_\alpha\, \dot{u}_\alpha. \tag{69}$$

The local conservation laws (63) and (64) become now balance laws. The balance laws of the canonical pseudomomentum and canonical energy for the elastic subsystem read

$$\mathrm{D}_t \mathcal{P}^{(\rm e)}_{k} - \mathrm{D}_i P^{(\rm e)}_{ki} = \mathcal{F}_k, \tag{70}$$

$$\mathrm{D}_t \mathcal{H}^{(\rm e)} - \mathrm{D}_i S^{(\rm e)}_{i} = \mathcal{W}, \tag{71}$$

where the configurational force and power densities are

$$\mathcal{F}_k = p_\alpha\, \varphi_{\alpha,k} - \sigma_{\alpha\beta} \phi_{\alpha\beta,k}, \tag{72}$$

$$\mathcal{W} = -p_\alpha\, \dot{\varphi}_\alpha + \sigma_{\alpha\beta}\, \dot{\phi}_{\alpha\beta}. \tag{73}$$

4.2 Dislocation Subsystem

Now we investigate the subsystem of dislocations which is similar to Maxwell's system of electromagnetic fields. For the dislocation subsystem we use the Lagrangian of dislocations

$$\mathcal{L}_{\rm di} = \frac{1}{2}\, D_{ij} I_{ij} - \frac{1}{4}\, H_{ijk} T_{ijk}. \tag{74}$$

So we obtain for the canonical currents of the dislocation part

$$P^{(\rm d)}_{ki} := -\mathcal{L}_{\rm di}\, \delta_{ki} - D_{\alpha i}\, \varphi_{\alpha,k} + H_{\alpha\beta i}\, \phi_{\alpha\beta,k}, \tag{75}$$

$$\mathcal{P}^{(\rm d)}_{k} := -D_{\alpha\beta}\, \phi_{\alpha\beta,k}, \tag{76}$$

$$S^{(\rm d)}_{i} := D_{\alpha i}\, \dot{\varphi}_\alpha - H_{\alpha\beta i}\, \dot{\phi}_{\alpha\beta}, \tag{77}$$

$$\mathcal{H}^{(\rm d)} := -\mathcal{L}_{\rm di} + D_{\alpha\beta}\, \dot{\phi}_{\alpha\beta}. \tag{78}$$

The balance laws of the canonical pseudomomentum and canonical energy for the dislocation subsystem read

$$\mathrm{D}_t \mathcal{P}_k^{(\mathrm{d})} - \mathrm{D}_i P_{ki}^{(\mathrm{d})} = -\mathcal{F}_k, \tag{79}$$

$$\mathrm{D}_t \mathcal{H}^{(\mathrm{d})} - \mathrm{D}_i S_i^{(\mathrm{d})} = -\mathcal{W}. \tag{80}$$

As source terms we obtain the configurational force and the power densities with opposite sign because the complete system is conserved. Also we observe that the configurational force (72) does not have the correct form of the Peach–Koehler force. Moreover, all the canonical currents are not gauge-invariant.

5 Gauge-Invariant Currents

We have seen that the canonical translational currents neither produce the correct configurational forces nor they are gauge-invariant. Therefore, we have to construct the gauge-invariant currents from the canonical ones. From the physical point of view, the gauge-invariant currents are the correct ones. From the mathematical point of view, the method behind to construct the gauge-invariant currents is the so-called Belinfante–Rosenfeld procedure (see, e.g., [10]). The result is given by

$$P_{ki}^{(\mathrm{g})} = P_{ki} + \mathrm{D}_t(D_{\alpha i}\, \phi_{\alpha k}) - \mathrm{D}_\beta(H_{\alpha\beta i}\, \phi_{\alpha k}), \tag{81}$$

$$\mathcal{P}_k^{(\mathrm{g})} = \mathcal{P}_k + \mathrm{D}_\beta(D_{\alpha\beta}\, \phi_{\alpha k}), \tag{82}$$

$$S_i^{(\mathrm{g})} = S_i - \mathrm{D}_t(D_{\alpha i}\, \varphi_\alpha) + \mathrm{D}_\beta(H_{\alpha\beta i}\, \varphi_\alpha), \tag{83}$$

$$\mathcal{H}^{(\mathrm{g})} = \mathcal{H} - \mathrm{D}_\beta(D_{\alpha\beta}\, \varphi_\alpha), \tag{84}$$

where the gauge-invariant currents are defined by

$$P_{ki}^{(\mathrm{g})} := -\mathcal{L}\,\delta_{ki} - \sigma_{\alpha i}\,\beta_{\alpha k} + D_{\alpha i}\, I_{\alpha k} - H_{\alpha\beta i}\, T_{\alpha\beta k}, \tag{85}$$

$$\mathcal{P}_k^{(\mathrm{g})} := -p_\alpha\, \beta_{\alpha k} + D_{\alpha\beta}\, T_{\alpha\beta k}, \tag{86}$$

$$S_i^{(\mathrm{g})} := \sigma_{\alpha i}\, v_\alpha - H_{\alpha\beta i}\, I_{\alpha\beta}, \tag{87}$$

$$\mathcal{H}^{(\mathrm{g})} := -\mathcal{L} + p_\alpha\, v_\alpha + D_{\alpha\beta}\, I_{\alpha\beta}. \tag{88}$$

The conservation laws of the gauge-invariant pseudomomentum and Hamiltonian are given by

$$\mathrm{D}_t \mathcal{P}_k^{(\mathrm{g})} - \mathrm{D}_i P_{ki}^{(\mathrm{g})} = 0, \tag{89}$$

$$\mathrm{D}_t \mathcal{H}^{(\mathrm{g})} - \mathrm{D}_i S_i^{(\mathrm{g})} = 0. \tag{90}$$

5.1 Elastic Subsystem

The gauge-invariant currents of the elastic system read

$$P_{ki}^{(\text{e-g})} := -\mathcal{L}_\text{e}\,\delta_{ki} - \sigma_{\alpha i}\,\beta_{\alpha k}, \tag{91}$$
$$\mathcal{P}_k^{(\text{e-g})} := -p_\alpha\,\beta_{\alpha k}, \tag{92}$$
$$S_i^{(\text{e-g})} := \sigma_{\alpha i}\,v_\alpha, \tag{93}$$
$$\mathcal{H}^{(\text{e-g})} := -\mathcal{L}_\text{e} + p_\alpha\,v_\alpha. \tag{94}$$

With Eqs. (91)–(94), the balance laws of the elastic system are

$$\mathrm{D}_t\,\mathcal{P}_k^{(\text{e-g})} - \mathrm{D}_i\,P_{ki}^{(\text{e-g})} = \mathcal{F}_k^{(\text{g})}, \tag{95}$$
$$\mathrm{D}_t\,\mathcal{H}^{(\text{e-g})} - \mathrm{D}_i\,S_i^{(\text{e-g})} = \mathcal{W}^{(\text{g})}, \tag{96}$$

where

$$\mathcal{F}_k^{(\text{g})} = -p_i\,I_{ik} + \sigma_{ij}\,T_{ijk}, \tag{97}$$
$$\mathcal{W}^{(\text{g})} = \sigma_{ij}\,I_{ij}, \tag{98}$$

are the gauge-invariant configurational force and power densities. In Eq. (97) we have two contributions to the configurational force density. The first term is the force density caused by the dislocation current density I_{ik} in presence of the physical momentum field p_i (see also [3,4,13]). The second term gives the force density caused by the dislocation density T_{ijk} in presence of the force stress field σ_{ij}. This is just the static Peach–Koehler force density [12]. Thus, Eq. (97) is the dynamical Peach–Koehler force density. Equation (98) is the interaction power density of a moving dislocation in an elastic medium [3, 13]. The dynamical Peach-Koehler force $\mathcal{F}^{(\text{PK})} = -\mathcal{F}^{(\text{g})}$ is analogous to the Lorentz force of electromagnetic fields: $\mathcal{F}_k^{(\text{em})} = \rho E_k + \epsilon_{klm}\,j_l\,B_m$. The configurational power density as energy exchange between the two systems is similar to the electric power density in the Maxwell theory: $\mathcal{W}^{(\text{em})} = j_k\,E_k$ (ρ – charge density, j_i – electric current, E_i – electric field, B_i – magnetic field). If we substitute the anisotropic constitutive relations (20) and (22) into (97) and (98), we obtain

$$\mathcal{F}_k^{(\text{g})} = -(M_{ij}v_j + F_{ijl}I_{jl})\,I_{ik} + (C_{ijlm}\beta_{lm} + B_{ijlmn}T_{lmn})\,T_{ijk}, \tag{99}$$
$$\mathcal{W}^{(\text{g})} = (C_{ijkl}\beta_{kl} + B_{ijklm}T_{klm})\,I_{ij}. \tag{100}$$

Thus, for the anisotropic case core-core interactions between the dislocation densities and the dislocation currents give contributions in the Peach–Koehler force (99) and interaction power density (100).

5.2 Dislocation Subsystem

The gauge-invariant currents of the dislocation part are given by

$$P^{(\text{d-g})}_{ki} := -\mathcal{L}_{\text{di}}\,\delta_{ki} + D_{\alpha i}\,I_{\alpha k} - H_{\alpha\beta i}\,T_{\alpha\beta k}, \tag{101}$$

$$\mathcal{P}^{(\text{d-g})}_{k} := D_{\alpha\beta}\,T_{\alpha\beta k}, \tag{102}$$

$$S^{(\text{d-g})}_{i} := -H_{\alpha\beta i}\,I_{\alpha\beta}, \tag{103}$$

$$\mathcal{H}^{(\text{d-g})} := -\mathcal{L}_{\text{di}} + D_{\alpha\beta}\,I_{\alpha\beta}. \tag{104}$$

If we use Eqs. (75)–(78) and (101)–(104) for the calculation of the canonical and gauge-invariant force and power, the same results as in Eqs. (95), (96) and (97), (98) appear, but with opposite signs:

$$\mathrm{D}_t\,\mathcal{P}^{(\text{d-g})}_{k} - \mathrm{D}_i\,P^{(\text{d-g})}_{ki} = -\mathcal{F}^{(\text{g})}_{k}, \tag{105}$$

$$\mathrm{D}_t\,\mathcal{H}^{(\text{d-g})} - \mathrm{D}_i\,S^{(\text{d-g})}_{i} = -\mathcal{W}^{(\text{g})}. \tag{106}$$

6 Conclusion

We have presented a translational gauge theory of dislocations with a Lagrangian $\mathcal{L} = \mathcal{L}(v_i, \beta_{ij}, I_{ij}, T_{ijk})$ given in terms of the physical state quantities: physical velocity vector, elastic distortion tensor, dislocation current tensor and dislocation density tensor. Such a theory of dislocations possesses asymmetric stresses. Ten material parameters enter the isotropic constitutive relations. We have shown that the Peach–Koehler force is the source for the gauge-invariant Eshelby stress tensor of the elastic and dislocation subsystems. The Peach–Koehler force is a material force due to the incompatibility tensors (dislocation current tensor, dislocation density tensor). From the mathematical point of view, the Peach–Koehler force is analogous to the Lorentz force in the Maxwell theory of electromagnetic fields. The gauge theory of dislocations is mathematically analogue to the Maxwell equations of electrodynamics. The gauge group of the dislocation theory is the three-dimensional translation group $T(3)$ and, on the other hand, the unitary group $U(1)$ is the gauge group of the Maxwell theory. Moreover, the Maxwell theory possesses a Lorentz symmetry with respect to the speed of light. Due to several speeds of sound, the dislocation theory does not have, in general, a Lorentz symmetry.

Acknowledgement

The authors have been supported by an Emmy-Noether grant of the Deutsche Forschungsgemeinschaft (Grant No. La1974/1-2).

References

1. Edelen, D.G.B. and Lagoudas, D.C. *Gauge Theory and Defects in Solids*. North-Holland, Amsterdam, 1988.
2. Kadić, A. and Edelen, D.G.B., *A Gauge Theory of Dislocations and Disclinations*, Lecture Notes in Physics, Vol. 174. Springer, Berlin, 1983.
3. Kluge, G., Über den Zusammenhang der allgemeinen Versetzungstheorie mit dem Cosserat-Kontinuum. *Wissen. Z. Techn. Hochschule Otto von Guericke Magdeburg* **13**, 1969, 377–380.
4. Kosevich, A.M., Crystal dislocations and the theory of elasticity. In: *Dislocations in Solids, Vol. 1*, F.R.N. Nabarro (Ed.). North-Holland, 1979, pp. 33–165.
5. Lazar, M., Dislocation theory as a 3-dimensional translation gauge theory. *Ann. Phys. (Leipzig)* **9**, 2000, 461–473.
6. Lazar, M., An elastoplastic theory of dislocations as a physical field theory with torsion. *J. Phys. A: Math. Gen.* **35**, 2002, 1983–2004.
7. Lazar, M., On conservation and balance laws in micromorphic elastodynamics. *J. Elast.* **88**, 2007, 63–78.
8. Lazar, M. and Anastassiadis, C., The gauge theory of dislocations: Conservation and balance laws. *Phil. Mag.* **88**, 2008, 1673–1699.
9. Lazar, M. and Anastassiadis, C., The gauge theory of dislocations: static solutions of screw and edge dislocations. *Phil. Mag.* **89**, 2009, 199–231.
10. Maugin, G.A., *Material Inhomogeneities in Elasticity*, Chapman and Hall, London, 1993.
11. Olver, P.J., *Applications of Lie Groups to Differential Equations*, Springer, New York, 1986.
12. Peach, M.O. and Koehler, J.S., Forces extended on dislocations and the stress fields produced by them. *Phys. Rev.* **80**, 1950, 436–439.
13. Schaefer, H., Maxwell-Gleichungen, Energiesatz und Lagrange-Dichte in der Kontinuumstheorie der Versetzungen. *Acta Mech.* **10**, 1970, 59–66.

Configurational Forces in Continuous Theories of Elastic Ferroelectrics

Vassilios K. Kalpakides and Antonios I. Arvanitakis

Abstract Domain walls in a ferroelectric crystal are considered as sharp interfaces, so their motion is governed by field equations, jump conditions and an appropriate kinetic relation between the domain wall velocity and the driving force. In this article, a regularized version of the sharp-interface theory in ferroelectrics is presented, by introducing a level set function that changes sign from domain to domain smoothly and thus eliminating discontinuities. It is proved that considering level set functions as constitutive variables in the energy functional, the driving forces that move domain walls are configurational forces obeying the canonical momentum equation. A new, recently proposed differential equation is used to describe the evolution of the level set function which keeps level set function closer to a signed distance function as possible. Theoretical considerations and numerical simulations show that configurational forces are closely related to the level set description of sharp interface theories in solids. Moreover, it is displayed that inhomogeneity forces drive the system successfully to the typical domain structure of elastic ferroelectrics.

1 Introduction

The electromechanical properties of ferroelectric materials are closely related to their microstructure, i.e. the domain structure. In a single crystal grain of a ferroelectric material there are domains of uniform polarization. The boundary between two distinct domains is called domain wall. A domain wall is an interface that separates phases of the material corresponding to different oriented polarization states.

In this article, at a first stage the sharp interface theory for elastic ferroelectrics is developed. Based on this, a theory of domain wall motion by introducing an ad-

---

Vassilios K. Kalpakides · Antonios I. Arvanitakis
Department of Materials Science and Engineering, University of Ioannina, GR-45110 Ioannina, Greece; e-mail: {vkalpak, aarvanit}@cc.uoi.gr

P. Steinmann (ed.), IUTAM Symposium on Progress in the Theory and Numerics of Configurational Mechanics, 229–238.

ditional level set function is presented. The domain wall is the zero level set of this function. In order to avoid discontinuities, domain wall is taken to be a transition layer using a small regularization parameter that controls domain wall thickness.

Considering level set functions as constitutive variables, it is proved that forces acting on domain walls are inhomogeneity forces and their expression enters the canonical momentum (pseudomomentum) equation. Soft lead zirconate titanate was chosen to perform computer simulations on ferroelectric domain structure.

2 Sharp Interfaces and Configurational Forces in Ferroelectrics

We assume an elastic ferroelectric body occupying an open set $\Omega \subset \Re^2$. The region Ω is separated by a domain wall S_t in two subregions Ω_t^- and Ω_t^+ corresponding to domains of different spontaneous polarization. The unit normal $\mathbf{n}$ to the domain wall is directed from subregion Ω_t^- to Ω_t^+.

The energy of the material in each region is assumed to be that of linear piezoelectric materials

$$W(\mathsf{e}, \mathbf{E}) = \frac{1}{2}(\mathsf{e} - \mathsf{e}^0) : [\mathsf{C}(\mathsf{e} - \mathsf{e}^0)] - (\mathsf{e} - \mathsf{e}^0) : \mathsf{d}^{\mathrm{T}}\mathbf{E} - \frac{1}{2}\mathbf{E} \cdot (\mathsf{k}\mathbf{E}) - \mathbf{P}^0 \cdot \mathbf{E}, \quad (1)$$

where C, d, k denote the elastic, piezoelectric and dielectric tensors of the material, respectively. Also, $\mathbf{P}^0$ is the spontaneous polarization vector and e^0 the spontaneous strain. The electric field $\mathbf{E}$ and strain e are given by the kinematical relations

$$\begin{aligned} \mathbf{E} &= -\nabla\phi, \\ \mathsf{e} &= \frac{1}{2}(\nabla\mathbf{u} + (\nabla\mathbf{u})^{\mathrm{T}}), \end{aligned} \quad (2)$$

where ϕ and $\mathbf{u}$ denote the electric potential and mechanical displacements, respectively. Two distinct energy densities are considered corresponding to the two disjoint regions Ω_t^- and Ω_t^+, i.e.

$$W = \begin{cases} W_1, & \text{in } \Omega_t^- \\ W_2, & \text{in } \Omega_t^+ \end{cases} \quad (3)$$

The branches of the energy correspond to different phases of the material associated with the orientation of the spontaneous polarization in each subregion. Within the smooth regions the standard equations and constitutive relations hold

$$\operatorname{div}\sigma = 0, \quad \operatorname{div}\mathbf{D} = 0, \quad \text{in} \quad \Omega_t^+ \cup \Omega_t^-, \quad (4)$$

$$\sigma = \frac{\partial W}{\partial \mathsf{e}}, \quad \mathbf{D} = -\frac{\partial W}{\partial \mathbf{E}}, \quad (5)$$

where σ is the Cauchy stress tensor and $\mathbf{D}$ the electric displacement vector. On domain wall S_t one has to deal with the jump conditions

$$[\![\sigma]\!] \cdot \mathbf{n} = 0, \quad [\![\mathbf{D}]\!] \cdot \mathbf{n} = 0, \tag{6}$$

where $[\![\bullet]\!] = \bullet^+ - \bullet^-$ is the jump of any field across the domain wall S_t.

The interface motion is determined by the normal velocity $V = V(\mathbf{x}, t) = \mathbf{v} \cdot \mathbf{n}$. This motion is accompanied by energy dissipation which can be computed as following [1, 7]

$$\Delta = \int_{S_t} f\, V ds \geq 0, \tag{7}$$

where $f = -(\mathbf{n} \cdot [\![\Sigma]\!]\, \mathbf{n})$, with $\Sigma = W\mathsf{I} - (\nabla \mathbf{u})\sigma + \mathbf{E} \otimes \mathbf{D}$ denoting the Eshelby stress tensor for electroelasticity.

Abeyaratne and Knowles [1] have proved that the solution uniqueness of such problems needs a kinetic relation

$$V = V(f, \mathbf{n}). \tag{8}$$

Finally, gathering all the above information one can solve numerically the problem of domain wall motion in a ferroelectric crystal. This approach is called local approach [5]. Approaches using configurational forces in interface motion problems can be found in literature (see for instance [2, 7]).

3 The Level-Set Method

The main idea behind level set method is based on the implicit interpretation of a surface (or curve) [8, 9]. A smooth, scalar function, $\psi(\mathbf{x}, t)$, $\mathbf{x} \in \Omega$ is introduced with the following property

$$\Omega_t^+ = \{\mathbf{x} \in \Omega \mid \psi(\mathbf{x}, t) > 0\}, \quad \Omega_t^- = \{\mathbf{x} \in \Omega \mid \psi(\mathbf{x}, t) < 0\},$$
$$S_t = \{\mathbf{x} \in \Omega \mid \psi(\mathbf{x}, t) = 0\}. \tag{9}$$

Notice that the interface S_t is the zero level set of the function ψ; for this reason ψ is referred to as level set function. A simple example of such a function is the signed distance function from the interface S_t. Assuming that the interface is moving with a given velocity field $\mathbf{v}(\mathbf{x}, t)$, the evolution equation for the level set function is of Hamilton–Jacobi type [8]

$$\frac{\partial \psi}{\partial t} - V|\nabla \psi| = 0. \tag{10}$$

One can set the signed distance function as initial condition for the above equation. Nevertheless, solving this initial value problem the level set function after some time steps will not remain a signed distance function leading to numerical instabilities. To remedy this, one has to solve an additional initial value problem after some time steps [8]. Alternatively, one can use a variational level set method [4, 11] given by the initial boundary value problem

$$\begin{aligned}\frac{\partial\psi}{\partial t} &= \mu\left(\Delta\psi - \kappa\right) + V|\nabla\psi|, \quad \text{in} \quad \Omega,\\ &\psi(\mathbf{x},0) = \psi_0(\mathbf{x}), \quad \forall\mathbf{x}\in\Omega,\\ &\nabla\psi\cdot\mathbf{n} = 0, \quad \text{on} \quad \partial\Omega,\end{aligned} \tag{11}$$

where μ is a penalizing parameter. The first term on the right-hand side of eq. (11) keeps the level set function close to a signed distance function, while the second term moves the level set according to a kinetic relation. Furthermore, the unit normal $\mathbf{n}$ and the curvature κ to the curve are given by

$$\mathbf{n} = \frac{\nabla\psi}{|\nabla\psi|}, \quad \kappa = \nabla\cdot\left(\frac{\nabla\psi}{|\nabla\psi|}\right), \tag{12}$$

respectively.

4 Continuous Energy and Configurational Forces

4.1 The Energy Function and the Level Set

For the needs of the level set method, one can write eq. (3) as an interpolation between the two branches W_1 and W_2:

$$W(\mathsf{e},\mathbf{E},h) = W_2(\mathsf{e},\mathbf{E}) + h(W_1(\mathsf{e},\mathbf{E}) - W_2(\mathsf{e},\mathbf{E})), \tag{13}$$

where $h = H(\psi)$ is the Heaviside function. Next one can introduce a continuous form of the energy by the use of a regularized version of the Heaviside function

$$H_\varepsilon(\psi) = \begin{cases} 1, & \psi > \varepsilon,\\ 0, & \psi < -\varepsilon,\\ \dfrac{1}{2}\left(1 + \dfrac{\psi}{\varepsilon} + \dfrac{1}{\pi}\sin\left(\dfrac{\pi\psi}{\varepsilon}\right)\right), & |\psi| \le \varepsilon, \end{cases} \tag{14}$$

where ε is a parameter defining the width of the layer $S_\varepsilon = \{\mathbf{x}\in\Omega \mid |\psi(\mathbf{x},t)| \le \varepsilon\}$. Then, the energy function and the constitutive relations become

$$\begin{aligned} W_\varepsilon(\mathsf{e},\mathbf{E},h_\varepsilon) &= W_2(\mathsf{e},\mathbf{E}) + h_\varepsilon(W_1(\mathsf{e},\mathbf{E}) - W_2(\mathsf{e},\mathbf{E})),\\ \sigma_\varepsilon &= \frac{\partial W_\varepsilon}{\partial\mathsf{e}} = \sigma_2 + h_\varepsilon(\sigma_1 - \sigma_2), \quad \sigma_i = \frac{\partial W_i}{\partial\mathsf{e}},\\ \mathbf{D}_\varepsilon &= -\frac{\partial W_\varepsilon}{\partial\mathbf{E}} = \mathbf{D}_2 + h_\varepsilon(\mathbf{D}_1 - \mathbf{D}_2), \quad \mathbf{D}_i = \frac{\partial W_i}{\partial\mathbf{E}}, \end{aligned} \tag{15}$$

where $h_\varepsilon = H_\varepsilon(\psi)$. Notice that the non-continuity of the energy function on the interface has been removed. Instead, it is now an inhomogeneous function in a narrow layer around the interface.

Estimating the dissipation rate, one can define the driving force that moves the level set function [3],

$$\Delta_\varepsilon = \int_{\partial\Omega} \boldsymbol{\sigma}\mathbf{n}\cdot\mathbf{v}ds - \int_{\partial\Omega} \mathbf{D}\cdot\mathbf{n}\dot{\varphi}dv - \frac{d}{dt}\int_{\Omega} W_\varepsilon dv \geq 0 \Rightarrow$$

$$\Delta_\varepsilon = -\int_{\Omega} \frac{\partial W_\varepsilon}{\partial\psi}\frac{\partial\psi}{\partial t}dv = -\int_{\Omega} \frac{\partial W_\varepsilon}{\partial h_\varepsilon}\delta_\varepsilon(\psi)V\,|\nabla\psi|dv, \tag{16}$$

where $\delta_\varepsilon(\psi) = H'_\varepsilon(\psi)$ is the regularized Dirac function. Taking the view of Hou et al. [3], we define the driving force as

$$f_\varepsilon = -\frac{\partial W_\varepsilon}{\partial h_\varepsilon}, \quad \text{in} \quad S_\varepsilon. \tag{17}$$

The last step is to determine an appropriate kinetic relation between the above driving force and the normal velocity component of the zero level set function. In this article the following expression is adopted

$$V = M_1 f_\varepsilon + M_2 f_\varepsilon |n_1 - n_2|, \tag{18}$$

where M_1, M_2 are the mobilities of the isotropic and anisotropic term, respectively and n_1, n_2 the components of $\mathbf{n}$ given by eq. (12). Thus, using eqs. (15) and (18), one can solve the field equations (4) and (11) accompanied by appropriate boundary conditions to determine the domain wall motion.

To reveal the role that the configurational forces might play in this problem, eq. (13) can be written in the following form

$$\bar{W}(\mathsf{e}, \mathbf{E}, \mathbf{x}) = W_\varepsilon(\mathsf{e}, \mathbf{E}, h_\varepsilon), \tag{19}$$

so as to underline that the energy function depends on $\mathbf{x}$ explicitly. Then, the equilibrium equation for configurational forces [6] holds

$$\operatorname{div}\Sigma + \mathbf{f}_{\mathrm{inh}} = 0 \tag{20}$$

where $\mathbf{f}_{\mathrm{inh}}$ represents the inhomogeneity forces. A simple calculation gives

$$\mathbf{f}_{\mathrm{inh}} = -\frac{\partial\bar{W}}{\partial\mathbf{x}} = -\frac{\partial W_\varepsilon}{\partial h_\varepsilon}H'_\varepsilon(\psi)\frac{\partial\psi}{\partial\mathbf{x}} = -\frac{\partial W_\varepsilon}{\partial h_\varepsilon}\delta_\varepsilon(\psi)\nabla\psi. \tag{21}$$

A direct conclusion is that the level set function introduces inhomogeneities in a material. Combining eqs. (17), (20) and (21) one can write

$$F_n = \int_\Omega f_\varepsilon \delta_\varepsilon(\psi)|\nabla\psi|dv = \int_\Omega (\mathbf{f}_{\text{inh}} \cdot \mathbf{n})dv = -\int_\Omega (\text{div}\,\Sigma \cdot \mathbf{n})dv, \tag{22}$$

where F_n is the resultant driving force in the layer. The above relation shows that the concept of material force can be naturally embedded in the framework of level set method. Furthermore, it provides an alternative way to compute the driving force by the use of the Eshelby stress tensor.

4.2 Two Level Set Functions

To study multi-phase problems, one has to introduce several level set functions and write an appropriate expression for the total energy. Motivated by observed domain patterns in a two dimensional ferroelectric sample, we introduce two distinct level set functions ψ_1, ψ_2. Thus, the body can be separated in four different phases corresponding to the four possible orientations of polarization in 2-d ferroelectrics with tetragonal crystal symmetry

$$\begin{aligned}
S_t^1 =\ & \{\mathbf{x} \in \Omega | \psi_1(\mathbf{x}, t) = 0\}, \quad S_t^2 = \{\mathbf{x} \in \Omega | \psi_2(\mathbf{x}, t) = 0\} \\
& \Omega_t^{++} = \{\mathbf{x} \in \Omega \mid \psi_1(\mathbf{x}, t) > 0 \ \text{and} \ \psi_2(\mathbf{x}, t) > 0\}, \\
& \Omega_t^{--} = \{\mathbf{x} \in \Omega | \ \psi(\mathbf{x}, t) < 0 \ \text{and} \ \psi_2(\mathbf{x}, t) < 0\} \\
& \Omega_t^{+-} = \{\mathbf{x} \in \Omega \mid \psi_1(\mathbf{x}, t) > 0 \ \text{and} \ \psi_2(\mathbf{x}, t) < 0\}, \\
& \Omega_t^{-+} = \{\mathbf{x} \in \Omega | \ \psi(\mathbf{x}, t) < 0 \ \text{and} \ \psi_2(\mathbf{x}, t) > 0\}.
\end{aligned} \tag{23}$$

The above regions correspond to polarization phases $\mathbf{P}^{++} = (0 \quad p_0)^{\mathrm{T}}$, $\mathbf{P}^{--} = (0 \quad -p_0)^{\mathrm{T}}$, $\mathbf{P}^{-+} = (p_0 \quad 0)^{\mathrm{T}}$, $\mathbf{P}^{+-} = (-p_0 \quad 0)^{\mathrm{T}}$. As before, a regularized energy containing the four phases can be formulated as

$$\begin{aligned}
W_\varepsilon(\mathsf{e}, \mathbf{E}, h_\varepsilon^1, h_\varepsilon^2) = (W_2 - W_1)h_\varepsilon^1 + (W_3 - W_1)h_\varepsilon^2 \\
+ (W_4 + W_1 - W_2 - W_3)h_\varepsilon^1 h_\varepsilon^2 + W_1,
\end{aligned} \tag{24}$$

where W_1, W_2, W_3, W_4 the energy expression of each region and $h_\varepsilon^1 = H_\varepsilon(\psi_1)$ and $h_\varepsilon^2 = H_\varepsilon(\psi_2)$.

Our next step is to determine the driving force that moves each level set function. Due to the non-homogeneous energy in eq. (24), one can write

$$\begin{aligned}
\frac{\partial \bar{W}}{\partial \mathbf{x}} &= \frac{\partial W_\varepsilon}{\partial h_\varepsilon^1}\delta_\varepsilon(\psi_1)\nabla\psi_1 + \frac{\partial W_\varepsilon}{\partial h_\varepsilon^2}\delta_\varepsilon(\psi_2)\nabla\psi_2 \\
&= \mathbf{f}_{\text{inh}}^1 + \mathbf{f}_{\text{inh}}^2, \quad f_\varepsilon^1 = -\frac{\partial W_\varepsilon}{\partial h_\varepsilon^1}, \quad f_\varepsilon^2 = -\frac{\partial W_\varepsilon}{\partial h_\varepsilon^2}.
\end{aligned} \tag{25}$$

Notice in eq. (25) the crucial role of Dirac function δ_ε which confines the configurational forces to act within the layer S_ε, where the inhomogeneities are present. The field equations and boundary conditions for the four phases problem are

$$\begin{aligned}
&\nabla\sigma_\varepsilon = 0, \quad \nabla\cdot\mathbf{D}_\varepsilon = 0, \quad \text{in} \quad \Omega \\
&\frac{\partial\psi_i}{\partial t} = \mu_i\left(\Delta\psi_i - \kappa_i\right) + V_i\left|\nabla\psi_i\right|, \quad \text{in} \quad \Omega, \quad i = 1,2 \\
&\sigma_\varepsilon\mathbf{n} = 0, \quad \mathbf{D}_\varepsilon\cdot\mathbf{n} = 0, \quad \nabla\psi_i\cdot\mathbf{n} = 0, \quad \text{on} \quad \partial\Omega,
\end{aligned} \tag{26}$$

where σ_ε, $\mathbf{D}_\varepsilon$ are taken by the constitutive eqs. (5). To complete this job, one has to choose appropriate kinetic relations. In this case the following isotropic forms

$$V_1 = M_1' f_\varepsilon^1, \quad V_2 = M_2' f_\varepsilon^2, \tag{27}$$

have been adopted.

5 Computational Results

In this section, the above system is solved numerically. To this end, we have made use of soft lead zirconate titanate the material parameters of which are taken [7],

$$\mathsf{C} = \begin{pmatrix} 12.6 & 5.3 & 0 \\ 5.3 & 11.7 & 0 \\ 0 & 0 & 3.53 \end{pmatrix} 10^{10}\,\text{Pa}, \quad \mathsf{d} = \begin{pmatrix} 0 & 0 & 17 \\ -6.5 & 23.3 & 0 \end{pmatrix} 10^{-2}\,\text{C/m}^2,$$

$$\mathsf{k} = \begin{pmatrix} 1.51 & 0 \\ 0 & 1.3 \end{pmatrix} 10^{-8}\,\text{C/Vm}, \quad \mathsf{e}^0 = \begin{pmatrix} -0.0039 \\ 0.0076 \\ 0 \end{pmatrix}, \quad \mathbf{P}^0 = \begin{pmatrix} 0 \\ 0.2 \end{pmatrix}\,\text{C/m}^2.$$

To describe the other three phases, apart from the modification of the polarization vector, one has to change the piezoelectric tensor and the spontaneous strain. The rest of the parameters used in this paper are: $\varepsilon \leq 0.2$, $\mu \simeq 10^{-6}\,1/\text{s}$. The field equations in each simulation were solved using finite element discretization in space (Lagrange quadratic elements) and finite differences in time (forward difference discretization).

In Figure 1, a domain of opposite polarization grows inside a mono-domain crystal under external electric field. The kinetic relation that has been used in this computation is given by eq. (18) with $M_1 = 2\cdot 10^{-4}\,\text{m}^3/\text{sN}$ and $M_2 = 0$. Notice that the domain grows vertically to reach the boundary. Then under the motion of 180° domain walls expands to cover the whole region of the sample.

In Figure 2, a domain of horizontal polarization grows in a crystal with vertical polarization under compressive stress. The kinetic relation used in this computation is given by eq. (18) with $M_1 = 2\cdot 10^{-4}\,\text{m}^3/\text{sN}$ and $M_2 = 20\cdot 10^{-4}\,\text{m}^3/\text{sN}$. This type of kinetic relation forces the domain to grow faster at 45° from the applied

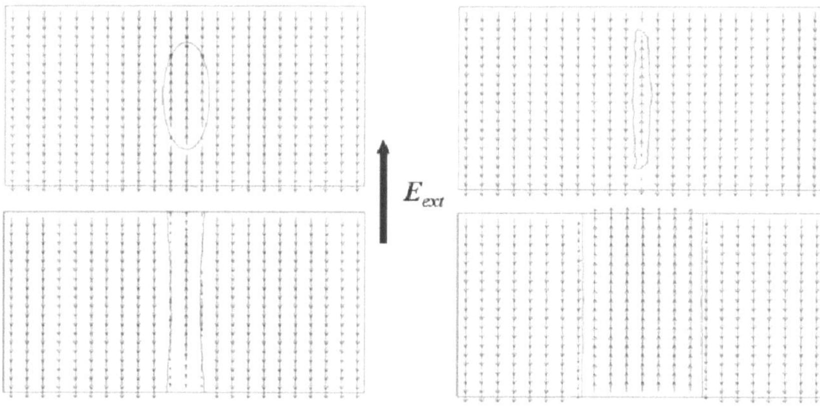

Fig. 1 Domain of opposite vertical polarization grows in a ferroelectric mono-domain crystal under external electric field. Zero level set represents 180° domain walls.

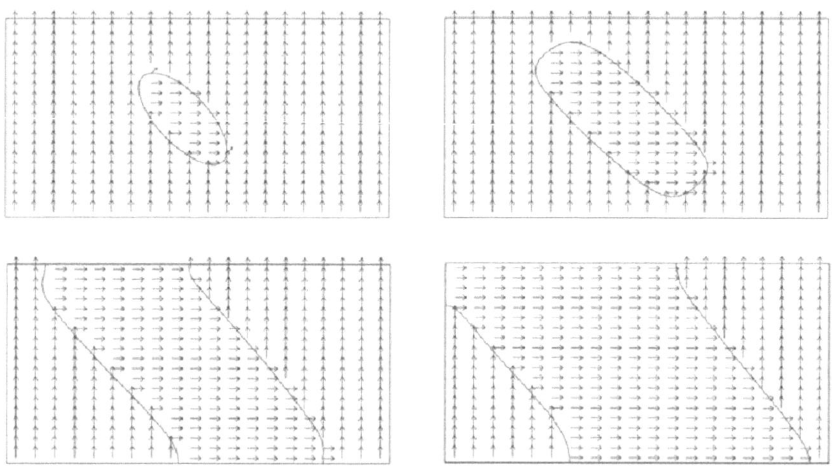

Fig. 2 Domain of horizontal polarization grows in a ferroelectric mono-domain crystal under compressive stress. Zero level set represents 90° domain walls.

stress axis, since such behavior is experimentally observed. The domain grows until it reaches the boundary, then under the motion of 90° domain walls expands up to occupy the whole crystal.

In Figure 3a, the profile of configurational forces on an 180° domain wall is displayed. As one can see, configurational forces take non zero values within the narrow layer representing the domain wall. In Figure 3b, the configurational traction norm is plotted at a specified height from one edge of the crystal up to the other at various instants. Note that the peak is on the zero level set and the inhomogeneity travels through the material up to the right end where it vanishes.

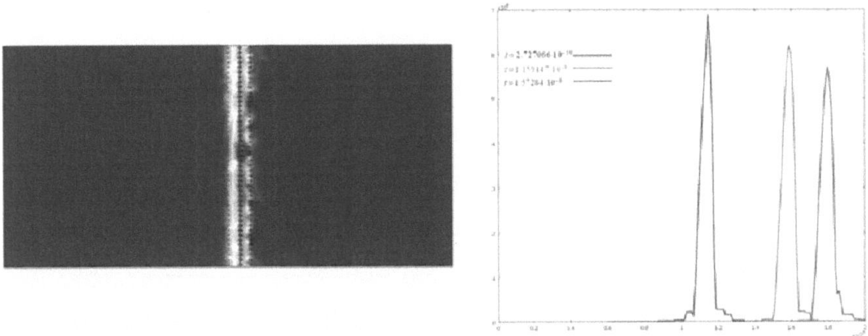

Fig. 3 Profile norm of material forces on a moving 180° domain wall and horizontal cross-section plot of the norm at different times.

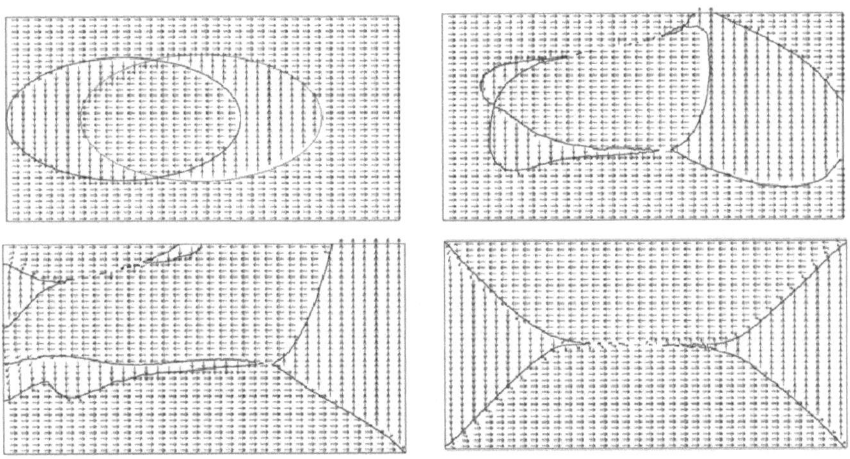

Fig. 4 The evolution of the level set functions in the case of a ferroelectric with four phases.

Also, we simulate the four phases of soft lead zirconate titanate as developed in Section 4.2. In Figure 4, the evolution of an arbitrary domain structure is illustrated. We choose two intersecting ellipses as initial values for the two level set functions and $M_1' = M_2' = 2 \cdot 10^{-4}\,\mathrm{m}^3/\mathrm{sN}$. The initial structure evolves until it reaches an equilibrium state. This vortex domain structure is quite typical to ferroelectrics and it is in agreement with experimental and theoretical results [10].

Last, in Figure 5 the electric energy density and the driving forces on each level set are plotted as functions of time. As expected, during domain evolution depicted in Figure 4, the energy of the material reaches a minimum and configurational forces tend to zero.

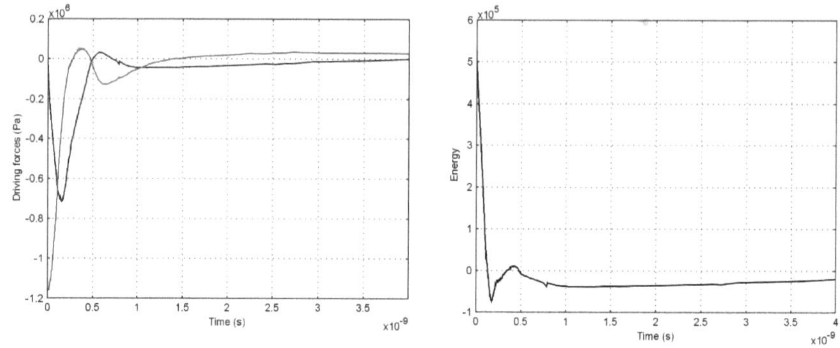

Fig. 5 Driving forces on level set functions and electric energy density with time.

Acknowledgement

Financial support from the State Scholarships Foundation of Greece (I.K.Y.) is gratefully acknowledged.

References

1. Abeyaratne, R. and Knowles, J.K., On the driving traction acting on a surface of strain discontinuity in a continuum. *J. Mech. Phys. Solids* **38**, 1990, 345–360.
2. Balassas, K.G. and Kalpakides, V.K., The equilibrium of material forces in a 1D phase transition problem. *Comput. Methods Appl. Engrg.* **196**, 2007, 2161–2172.
3. Hou. T., Rosakis, P. and LeFloch, P.G., A level-set approach to the computation of twinning and phase-transition dynamics. *J. Comput. Phys.* **150**, 1999, 302–331.
4. Li, C., Xu, C., Gui, C. et al., Level set evolution without re-initialization: A new variational formulation. In *Proceedings of the 2005 IEEE Computer Society Conference on Computer Vision and Pattern Recognition (CVPR'05)*, 2005.
5. Loge, R.E. and Suo, Z., Nonequilibrium thermodynamics of ferroelectric domain evolution. *Acta Mater.* **44**, 1996, 3429–3438.
6. Maugin, G.A., *Material Inhomogeneities in Elasticity*. Chapman and Hall, London, 1993.
7. Mueller, R., Gross, D. and Lupascu, D.C., Driving forces on domain walls in ferroelectric materials. *Comput. Mater. Sci.* **35**, 2006, 42–52.
8. Osher, S. and Fedkiw, R., *Level Set Methods and Dynamic Implicit Surfaces*. Springer, New York, 2002.
9. Osher, S. and Sethian, J.A., Front propagating with curvature-dependent speed: Algorithms based on Hamilton–Jacobi formulations. *J. Comput. Phys.* **79**, 1988, 22–49.
10. Schrade, D., Mueller, R., Xu, B.X. et al., Domain evolution in ferroelectric materials: A continuum phase field model and finite element implementation. *Comput. Methods Appl. Engrg.* **196**, 2007, 4365–4374.
11. Zhao, H.K., Chan, T., Merriman, B. et al., A variational level set approach to multiphase motion. *J. Comput. Phys.* **127**, 1996, 179–195.

A Variationally Consistent Approach for Crack Propagation Based on Configurational Forces

Jörn Mosler

Abstract This paper is concerned with a variationally consistent approach suitable for the analysis of cracking in brittle materials. In line with the pioneering works by Griffith, it is assumed that a crack propagates, if this is energetically favorable. However, in order to bypass the well-known defects of Griffith's original idea such as the requirement of a pre-existing crack, a modified energy-based criterion is proposed. In contrast to Griffith and similar to Francfort and Marigo, the novel cracking model is based on a finite crack extension. More precisely, new crack surfaces form, if this leads to a reduction in energy within a finite (but not global) neighborhood. The features of the advocated model are critically analyzed.

1 Introduction

Since the pioneering work [1] by Griffith has been published, energy-based criteria have been frequently applied to the analysis of crack propagation. The underlying mathematically and physically sound idea is that a crack propagates, if this is energetically favorable. For the derivation of this strategy, Griffith analyzed the force F associated with a (material) variation of a crack tip and postulated that a new crack segment would form, if F reached a critical threshold, i.e., $F > F_{\text{crit}}$, with F_{crit} denoting a material parameter. Clearly, F represents a material or configurational force, cf. [2]. Unfortunately, Griffith's model shows some problems.

Griffith's model was critically analyzed in [3, 4]. Francfort and Marigo showed that Griffith's local criterion cannot predict crack initiation, and in its original version, it does neither provide any information concerning the length nor the direction of the new crack segment. Based on this observation, Francfort and Marigo advocated a global criterion. Conceptually identical to the ideas proposed by Griffith, they

Jörn Mosler
GKSS Research Centre, Institute for Materials Research, Materials Mechanics, D-21502 Geesthacht, Germany; e-mail: joern.mosler@gkss.de

P. Steinmann (ed.), IUTAM Symposium on Progress in the Theory and Numerics of Configurational Mechanics, 239–247.

considered a minimization problem of the type

$$\inf_{\Gamma_c, \boldsymbol{\varphi}} I(\Gamma_c, \boldsymbol{\varphi}), \quad \text{with} \quad I(\Gamma_c, \boldsymbol{\varphi}) = \int_{\Omega} \Psi_{\text{bulk}} \, dV + \int_{\Gamma_c} \Psi_s \, dA. \tag{1}$$

Here, Ψ_{bulk} and Ψ_s are the Helmholtz energy of the bulk Ω and the surface energy associated with the formation of new crack surfaces, $\boldsymbol{\varphi}$ denotes the deformation mapping and Γ_c is the crack surface. Obviously, the topology of the crack surface Γ_c is usually unknown. According to Eq. (1), crack propagation is understood as a competition between surface energy and energy released in the bulk. Although the global criterion (1) eliminates the aforementioned problems associated with Griffith's theory, it induces some new difficulties. From a mathematical point of view, Eq. (1) represents a so-called free discontinuity problem. The computation of the solution of such a problem is far from being straightforward. Even more importantly, in additional to these technical difficulties, the global energy minimization problem (1) may lead to non-physical minimizers. Such a pathological example was recently given by Larsen [5]. Larsen analyzed a brittle bar (in this case, $\Psi_s = \text{const}$ denotes the fracture energy of the considered material) subjected to a one-dimensional (homogeneous) tension stress state. Clearly, if this bar cracks, the energy necessary for forming the crack Γ_c equals $\Psi_s \int_{\Gamma_c} dA$. As a consequence, a crack will develop, if the Helmholtz energy of the bulk material reaches the critical threshold

$$\Psi_{\text{bulk}} = \frac{\Psi_s \int\limits_{\Gamma_c} dA}{\int\limits_{\Omega} dV}. \tag{2}$$

Here, it is assumed that Ψ_{bulk} is spatially constant. According to Eq. (2), for a prismatic bar with length l, Eq. (2) results in

$$\Psi_{\text{bulk}} = \Psi_s / l \tag{3}$$

As a result, crack initiation depends strongly on the length of the bar. Clearly, even if stochastic effects are taken into account, this dependency is too pronounced in Eq. (3) and thus, the model (1) is not in line with experimental observations.

Griffith's local criterion and the global counterpart proposed by Francfort and Marigo can be recast into the unified framework

$$\inf_{\Gamma_c, \boldsymbol{\varphi}} \tilde{I}(\Gamma_c, \boldsymbol{\varphi}), \quad \text{with} \quad \tilde{I}(\Gamma_c, \boldsymbol{\varphi}) = \int_{E \subset \Omega} \Psi_{\text{bulk}} \, dV + \int_{\Gamma_c \subset E} \Psi_s \, dA. \tag{4}$$

By setting $E = \Omega$ the approach advocated by Francfort and Marigo is obtained, while the limiting case $\text{diam}(E) \to 0$ corresponds to Griffith's method (diam denotes the diameter).

In this contribution, the truly non-local minimization problem (4) with

$$0 < \mathrm{diam}(E) < \mathrm{diam}(\Omega) \tag{5}$$

is analyzed. As a prototype, the model is implemented by employing the so-called Strong Discontinuity Approach (SDA), cf. [6–9]. In this contribution, it will be shown that the resulting finite element formulation allows to model crack initiation consistently and that the approach avoids the non-physical minimizers predicted by the global model proposed by Francfort and Marigo [3, 4].

2 The Strong Discontinuity Approach – Fundamentals

In this section, the kinematics associated with the Strong Discontinuity Approach (SDA) as introduced in [6] is briefly presented. For an overview, the interested reader is referred to [9, 10].

First, the general case dealing with mixed-mode failure is discussed in Section 2.1. Subsequently, the kinematics corresponding to the formation of slip bands occurring in ductile materials is addressed in Section 2.2.

2.1 Kinematics Associated with Mixed Mode Brittle Failure

According to Simo et al. [6] and without going too much into detail, the discontinuous deformation mapping (more precisely, the displacement field) characterizing the SDA is given by

$$\boldsymbol{u} = \hat{\boldsymbol{u}} + [\![\boldsymbol{u}]\!]\ (H_s - \varphi), \quad \text{with} \quad \hat{\boldsymbol{u}} \in \mathcal{C}^\infty(\Omega, \mathbb{R}^3),\ \varphi \in \mathcal{C}^\infty(\Omega, \mathbb{R}). \tag{6}$$

Here and henceforth, $\hat{\boldsymbol{u}}$, $[\![\boldsymbol{u}]\!]$, H_s and φ represent a continuous displacement field, the displacement discontinuity, the Heaviside function ($H_s = 1, \forall \boldsymbol{X} \in \overline{\Omega^+}$ and $H_s = 0, \forall \boldsymbol{X} \in \Omega^-$, see Figure 1) and a scalar-valued ramp function (not to be confused with the vector-valued deformation mapping), respectively. The smooth ramp function φ allows to prescribe the Dirichlet boundary conditions in terms of $\hat{\boldsymbol{u}}$ (see [7, 11]).

Applying Eq. (6) to the finite element method, the displacement field within the considered element e is approximated by using standard interpolation functions N_i. More precisely,

$$\hat{\boldsymbol{u}} = \sum_{i=1}^{n} N_i\ \hat{\boldsymbol{u}}_i^{(e)} \in \mathcal{C}^\infty(\Omega^{(e)}, \mathbb{R}^3) \tag{7}$$

and

$$\varphi = \sum_{i=1}^{n_{\overline{\Omega^+}}} N_i \in \mathcal{C}^\infty(\Omega^{(e)}, \mathbb{R}). \tag{8}$$

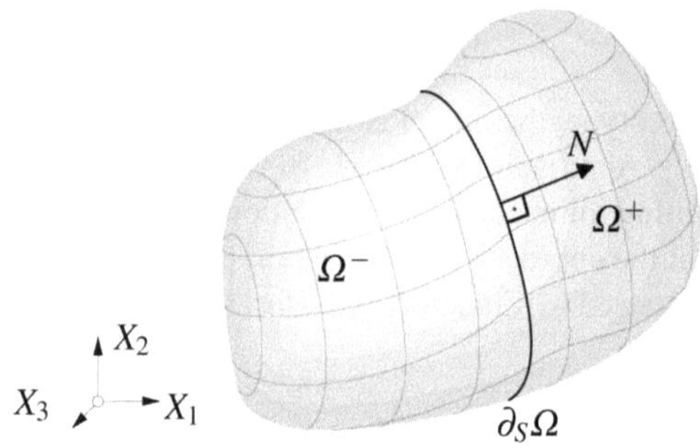

Fig. 1 Body $\Omega \subset \mathbb{R}^3$ separated into two parts Ω^- and Ω^+ by a two-dimensional submanifold $\partial_s \Omega$ of class $\mathcal{C}^1$.

In Eq. (8), the sum over all shape functions corresponding to nodes belonging to the closure of Ω^+ is computed. Further details may be found in [9].

With approximation (7) and (8), the deformation gradient $\boldsymbol{F} = \mathrm{GRAD}\boldsymbol{\varphi}$ can be computed. It results in

$$\boldsymbol{F} = \mathbf{1} + \mathrm{GRAD}\hat{\boldsymbol{u}} + [\![\boldsymbol{u}]\!] \otimes \boldsymbol{N}\, \delta_s - [\![\boldsymbol{u}]\!] \otimes \mathrm{GRAD}\varphi. \tag{9}$$

In Eq. (9), a spatially constant displacement discontinuity $\mathrm{GRAD}\,[\![\boldsymbol{u}]\!] = \mathbf{0}$ is assumed and the dirac-delta distribution denoted as δ_s is introduced. According to Eq. (9), the regularly distributed part of the deformation gradient governing the material response in $\Omega^\pm := \Omega^- \cup \Omega^+$ reads

$$\boldsymbol{F} = \hat{\boldsymbol{F}} - [\![\boldsymbol{u}]\!] \otimes \mathrm{GRAD}\varphi \quad \text{with} \quad \hat{\boldsymbol{F}} := \mathbf{1} + \mathrm{GRAD}\hat{\boldsymbol{u}}. \tag{10}$$

2.2 Kinematics Associated with Slip Bands

The kinematics as defined in the previous subsection is associated with the most general case ranging from mode-I to mode-II and mode-III failure. However, sometimes, e.g., for the analysis of slip bands, it is more convenient to modify the kinematics a priori according to the considered physical phenomenon.

In this section, the deformation characterizing slip bands is analyzed. Hence, the displacement jump $[\![\boldsymbol{u}]\!]$ is orthogonal to the spatial normal vector $\boldsymbol{n} = \boldsymbol{F}^{-T} \cdot \boldsymbol{N}/\|\boldsymbol{F}^{-T} \cdot \boldsymbol{N}\|$ (the push forward of $\boldsymbol{N}$), i.e.,

$$\boldsymbol{n} \cdot [\![\boldsymbol{u}]\!] = 0. \tag{11}$$

Introducing the so-called material displacement jump $\boldsymbol{J} := \boldsymbol{F}^{-1} \cdot [\![\boldsymbol{u}]\!]$, Eq. (11) can be re-written as

$$\boldsymbol{J} \cdot \boldsymbol{N} = 0 \tag{12}$$

and Eq. (10) results in

$$F = \hat{F} - F \cdot J \otimes \mathrm{GRAD}\varphi. \tag{13}$$

Alternatively, the additive decomposition (13) can be recast into the equivalent multiplicative counterpart

$$F = \hat{F} \cdot F^J, \quad \text{with} \quad F^J := \mathbf{1} - J \otimes \mathrm{GRAD}\varphi. \tag{14}$$

Here, the identity $F \cdot J = \hat{F} \cdot J$ is applied. It bears emphasis that this identity is only fulfilled in case of mode-II or mode-III failure, i.e., if $J \cdot N = 0$. Interestingly, the multiplicative decomposition (14) is formally identical to that used in finite strain plasticity (more precisely, crystal plasticity). Further details concerning this analogy may be found in [8,9].

3 A Non-Local Crack Initiation Criterion

In this section, a variationally consistent formulation suitable for the analysis of deformation-driven, localized material defects such as cracks or shear bands is proposed. For that purpose, the underlying variational framework is introduced in Section 3.1. Subsequently, both limiting cases of material failure are carefully analyzed. While Section 3.2 deals with brittle fracture, the opposite, namely a ductile slip band, is addressed in Section 3.3.

3.1 Fundamentals

It is well known that the Strong Discontinuity Approach (SDA) in its original form is based on a Petrov–Galerkin discretization, cf. [9]. Hence, it is not variationally consistent. More precisely, the SDA cannot be derived from a potential in general. However, if the failure surface (crack or shear band, see Figure 1) is parallel to one of the boundaries of the cut finite element, a symmetric scheme is obtained, see [9]. Consequently and for the sake of simplicity, the aforementioned condition is assumed to hold true. Clearly, within a (standard) SDA-based finite element analysis this condition is usually too restrictive in practice. Therefore, a combination with adaptive strategies such as r-adaptivity is required, cf. [12, 13].

In line with Griffith's local criterion and Francfort's and Marigo's global counterpart, we advocate the non-local minimization principle

$$\inf_{\Gamma_c, \boldsymbol{\varphi}} \tilde{I}(\Gamma_c, \boldsymbol{\varphi}), \quad \text{with} \quad \tilde{I}(\Gamma_c, \boldsymbol{\varphi}) = \int_{E \subset \Omega} \Psi_{\text{bulk}} \, \mathrm{d}V + \int_{\Gamma_c \subset E} \Psi_s \, \mathrm{d}A. \tag{15}$$

with Ψ_{bulk} and Ψ_s denoting the bulk energy and the surface energy of the considered body. The neighborhood $E \subset \Omega$ is defined as the domain of the respective finite element e. Alternatively, assuming that the crack is aligned with the facets of the finite elements (The topology of the crack is to some extend known in advance and it is approximated by means of a finite space), Eq. (15) can be re-written as

$$\inf_{\hat{\boldsymbol{u}}, [\![\boldsymbol{u}]\!]} \tilde{I}(\hat{\boldsymbol{u}}, [\![\boldsymbol{u}]\!]), \quad \text{with} \quad \tilde{I}(\hat{\boldsymbol{u}}, [\![\boldsymbol{u}]\!]) = \int_{E \subset \Omega} \Psi_{\text{bulk}}(\hat{\boldsymbol{u}}, [\![\boldsymbol{u}]\!]) \, \mathrm{d}V + \int_{\Gamma_c \subset E} \Psi_s([\![\boldsymbol{u}]\!]) \, \mathrm{d}A. \tag{16}$$

Suppose the continuous part $\hat{\boldsymbol{u}}$ as well as the discontinuous part $[\![\boldsymbol{u}]\!]$ of the displacement field have been computed by means of minimization problem (16) (it will be shown later that this leads to physically sound solutions). The respective solution is denoted as $\hat{\boldsymbol{u}}_n$, $[\![\boldsymbol{u}]\!]_n$. Based on this solution and the respective boundary conditions at time t_{n+1}, the goal is to compute the updated variables $\hat{\boldsymbol{u}}_{n+1}$ and $[\![\boldsymbol{u}]\!]_{n+1}$. Clearly, if no new crack segments form, the new solution can also be computed by relaxing the energy with respect to $\hat{\boldsymbol{u}}_{n+1}$ and $[\![\boldsymbol{u}]\!]_{n+1}$ (and some additional restrictions). However, first it has to be checked, if a new crack forms. For that purpose, the energy associated with an uncracked body

$$I^e_{\text{cont}} := \int_{\Omega^e} \Psi_{\text{bulk}}(\hat{\boldsymbol{F}}) \, \mathrm{d}V \tag{17}$$

is compared to that of the cracked counterpart, i.e.,

$$I^e_{\text{disc}} := \inf_{[\![\boldsymbol{u}]\!]} \left\{ \int_{\Omega^e} \Psi_{\text{bulk}}(\hat{\boldsymbol{F}}, [\![\boldsymbol{u}]\!]) \, \mathrm{d}V + \int_{\Gamma_c \subset \Omega^e} \Psi_s([\![\boldsymbol{u}]\!]) \, \mathrm{d}A \right\} \Bigg|_{\hat{\boldsymbol{F}}=\text{const}} . \tag{18}$$

Since within the SDA the displacement discontinuity is modeled in an element-wise fashion, the computation of the aforementioned energies can conveniently be restricted to the considered finite element e. It bears emphasis that similar to computational plasticity (cf., the return-mapping scheme), the aforementioned crack initiation criterion is checked several times within a considered loading increment (the same holds for the yield function in plasticity theory). By doing so, it is guaranteed that crack propagation is governed by means of a physically sound state (converged).

Remark 1. The crack initiation criterion (17) and (18) can only be applied to fully open cracks (or shear bands). If micro-defects characterized by non-vanishing traction vectors are to be modeled, dissipation has to be added to Eq. (18).

Remark 2. The crack initiation criterion (17) and (18) shows a similar size effect as the model proposed by Francfort and Marigo [3]. However, without going too much into detail, this effect can be eliminated by replacing Eq. (18) by the more general minimization principle

$$I^e_{\text{disc}} := \inf_{[\![\boldsymbol{u}]\!]} \left\{ \int_{\Omega^e} \Psi_{\text{bulk}}(\hat{\boldsymbol{F}}, [\![\boldsymbol{u}]\!]) \, \mathrm{d}V + l_c \int_{\Gamma_c \subset \Omega^e} \tilde{\Psi}_s([\![\boldsymbol{u}]\!]) \, \mathrm{d}A \right\} \Bigg|_{\hat{\boldsymbol{F}}=\text{const}} . \tag{19}$$

Here, $l_c = 1/\|\mathrm{GRAD}\varphi\|$ is the characteristic length of the considered finite element, cf. [14, 15]. It can be shown in a relatively straightforward manner that the modified crack initiation criterion is purely local in nature (stress based). This is a direct consequence of the locally incompatibly enhanced displacement field associated with the SDA.

3.2 Mixed Mode Brittle Failure

In this subsection, the variational framework as briefly introduced before is applied to the analysis of brittle fracture. Hence, the surface energy Ψ_s is constant and furthermore, it is identical to the fracture energy of the respective solid. To prove that the proposed energy-driven method leads to physically sound results, the stationarity condition corresponding to a fully open macro-crack is analyzed. Obviously, in this case, the stress vector $\boldsymbol{T} := \boldsymbol{P} \cdot \boldsymbol{N}$ acting at such a crack is supposed to vanish. Here and henceforth, $\boldsymbol{P}$ denotes the first Piola–Kirchhoff stress tensor.

The condition $\boldsymbol{T} = \boldsymbol{0}$ characterizing brittle fracture can be checked in a straightforward manner by computing the stationarity condition associated with Eq. (18). Applying Eq. (10), it results in

$$\frac{\partial I^e_{\mathrm{disc}}}{\partial [\![\boldsymbol{u}]\!]} = \int\limits_{\Omega^e} \frac{\partial \Psi_{\mathrm{bulk}}}{\partial [\![\boldsymbol{u}]\!]} \, \mathrm{d}V = -\|\mathrm{GRAD}\varphi\| \int\limits_{\Omega^e} \boldsymbol{T} \, \mathrm{d}V = \boldsymbol{0}. \tag{20}$$

Here, the definition of the first Piola–Kirchhoff stress tensor $\boldsymbol{P} := \partial_{\boldsymbol{F}} \Psi_{\mathrm{bulk}}$, together with the identity

$$\mathrm{GRAD}\varphi = \boldsymbol{N} \, \|\mathrm{GRAD}\boldsymbol{\varphi}\|, \tag{21}$$

have been used. Eq. (21) is a direct consequence following from the assumption that the crack or the shear band surface is aligned with one of the facets of the finite element. Furthermore, a linearly varying ramp function φ is considered, i.e., $\varphi \in \mathbb{P}_1$. As evident from Eq. (20), the model indeed predicts a stress-free macro-crack (in an average sense).

3.3 Slip Bands

The applicability of the proposed variational method to the analysis of shear bands can be proven in the same manner as shown for brittle cracks. For the sake of simplicity, it is again assumed that $\Psi_s = \mathrm{const.}$ However, the more general case, i.e., $\Psi_s = \Psi_s(\boldsymbol{J})$ does not rise any new problems.

In case of slip bands, it is more convenient to start from kinematics (14), i.e., the multiplicative decomposition of the deformation gradient. With this split, the stationarity condition reads

$$\frac{\partial I^e_{\text{disc}}}{\partial \boldsymbol{J}} \cdot \delta \boldsymbol{J} = \int_{\Omega^e} \frac{\partial \Psi_{\text{bulk}}}{\partial \boldsymbol{J}} \cdot \delta \boldsymbol{J} \, \mathrm{d}V = -\|\text{GRAD}\varphi\| \int_{\Omega^e} \left\{ \left[\hat{\boldsymbol{F}}^T \cdot \boldsymbol{P} \right] \cdot \boldsymbol{N} \right\} \cdot \delta \boldsymbol{J} \, \mathrm{d}V, \quad (22)$$

with $\delta \boldsymbol{J}$ denoting the variation of $\boldsymbol{J}$. Hence, the stress vector $\bar{\boldsymbol{T}} := [\hat{\boldsymbol{F}}^T \cdot \boldsymbol{P}] \cdot \boldsymbol{N}$ (intermediate configuration) has to vanish for any tangential direction (in case of slip bands, $\delta \boldsymbol{J}$ belongs to the tangent plane). Consequently, the proposed variational model enforces the constraints associated with shear bands naturally (in an average sense).

4 Conclusions

In this paper, a fully variational framework suitable for the analysis of localized material failure such as cracking in brittle materials or shear bands in ductile metals has been developed. Based on the (embedded) Strong Discontinuity Approach (SDA), the final failure kinematics of solids characterized by a discontinuous deformation mapping is approximated. In contrast to previous, conventional models, the unknown deformation including the displacement discontinuity follows naturally by minimizing the energy of the respective solid. Clearly, besides the physical interpretation of cracking as an energy minimizer, the variational formulation shows several significant advantages compared to classical methods. For instance, the proposed minimum principle opens up the possibility of deriving error estimates necessary for adaptive finite element analyses.

The presented energy-driven approach is in line with Griffith's classical local criterion and Marigo's and Francfort's global extension. More precisely, the energy within a non-local neighborhood, i.e., the domain of the respective finite element, is analyzed. Interestingly, it turns out that the final model does not involve any length scale concerning the spatial discretization. Furthermore, by incorporating the characteristic length of the considered finite element into the minimization principle, a purely local crack initiation criterion is obtained (stress based). This reflects the local incompatibility of the enhanced displacement field associated with the adopted Strong Discontinuity Approach (SDA). For models showing a compatible discontinuous part of the displacement, the proposed minimization principles have to be carfully analyzed in the future.

References

1. Griffith A., The phenomena of rupture and flow in solids. *Phil. Trans. Roy. Soc. London* **CCXXI-A**, 1920, 163–198.
2. Maugin, G.A., *Configurational Mechanics of Materials*. Springer Wien, 2001.
3. Francfort, G.A. and Marigo, J.-J., Revisiting brittle fracture as an energy minimization problem. *Journal of the Mechanics and Physics of Solids* **46**(8), 1998, 1319–1342.

4. Bourdin, B., Francfort, G.A. and Marigo, J.-J., Numerical experiments in revisited brittle fracture. *Journal of the Mechanics and Physics of Solids* **48**(4), 2000, 797–826.
5. Larsen, C., Quasi-static evolution in brittle fracture based on local minimization. In *GAMM Annual Meeting*, 2007.
6. Simo, J.C., Oliver, J. and Armero, F., An analysis of strong discontinuities induced by strain softening in rate-independent inelastic solids. *Computational Mechanics* **12**, 1993, 277–296.
7. Oliver, J., Modelling strong discontinuities in solid mechanics via strain softening constitutive equations. Part 1: Fundamentals. Part 2: Numerical simulations. *International Journal for Numerical Methods in Engineering* **39**, 1996, 3575–3623.
8. Mosler, J., Modeling strong discontinuities at finite strains – A novel numerical implementation. *Computer Methods in Applied Mechanics and Engineering* **195**(33–36), 2006, 4396–4419.
9. Mosler, J., On the numerical modeling of localized material failure at finite strains by means of variational mesh adaption and cohesive elements. Habilitation, Ruhr University Bochum, Germany, 2007.
10. Mosler, J., On the modeling of highly localized deformations induced by material failure: The strong discontinuity approach. *Archives of Computational Methods in Engineering* **11**(4), 2004, 389–446.
11. Simo, J.C. and Oliver, J., A new approach to the analysis and simulation of strain softening in solids. In Z.P. Bažant, Z. Bittnar, M. Jirásek, and J. Mazars (Eds.), *Fracture and Damage in Quasibrittle Structures*. E. &F.N. Spon, London, 1994, pp. 25–39.
12. Mosler, J. and Ortiz, M., On the numerical implementation of variational arbitrary Lagrangian–Eulerian (VALE) formulations. *International Journal for Numerical Methods in Engineering* **67**, 2006, 1272–1289.
13. Miehe, C. and Gürses, E., A robust algorithm for configurational-force-driven brittle crack propagation with r-adaptive mesh alignment. *International Journal for Numerical Methods in Engineering* **72**, 2007, 127–155.
14. Oliver, J., A consistent characteristic length for smeared cracking models. *International Journal for Numerical Methods in Engineering* **28**, 1989, 461–474.
15. Mosler, J. and Meschke, G., Embedded cracks vs. smeared crack models: A comparison of elementwise discontinuous crack path approaches with emphasis on mesh bias. *Computer Methods in Applied Mechanics and Engineering* **193**(30–32), 2004, 3351–3375.

Computational Homogenization of Defect Driving Forces

Sarah Ricker, Julia Mergheim and Paul Steinmann

Abstract Due to the fact that many engineering materials and also biological tissues possess an underlying (heterogeneous) micro-structure it is not sufficient to simulate these materials by pre-assumed overall constitutive assumptions. Therefore, we apply a homogenization scheme, which determines the macroscopic material behavior based on analysis of the underlying micro-structure. In the work at hand focus is put on the extension of the classical computational homogenization scheme towards the homogenization of material forces. Therefore, volume forces have to incorporated which may emerge due to inhomogeneities in the material. With assistance of this material formulation and the equivalence of the J-integral and the material force at a crack tip, studies on the influence of the micro-structure onto the macroscopic crack-propagation are carried out.

1 Introduction

The main goal of the current work is the combination of the concept of computational homogenization with the concept of configurational mechanics. Homogenization schemes which incorporate simulations on different length scales play an important role in the simulation of heterogeneous materials. Apparent classes of such materials in the field of engineering applications are given, e.g., by composite materials, metal foams which are often used in lightweight construction due to their strength–weight ratio, or different kinds of alloys. Another application is the prediction of the mechanical behavior of biological tissues like bones or pulmonary

Sarah Ricker · Julia Mergheim
Chair of Applied Mechanics, University of Kaiserslautern, P.O. Box 3049, 67653 Kaiserslautern, Germany; e-mail: {sricker, mergheim}@rhrk.uni-kl.de

Paul Steinmann
Chair of Applied Mechanics, University of Erlangen-Nuremberg, Egerlandstr. 5, 91058 Erlangen, Germany; e-mail: steinmann@ltm.uni.erlangen.de

P. Steinmann (ed.), IUTAM Symposium on Progress in the Theory and Numerics of Configurational Mechanics, 249–259.

tissue, which is of great interest in biomechanical simulation. In the computational homogenization scheme applied here, the macroscopic response at a certain material point is determined by analysis of its underlying micro-structure. Therefore in each macroscopic simulation point of interest the microscopic setting is attached via a so-called representative volume element (RVE). On the macro-level no constitutive assumption is stated because the required quantities, which are the stress and the tangent modulus in the applied deformation-driven scheme, emerge from an averaging over the RVE. Due to this nested solution scheme, wherein for each macroscopic simulation point a finite-element simulation is performed on the microscopic RVE, this particular approach is denoted as FE^2 method in the work of Feyel and Chaboche [3] or Feyel [4]. For further reading on this computational homogenization scheme the reader is referred, e.g., to the work of Kouznetsova et al. [6] and Miehe [8, 9].

The fundamental feature of this work is the extension of the classical computational homogenization towards the homogenization of material or configurational forces. Therefore, in contrast to state-of-the-art homogenization schemes, volume forces have to be taken into account, which in the material motion problem occur due to, e.g., inhomogeneous materials. The material formulation in the sense of Eshelby, which is inverse to the spatial motion problem, is motivated by the following description of Eshelby in [2]:

> ... the total energy of a system ... is a function of the set of parameters necessary to specify the configuration of the imperfections. The negative gradient of the total energy with respect to the position of an imperfection may conveniently be called the force on it. This force, in a sense fictitious, is introduced to give a picturesque description of energy changes, and must not be confused with the ordinary surface and body forces acting on the material.

Thus, it can be summarized that in contrast to classical spatial forces in the sense of Newton which are linked to the variation of spatial positions of physical particles with respect to the ambient space, the material forces in the scnse of Eshelby are linked to the variations of material positions of physical particles with respect to the ambient material. Therefore, this material formulation exhibits a wide range of applications in the field of defect mechanics. It can be shown that the material force at a crack-tip corresponds to the J-integral, introduced by Rice [11], and thus yields a criterion whether a crack propagates or not. Further details describing the material motion problem and its application to fracture mechanics can be found in the textbook by Maugin [7] and, e.g., in the work of Steinmann [12], Steinmann et al. [13], Denzer et al. [1] among many other publications in this field.

The current work is structured as follows: In Section 2 the governing equations for the spatial and the material motion problem, which are given by the kinematics and the balance equations, are reviewed. In Section 3 the averages of the variables of interest are summarized for the spatial and the material motion problem. Then, in Section 4 the essential theorems and the boundary conditions, which are necessary to perform the scale-transitions, are elaborated. The influence of different microstructures onto the macroscopic behavior as well for the spatial as for the material motion problem is studied by numerical examples in Section 5. The work is closed by the conclusions given in Section 6.

2 Governing Equations

In the following the governing equations for the spatial as well as for the material motion problem are reviewed. No distinction is made between the macro- and the micro-level, due to the fact that the kinematics and the balance equations are valid for both levels. Please note that in the homogenization scheme the constitutive assumption is only stated at the micro-level, because the macroscopic behavior is obtained from the analysis of the microscopic RVE.

2.1 Governing Equations for the Spatial Motion Problem

In the spatial motion problem a body $\mathscr{B}_0$ with material points $\mathbf{X}$ is considered. These points are mapped to the spatial configuration via the deformation map $\boldsymbol{\varphi}$, whose gradient yields the deformation gradient tensor:

$$\mathbf{x} = \boldsymbol{\varphi}(\mathbf{X}) \quad \text{and} \quad \mathbf{F} = \frac{\partial \boldsymbol{\varphi}}{\partial \mathbf{X}}. \tag{1}$$

The spatial motion Piola stress tensor and the body forces are given in terms of the energy functional U_0

$$\mathbf{P} = \frac{\partial U_0}{\partial \mathbf{F}} \quad \text{and} \quad \mathbf{b}_0 = -\frac{\partial U_0}{\partial \mathbf{x}}. \tag{2}$$

The connection between the Piola stress and the Cauchy stress is given by the following push-forward operation

$$\boldsymbol{\sigma} = \det(\mathbf{F})^{-1}\, \mathbf{P} \cdot \mathbf{F}^t. \tag{3}$$

This set of equations is completed by the quasi-static balance of momentum which reads

$$-\operatorname{Div} \mathbf{P} = \mathbf{b}_0 \quad \Rightarrow \quad -\operatorname{div} \boldsymbol{\sigma} = \mathbf{b}_t = \det(\mathbf{F})^{-1}\, \mathbf{b}_0. \tag{4}$$

Thereby, $\operatorname{Div}(\bullet)$ and $\operatorname{div}(\bullet)$ denote the divergence operators with respect to the material coordinates $\mathbf{X}$ and the spatial coordinates $\mathbf{x}$.

2.2 Governing Equations for the Material Motion Problem

In the material motion problem a body $\mathcal{B}_t$ with spatial points $\mathbf{x}$ is considered. In analogy with the spatial motion problem we define the deformation map and the deformation gradient as

$$\mathbf{X} = \Phi(\mathbf{x}) \quad \text{and} \quad \mathbf{f} = \frac{\partial \Phi}{\partial \mathbf{x}}. \tag{5}$$

The material motion two-point stress and the body forces are given in terms of the energy functional U_t

$$\mathbf{p} = \frac{\partial U_t}{\partial \mathbf{f}} \quad \text{and} \quad \mathbf{B}_t = -\frac{\partial U_t}{\partial \mathbf{X}}, \tag{6}$$

wherein U_t is obtained via $U_t = \det(\mathbf{f})\, U_0$. The connection between the material two-point stress and the Eshelby stress is given by the following pull-back operation

$$\boldsymbol{\Sigma} = \det(\mathbf{f})^{-1}\, \mathbf{p} \cdot \mathbf{f}^t. \tag{7}$$

This set of equations is completed by the quasi-static balance of momentum which reads

$$-\operatorname{div} \mathbf{p} = \mathbf{B}_t \quad \Rightarrow \quad -\operatorname{Div} \boldsymbol{\Sigma} = \mathbf{B}_0 = \det(\mathbf{f})^{-1}\, \mathbf{B}_t. \tag{8}$$

3 Homogenization

In order to determine the averaged or homogenized macroscopic quantities an averaging over the micro-domain has to be established. Therefore, in our deformation-driven framework we apply the macroscopic deformation gradient tensor $\mathbf{F}_M$ via admissible boundary conditions onto the RVE. After solving the microscopic boundary value problem within a finite-element scheme the homogenization procedure is carried out. The homogenized quantities are not simply defined as the volume averages over their microscopic counterparts, but as averages over boundary terms of the RVE. This modus operandi has already been proposed by Hill in [5]:

> Macro-variables intended for constitutive laws should thus be capable of definition in terms of surface data alone, either directly or indirectly. It is not necessary, by any means, that macro-variables so defined should be unweighted volume averages of their microscopic counterparts. [...] Accordingly, we approach the construction of macro-variables by first identifying some relevant averages that depend uniquely on surface data.

Note that in the following macro-variables are denoted by $(\bullet)_M$ and averaged variables by $(\bar{\bullet})$. Non-indexed variables belong to the micro-level.

3.1 Averaged Variables for the Spatial Motion Problem

In this section we summarize the desired homogenized variables for the spatial motion problem as averages over the boundary of the RVE. Firstly, we need the average of the deformation gradient tensor, which is defined by:

$$\bar{\mathbf{F}} := \frac{1}{V_0} \int_{\partial \mathcal{V}_0} \boldsymbol{\varphi} \otimes \mathbf{N} \, \mathrm{d}A \tag{9}$$

where V_0 denotes the volume of the RVE in the material configuration, $\boldsymbol{\varphi}$ the spatial micro-deformation mapping and $\mathbf{N}$ the normal vector of the boundary of the RVE in the material configuration. Furthermore, the averaged Cauchy stress is defined as the average of the product of the traction vector $\mathbf{t}_0$ and the deformation mapping vector $\boldsymbol{\varphi}$

$$\bar{\boldsymbol{\sigma}} := \frac{1}{V_t} \int_{\partial \mathcal{V}_t} \mathbf{t}_t \otimes \boldsymbol{\varphi} \, \mathrm{d}a \tag{10}$$

where V_t denotes the volume of the RVE in the spatial setting. For further insight into the averaging of variables, the reader is referred to the textbook by Nemat-Nasser and Hori [10].

3.2 Averaged Variables for the Material Motion Problem

In analogy with the spatial computational homogenization scheme, in the material homogenization scheme averages of the microscopic material quantities have to be defined. Firstly, we consider the average of the material deformation gradient tensor, which reads

$$\bar{\mathbf{f}} := \frac{1}{V_t} \int_{\partial \mathcal{V}_t} \Phi \otimes \mathbf{n} \, \mathrm{d}a. \tag{11}$$

Therein, Φ denotes the microscopic material deformation map, i.e., $\Phi = \boldsymbol{\varphi}^{-1}$ holds, and $\mathbf{n}$ identifies the normal vector on the boundary of the RVE in the spatial configuration. Furthermore, the material Eshelby stress $\bar{\boldsymbol{\Sigma}}$ is obtained via

$$\bar{\boldsymbol{\Sigma}} := \frac{1}{V_0} \left[\int_{\partial \mathcal{V}_0} \mathbf{T}_0 \otimes \boldsymbol{\Phi} \, \mathrm{d}A + \int_{\mathcal{S}_0} \mathbf{T}_0^s \otimes \boldsymbol{\Phi} \, \mathrm{d}A \right]. \tag{12}$$

The material traction vector on the boundary is denoted with $\mathbf{T}_0$ and is obtained through $\mathbf{T}_0 := \boldsymbol{\Sigma} \cdot \mathbf{N}$. The tractions at the interface $\mathcal{S}_0$ are defined as $\mathbf{T}_0^s = (\mathbf{T}_0)_1 + (\mathbf{T}_0)_2 = \boldsymbol{\Sigma}_1 \cdot \mathbf{N}_1 + \boldsymbol{\Sigma}_2 \cdot \mathbf{N}_2$. The interface integral needs to be incorporated due to the fact that the configurational tractions on opposite sides of the interface are not in equilibrium.

4 Scale-Transition and Boundary Conditions

In order to connect the macroscopic simulation points and the underlying micro-RVE a scale-transition has to be performed. In case of the applied deformation-driven scheme, this macro-micro transition is performed by solving a boundary value problem on the micro-scale, driven by the macroscopic deformation gradient. After solving this particular microscopic boundary value problem the micro-macro transition is performed via averaging the desired variables according to Section 3.

4.1 Spatial Motion Problem

The microscopic deformation map for the spatial motion problem is given in terms of the macroscopic deformation map $\boldsymbol{\varphi}_M$, the deformation gradient tensor $\mathbf{F}_M$ and the spatial fluctuations $\mathbf{w}$

$$\boldsymbol{\varphi}(\mathbf{X}) = \boldsymbol{\varphi}_M + \mathbf{F}_M \cdot \mathbf{X} + \mathbf{w}(\mathbf{X}). \tag{13}$$

In order to perform the scale-transition in a consistent manner the Average-Strain Theorem, which claims the equivalence of the macroscopic and averaged deformation gradient, has to be fulfilled:

$$\bar{\mathbf{F}} \equiv \mathbf{F}_M. \tag{14}$$

This constraint yields two types of admissible boundary conditions, namely displacement boundary conditions

$$\mathbf{w} = \mathbf{0} \text{ on } \partial\mathcal{V}_0 \tag{15}$$

and periodic displacement boundary conditions

$$\mathbf{w}^+ = \mathbf{w}^- \quad \text{on} \quad \partial\mathcal{V}_0 = \partial\mathcal{V}_0^+ \cup \partial\mathcal{V}_0^-. \tag{16}$$

Please note that in case of periodic boundary conditions it follows directly from the microscopic equilibrium that the tractions behave anti-periodically, i.e., $\mathbf{t}_0^+ = -\mathbf{t}_0^-$.

After defining the boundary conditions for the scale-transition focus is put onto the variation of averaged work. The Hill–Mandel condition given in equation (17) states that the averaged variation of microscopic work $D_\delta \bar{U}_0$ has to be equivalent to the variation of macroscopic work $D_\delta U_{0M}$ at fixed coordinates $\mathbf{X}$.

$$D_\delta U_{0M} = D_\delta \bar{U}_0$$
$$\Rightarrow \left[D_\delta \mathbf{F}_M \cdot \mathbf{F}_M^{-1} \right] : \boldsymbol{\sigma}_M - D_\delta \boldsymbol{\varphi}_M \cdot \mathbf{b}_{tM} = \frac{1}{V_t} \int_{\partial \mathcal{V}_t} D_\delta \boldsymbol{\varphi} \cdot \mathbf{t}_t \, da. \tag{17}$$

Inserting the boundary conditions given in equations (15) and (16) into the Hill–Mandel (17) condition yields the following equivalences between the averaged and the macroscopic variables:

$$\boldsymbol{\sigma}_M = \bar{\boldsymbol{\sigma}} \quad \text{and} \quad \mathbf{b}_{tM} = \mathbf{0}. \tag{18}$$

4.2 Material Motion Problem

The microscopic deformation map for the material motion problem is here computed from a pull-back and a composition with the spatial deformation map as

$$\mathbf{X} = \boldsymbol{\Phi} = -\mathbf{f}_M \cdot \boldsymbol{\varphi}_M + \mathbf{f}_M \cdot \mathbf{x} - \mathbf{f}_M \cdot \mathbf{w}(\boldsymbol{\varphi}^{-1}(\mathbf{x})). \tag{19}$$

Thus, the variation of the material deformation map at fixed coordinates $\mathbf{x}$ reads

$$d_\delta \boldsymbol{\Phi} = d_\delta \mathbf{f}_M \cdot \left[-\boldsymbol{\varphi}_M + \mathbf{x} - \mathbf{w} \right] = d_\delta \mathbf{f}_M \cdot \mathbf{F}_M \cdot \mathbf{X}. \tag{20}$$

By assistance of the average strain theorem for the spatial motion problem one can show that the average strain theorem is also valid for the material motion problem, i.e.,

$$\bar{\mathbf{f}} \equiv \mathbf{f}_M. \tag{21}$$

In order to perform the scale-transition in a consistent manner, the Hill–Mandel condition adapted to the material motion problem has to be fulfilled

$$d_\delta U_{tM} = d_\delta \bar{U}_t$$
$$\Rightarrow [d_\delta \mathbf{f}_M \cdot \mathbf{F}_M] : \boldsymbol{\Sigma}_M - d_\delta \boldsymbol{\Phi}_M \cdot \mathbf{B}_{0M} = \frac{1}{V_0} \left[\int_{\partial \mathcal{V}_0} d_\delta \boldsymbol{\Phi} \cdot \mathbf{T}_0 \, dA + \int_{\mathcal{S}_0} d_\delta \boldsymbol{\Phi} \cdot \mathbf{T}_0^s \, dA \right]. \tag{22}$$

Inserting the expression for the variation of the microscopic deformation map given in equation (20) and comparing the coefficients yields following equivalences

$$\boldsymbol{\Sigma}_M = \bar{\boldsymbol{\Sigma}} \quad \text{and} \quad \mathbf{B}_{0M} = \mathbf{0}. \tag{23}$$

From $\mathbf{B}_{0M} = \mathbf{0}$ we can conclude that the inhomogeneities at the micro-level, captured by $\mathbf{B}_0$, as well as inhomogeneities at the macro-level due to different attached RVEs are completely contained in the format of the Eshelby stress $\boldsymbol{\Sigma}_M$ at the macro-level.

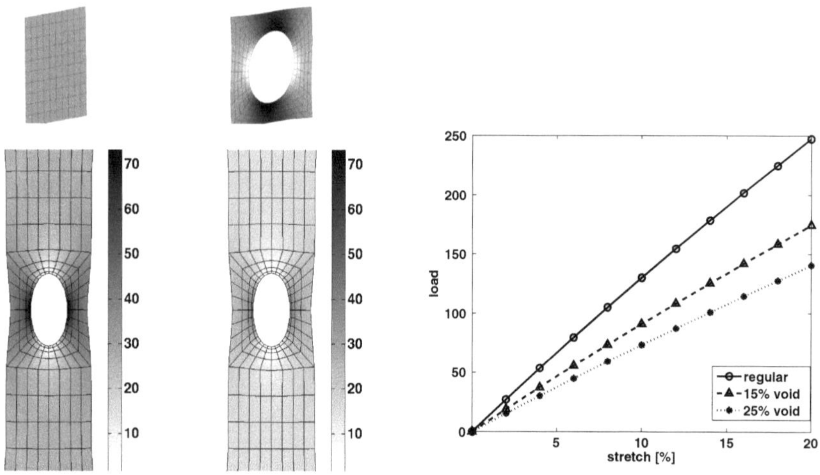

Fig. 1 Macroscopic and microscopic longitudinal Cauchy stress for macroscopic specimen with centered hole under tensile loading.

5 Examples

In this section numerical examples for the multi-scale computation of the spatial and the material motion problem will be presented.

The influence of different kinds of micro-structures on the spatial as well as on the material response of a macroscopic specimen is studied. All examples are calculated with periodic boundary conditions and Neo-Hookean type constitutive law on the micro-level.

In a first example a macroscopic specimen with a centered hole under tensile loading is considered – see Figure 1. A homogeneous micro-structure is compared to a micro-structure with a void. Figure 1 displays the longitudinal Cauchy-stresses on the macro-scale as well as the microscopic stress in particular RVEs, located on the left side of the macroscopic hole. As expected, both the stress plot and the load-displacement diagram reveal that the macro-specimen is softer for a higher percentage of micro-voids. In the stress plot this can be identified by the decreasing stress peak near the boundary of the hole.

Secondly, focus is put on the same macro-structure, but this time the material nodal forces are of interest. Figure 2 monitors the material forces on the macro and the micro nodes for different micro-structures – a homogeneous structure, a structure with 15% voids and a structure with 25% voids. Again, the displayed microscopic RVEs are located near the macroscopic hole. In the homogeneous micro-structure no material nodal forces are present. In the micro-structures containing voids material nodal forces occur on the boundary of the hole. The softening effect of the micro-voids onto the macroscopic material behavior can be observed in

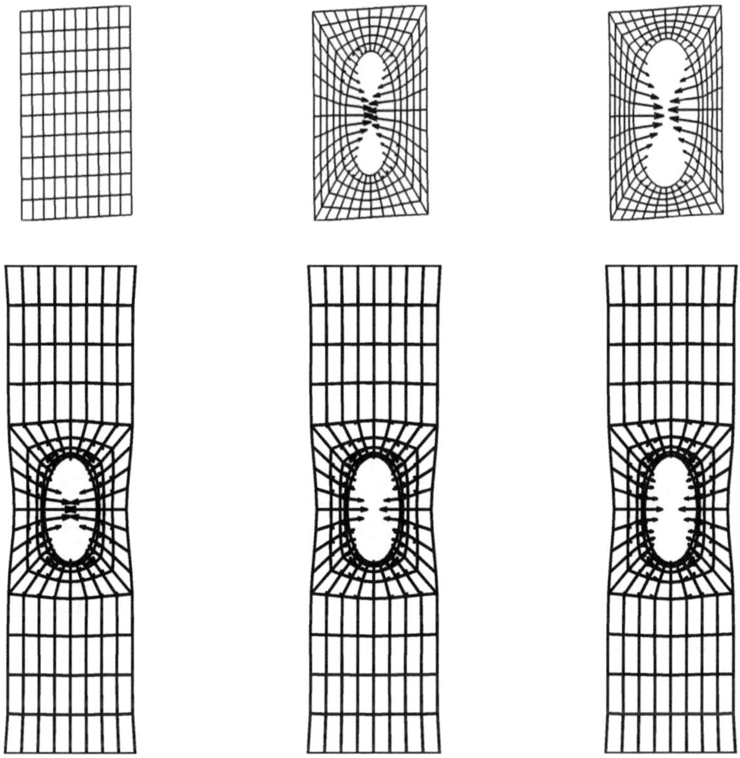

Fig. 2 Macroscopic and microscopic material nodal forces for macroscopic specimen with a centered hole under tensile loading.

the decrease of the macroscopic material forces at the boundary of the macroscopic hole.

In a last example the influence of different micro-structures on the material force at a crack tip is studied. It can be shown that the material force at the crack tip corresponds to the J-integral and thus yields a criterion whether a crack propagates or not. In Figure 3 we consider three different RVEs, each representing a micro-structure with two material layers and a straight interface. For all three RVEs the average of the Young's moduli is the same, but the ratio of the moduli differs. From Figure 3 it can be observed that the bigger the difference between the stiffnesses of the two materials becomes, the smaller is the macroscopic material force. Thus, we have shown that the present approach captures the influence of the heterogeneous micro-structure on the macroscopic J-integral.

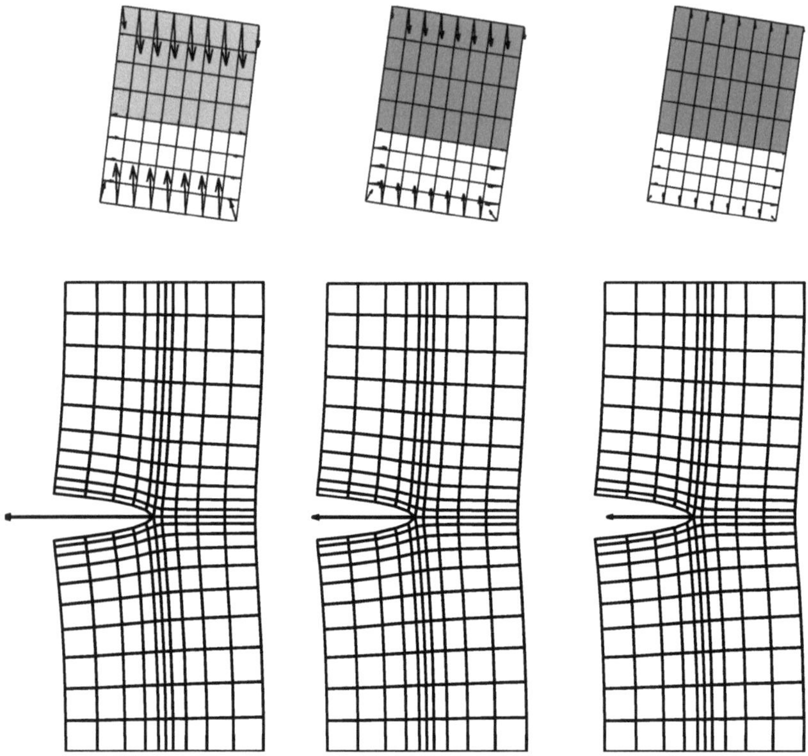

Fig. 3 Macroscopic material force at crack tip for different ratios of microscopic Young's moduli – left to right ratio of 1:2; 1:10 and 1:20.

6 Conclusions

In the current work we have extended the classical computational homogenization scheme towards the homogenization of the material motion problem. Therefore, a formulation containing the homogenization of volume forces, which may occur in the material motion problem due to inhomogeneities on the micro-structure, has been elaborated. Furthermore, it has been shown that in the homogenization of the Eshelby stress interface terms have to be taken into account, because material tractions at the interface are not vanishing in general. Based on the spatial average strain theorem admissible boundary conditions have been developed which fulfill the Hill–Mandel condition. In a next step the microscopic deformation map for the material motion problem has been introduced which provides the basis for the material analogon of the average strain theorem. Then by assistance of a Hill–Mandel type condition the equivalence between the averaged and the macroscopic material stress has been elaborated. To complete these studies the influence of the micro-structure onto macroscopic behavior, like the deformation, the stresses or the material force at a crack-tip, has been monitored by numerical examples. Thus, the presented approach captures the influence of the microscopic setting for the spatial and the material motion problem.

References

1. Denzer, R., Barth, F.J. and Steinmann, P., Studies in elastic fracture mechanics based on the material force method. *Int. J. Num. Meth. Eng.* **58**, 2003, 1817–1835.
2. Eshelby, J.D., The force on an elastic singularity. *Phil. Trans. Royal Soc. A* **244**, 1951, 87–112.
3. Feyel, F. and Chaboche, J.-L., FE^2 multi-scale approach for modelling the elastoplastic behavior of long fibre SiC/Ti composite materials. *Comput. Meth. Appl. Mech. Eng.* **183**, 2000, 309–330.
4. Feyel, F., A multilevel finite element method (FE^2) to describe the response of highly nonlinear structures using generalized continua. *Comput. Meth. Appl. Mech. Eng.* **192**, 2003, 3233–3244.
5. Hill, R., On constitutive macro-variables for heterogeneous solids at finite strain. *Proc. Royal Soc. A* **326**, 1972, 131–147.
6. Kouznetsova, V., Brekelmans, W.A.M. and Baaijens, F.P.T., An approach to micro-macro modeling of heterogeneous materials. *Comput. Mech.* **27**, 2001, 27–48.
7. Maugin, G.A., *Material Inhomogeneities in Elasticity*. Chapman and Hall, London, 1993.
8. Miehe, C., Strain-driven homogenization of inelastic microstructures and composites based on an incremental variational formulation. *Int. J. Numer. Meth. Eng.* **55**, 2002, 1285–1322.
9. Miehe, C., Computational micro-to-macro transitions for discretized micro-structures of heterogeneous materials at finite strains based on the minimization of averaged incremental energy. *Comput. Meth. Appl. Mech. Eng.* **192**, 2003, 559–291.
10. Nemat-Nasser, S. and Hori, M., *Micromechanics: Overall Properties of Heterogeneous Materials*. North-Holland, Amsterdam, 1993.
11. Rice, J.R., A path independent integral and the approximate analysis of strain concentration by notches and cracks. *J. Appl. Mech.* **35**, 1968, 379–386.
12. Steinmann, P., Application of material forces to hyperelastostatic fracture mechanics. I. Continuum mechanical setting. *Int. J. Solid Struct.* **37**, 2000, 7371–7391.
13. Steinmann, P., Ackermann, D. and Barth, F.J., Application of material forces to hyperelastostatic fracture mechanics. II. Computational setting. *Int. J. Solid Struct.* **38**, 2001, 5509–5526.

On the Computation of Configurational Forces in Anisotropic Hyperelastic Solids

V. Ebbing, J. Schröder, P. Steinmann and P. Neff

Abstract In the field of configurational mechanics we study energetic changes associated to variations of material configurations. The central part in this notion is the energy momentum tensor, also known as the Eshelby stress tensor, which enters the configurational force balance. The mechanics of material forces has been successfully applied to a variety of important fields in applied mechanics such as e.g. the evolution of interfaces, growth in biomechanical systems, the kinetics of dislocations, fracture mechanics, morphology/structure optimization in heterogeneous microstructures.

In most contributions configurational forces are computed for isotropic bodies; in this presentation we consider the influence of anisotropy. As a specific model problem we consider a single-edged-tension specimen, where we analyze the sensitivity of the configurational forces in amplitude and orientation with respect to changing main axes of anisotropy of a hyperelastic material. These configurational forces can be interpreted as driving forces on the crack tips of the considered boundary value problem, which are directly related to the classical J-integral in fracture mechanics. In order to guarantee the existence of minimizers we use an anisotropic polyconvex energy. Here, we focus on a transversely isotropic constitutive law formulated in the framework of invariant theory.

J. Schröder · V. Ebbing
Institut für Mechanik, Fakultät für Ingenieurwissenschaften, Abteilung Bauingenieurwesen, Universität Duisburg-Essen, Universitätsstr. 15, 45117 Essen, Germany;
e-mail: j.schroeder@uni-due.de

P. Steinmann
Lehrstuhl für Technische Mechanik, Universität Erlangen-Nürnberg, Egerlandstr. 5, 91058 Erlangen, Germany; e-mail: p.steinmann@ltm.uni-erlangen.de

P. Neff
Fachbereich Mathematik, Lehrstuhl für Nichtlineare Analysis, Universität Duisburg-Essen, Universitätsstr. 2, 45117 Essen, Germany; e-mail: patrizio.neff@uni-due.de

P. Steinmann (ed.), IUTAM Symposium on Progress in the Theory and Numerics of Configurational Mechanics, 261–270.

1 Introduction

A scheme of configurational forces entering computational mechanics was first mentioned in [7]. A prominent application is in the field of fracture mechanics [27, 28]. Here the material forces are directly related to the classical J-integral, see e.g. [10, 13]. Besides the application of this concept in physical problems as mentioned above it can also be used as an error indicator in adaptive Finite-Element-Method, see e.g. [18–21, 31]. For an overview of recent developments in this rapidely expanding branch of continuum mechanics we refer to [29, 30].

In the following we restrict our discussion on hyperelasticity, where we assume the existence of an energy function W formulated in terms of the deformation gradient $\mathbf{F}$. The existence of minimizers of boundary value problems is guaranteed if the functional to be minimized is sequentially weakly lower semicontinuous (s.w.l.s.) and coercive. The concept of quasiconvexity, introduced by Morrey [17], ensures the s.w.l.s. condition if further growth conditions are satisfied. The polyconvexity condition, introduced by Ball [3], directly implies s.w.l.s., quasiconvexity and rank-one convexity. For the proof of polyconvexity of the free energy function $W = W(\mathbf{F})$, we have to prove the convexity of the free energy with respect to the argument $(\{\mathbf{F}, \operatorname{Cof}\mathbf{F}, \det\mathbf{F}\}) \in \mathbb{R}^{19}$. This local convexity condition is much simpler to handle than the non-local integral inequality of the quasiconvexity condition. Investigations concerning the relation of polyconvexity to the notion of material stability, in the sense of Legendre–Hadamard ellipticity, are documented in [25], in this context see also [14].

For isotropic materials there exist some models, e.g., the Ogden, Mooney–Rivlin and Neo-Hooke-type free energy functions, which satisfy the polyconvexity condition. In the case of anisotropy only a few years ago the construction of polyconvex energy functions was a question yet to be answered. In 2002 Ball still noticed in *Some open problems in elasticity* [2]: *Are there ways of verifying polyconvexity and quasiconvexity for a useful class of anisotropic stored-energy functions?*

Almost at the same time the first anisotropic polyconvex energy functions in case of transverse isotropy were proposed by Schröder and Neff [23]; an extension of this work as well as the general formulation of orthotropic polyconvex functions is discussed in [24]. A more general concept for the construction of triclinic, monoclinic, orthotropic and transversely isotropic polyconvex energy functions is given in [26]. In order to account for the condition of a stress-free reference configuration a priori, an attractive approach is given in [4, 11]. Applications of anisotropic polyconvex models to biomechanics can be found in [5, 9, 12, 16].

2 Continuum Mechanical Preliminaries.

The body of interest in the material configuration is denoted by $\mathcal{B}_0 \subset \mathbb{R}^3$, parametrized in $\mathbf{X}$, and the spatial configuration by $\mathcal{B}_t \subset \mathbb{R}^3$, parametrized in $\mathbf{x}$.

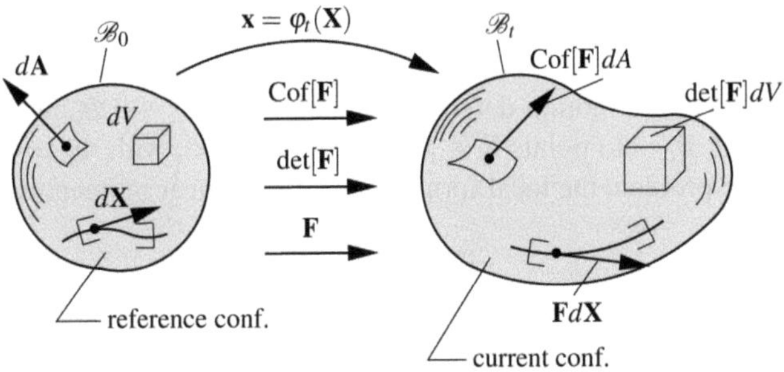

Fig. 1 Material (reference) and spatial (current) configuration of the considered body.

The nonlinear deformation map $\varphi_t : \mathcal{B}_0 \to \mathcal{B}_t$ at time $t \in \mathbb{R}_+$ maps points $\mathbf{X} \in \mathcal{B}_0$ onto points $\mathbf{x} \in \mathcal{B}_t$. The deformation gradient $\mathbf{F}$ is defined by

$$\mathbf{F}(\mathbf{X}) := \mathrm{Grad}\varphi_t(\mathbf{X}), \tag{1}$$

with the Jacobian $J(\mathbf{X}) := \det \mathbf{F}(\mathbf{X}) > 0$. In the sequel we will use polynomial invariants governed by the mappings of the infinitesimal referential line $d\mathbf{X}$, area $d\mathbf{A} = \mathbf{N}dA$ and volume elements dV, respectively. These quantities, defined w.r.t. the reference configuration, are mapped to their spatial counterparts $d\mathbf{x}$, $d\mathbf{a} = \mathbf{n}da$ and dv via

$$d\mathbf{x} = \mathbf{F}d\mathbf{X}, \quad \mathbf{n}da = \mathrm{Cof}\,[\mathbf{F}]\mathbf{N}dA \quad \text{and} \quad dv = \det\,[\mathbf{F}]\,dV, \tag{2}$$

see Figure 1. Equation $(2)_2$ is the well-known Nanson's formula. Furthermore, as already mentioned the argument $(\mathbf{F}, \mathrm{Cof}\,\mathbf{F}, \det \mathbf{F})$ plays an important role in the definition of polyconvexity. The local form of the balance of linear momentum reads

$$-\mathrm{Div}\mathbf{P} = \bar{\mathbf{f}}, \tag{3}$$

where $\bar{\mathbf{f}}$ denotes the body forces. The Piola stresses $\mathbf{P}$ are derived from an energy density $W(\mathbf{F}; \mathbf{X})$, defined per unit volume in the reference configuration, by the partial derivative $\mathbf{P} = \partial_F W$. Multiplying (3) by a suitable vector-valued test function $\delta\mathbf{u}$ and integrating over the reference configuration $\mathcal{B}_0$ yields the weak form of the balance of linear momentum $G = 0$, with

$$G = \int_{\mathcal{B}_0} \mathrm{Grad}[\delta\mathbf{u}] : \mathbf{P}\,dV - \int_{\partial\mathcal{B}_0} \delta\mathbf{u} \cdot \bar{\mathbf{t}}_0\,dA - \int_{\mathcal{B}_0} \delta\mathbf{u} \cdot \bar{\mathbf{f}}\,dV, \tag{4}$$

with the traction vector $\bar{\mathbf{t}}_0 = \mathbf{P}\mathbf{N}$ and the outward unit normal $\mathbf{N}$.

3 Balance of Configurational Forces

The nonlinear material motion deformation map $\Phi_t : \mathcal{B}_t \to \mathcal{B}_0$ at time $t \in \mathbb{R}_+$ maps points $\mathbf{x} \in \mathcal{B}_t$ onto points $\mathbf{X} \in \mathcal{B}_0$, i.e. $\mathbf{X} = \Phi_t(\mathbf{x})$, with $\Phi_t = \varphi_t^{-1}$. For the material motion problem the local form of balance of linear momentum reads

$$-\text{Div}\Sigma - \bar{\mathsf{F}} = \mathbf{0}, \tag{5}$$

where Σ and $\bar{\mathsf{F}}$ denote the Eshelby stress tensor and the configurational body forces, respectively. These quantities can be derived from the energy density $W(\mathbf{F};\mathbf{X})$ as follows

$$\Sigma = W\,\mathbf{1} - \mathbf{F}^T\,\mathbf{P} \qquad \bar{\mathsf{F}} = -\partial_X W - \mathbf{F}^T\,\bar{\mathbf{f}}. \tag{6}$$

Note that $\bar{\mathsf{F}}$ captures essentially possible inhomogeneities in the material properties. Multiplying (5) by a test function $\delta\mathbf{U}$, under consideration of the necessary boundary conditions, leads to the weak form of the balance of material momentum of the material motion problem (sometimes also denoted as balance of pseudo or configurational momentum), $\tilde{G} = 0$, with

$$\tilde{G} = \int_{\mathcal{B}_0} \text{Grad}[\delta\mathbf{U}] : \Sigma\, dV - \int_{\partial\mathcal{B}_0} \delta\mathbf{U} \cdot \Sigma\,\mathbf{N}\, dA. \tag{7}$$

Here we neglected configurational body forces for the sake of simplicity since our numerical example in Section 5 will exclusively concentrate on homogeneous material properties and vanishing (spatial) body forces. Note that this case corresponds to divergence free (solenoidal) Eshelby stresses and thus to path-independent J-integrals, an attractive property in fracture mechanics. Next, the test function is selected from a proper finite dimensional test space $\mathcal{V}_0^h$ that is typically spanned by polynomial basis functions N_K as already used in the finite element discretization of the spatial deformation problem. Finally, the discrete configurational (nodal) material force at node point K is computed by

$$\mathbf{f}_K^{\text{config}} = \int_{\mathcal{B}_0^h} [\Sigma^h \cdot \text{ Grad N}_\text{K}]\, dV. \tag{8}$$

These vectorial quantities are power conjugated to configurational changes of the material node point positions; they are easily calculated in a *post processing* step, the so-called *Material Force Method*, once the solution to the spatial deformation problem has been computed, see [1]. Here the sign convention associates a configurational change of the material node point position in the direction of the discrete material node point force with an energy increase of the system. Thus in order to release energy, that can then potentially be used for other physical processes like, e.g., the creation of new surfaces or the motion of defects, material node point positions have to move opposite to the material node point forces; compare our results in Section 5.

4 Polyconvex Anisotropic Energy Functions

Concerning the construction of the constitutive equations in anisotropic finite elasticity the principle of objectivity and the principle of material symmetry play an important role. The principle of objecticity is automatically fulfilled since we use the reduced constitutive equations formulated in terms of the right Cauchy–Green tensor $\mathbf{C} := \mathbf{F}^T\mathbf{F}$, i.e., $\psi(\mathbf{C}) = W(\mathbf{F})$. The principle of material symmetry enforces the invariance of the constitutive equations with respect to the transformations $\mathbf{Q} \in \mathcal{G} \subset \mathrm{O}(3)$ of the material symmetry group $\mathcal{G}$, which is a subgroup of the full orthogonal group O(3), i.e.,

$$\psi(\mathbf{C}) = \psi(\mathbf{QCQ}^T) \quad \forall \mathbf{Q} \in \mathcal{G}. \tag{9}$$

The anisotropy of the material can be described by structural tensors $\mathbf{M}$, which are invariant with respect to transformations of the underlying material symmetry group:

$$\mathbf{M} = \mathbf{QMQ}^T \quad \forall \mathbf{Q} \in \mathcal{G} \subset \mathrm{O}(3), \tag{10}$$

see [6, 15]. Inserting the structural tensor as a further tensorial argument into the free energy function (9), then

$$\psi(\mathbf{C}, \mathbf{M}) = \psi(\mathbf{QCQ}^T, \mathbf{QMQ}^T) \quad \forall \mathbf{Q} \in \mathrm{O}(3), \tag{11}$$

holds. This is the definition of an isotropic tensor function, therefore the scalar-valued function (11) can be formulated in terms of the principal invariants

$$I_1 := \mathrm{tr}\,\mathbf{C}, \; I_2 := \mathrm{tr}\,[\mathrm{Cof}\,\mathbf{C}], \; I_3 := \det \mathbf{C}, \tag{12}$$

and the mixed invariants

$$J_4 := \mathrm{tr}\,[\mathbf{CM}], \; J_5 := \mathrm{tr}\,[\mathrm{Cof}\,[\mathbf{C}]\mathbf{M}]. \tag{13}$$

In the following we focus on transversely isotropic materials and introduce a structural tensor defined as

$$\mathbf{M} = \mathbf{a} \otimes \mathbf{a}, \tag{14}$$

where $\mathbf{a}$ characterizes the preferred direction of the material, with the normalization condition $\|\mathbf{a}\| = 1$. $\mathbf{M}$ preserves the transversely isotropic material symmetry group $\mathcal{G}_{ti} := \{\pm\mathbf{1};\; \mathbf{Q}(\alpha, \mathbf{a}) \mid 0 < \alpha < 2\,\pi\}$, where $\mathbf{Q}(\alpha, \mathbf{a})$ denotes all rotations about the $\mathbf{a}$-axis.

4.1 Special Anisotropic Polyconvex Energy

The material model is assumed to be additively decomposed as follows

$$\psi^{ti} = \hat{\psi}^{\text{iso}} + \hat{\psi}^{ti}, \tag{15}$$

where the isotropic function $\hat{\psi}^{\text{iso}}$ is chosen to be a function depending on the principal invariants (12). In detail, we consider the compressible Mooney–Rivlin model

$$\hat{\psi}^{\text{iso}} = \alpha_1\, I_1 + \alpha_2\, I_2 + \delta_1\, I_3 - (2\,\alpha_1 + 4\,\alpha_2 + 2\delta_1)\ln(\sqrt{I_3}), \tag{16}$$

which satisfies the polyconvexity condition if $\alpha_1,\ \alpha_2,\ \delta_1 \geq 0$ holds, see e.g. [8]. The resulting isotropic second Piola–Kirchhoff stresses $\mathbf{S}^{\text{iso}}$ satisfy automatically the stress-free reference configuration condition

$$\mathbf{S}^{\text{iso}}(\mathbf{C} = \mathbf{1}) := 2\partial_C \hat{\psi}^{\text{iso}}(\mathbf{C} = \mathbf{1}) = \mathbf{0}. \tag{17}$$

For the anisotropic part $\hat{\psi}^{ti}$ we use the simple polyconvex function

$$\hat{\psi}^{ti} = \left[f_3(I_3) + f_4(J_4) + f_5(J_5)\right]. \tag{18}$$

The functions $f_4(J_4)$ and $f_5(J_5)$ have the explicit form

$$f_4 = \frac{\xi}{\alpha + 1} J_4^{\alpha+1}, \quad f_5 = \frac{\xi}{\beta + 1} J_5^{\beta+1}, \quad \forall\, \xi,\, \alpha,\, \beta \geq 0 \tag{19}$$

and are taken from [26].The proof of the polyconvexity of the powers of J_4 and J_5 is given in [23,24]. For an a priori fulfillment of the stress-free reference configuration condition of the anisotropic part of the stresses at natural state, i.e.,

$$\mathbf{S}^{ti}(\mathbf{C} = \mathbf{1}) := 2\partial_C \hat{\psi}^{ti}(\mathbf{C} = \mathbf{1}) = \mathbf{0}, \tag{20}$$

we consider additionally an isotropic function in terms of I_3 in (18), proposed in this way in [11]. In detail, the function $f_3(I_3)$ appears in the form

$$f_3 = \frac{\xi}{\gamma} I_3^{-\gamma}, \quad \forall\, \xi \geq 0,\ \gamma \geq -\frac{1}{2}. \tag{21}$$

The explixit expression of the anisotropic part is then given by

$$\hat{\psi}^{ti} = \xi \left(\frac{1}{\alpha + 1} (J_4)^{\alpha+1} + \frac{1}{\beta + 1} (J_5)^{\beta+1} + \frac{1}{\gamma} (I_3)^{-\gamma} \right). \tag{22}$$

The Eshelby stress tensor, defined in (6), is finally calculated by

$$\Sigma = \psi_0^{ti} \mathbf{1} - \mathbf{C}\,\mathbf{S}, \tag{23}$$

where we have introduced the abbreviation

$$\psi_0^{ti} = \psi^{ti} - \psi^{ti}(\mathbf{C} = \mathbf{1}), \tag{24}$$

in order to satisfy the condition $\Sigma = \mathbf{0}$ in the unloaded reference configuration.

5 Numerical Example: Cracked SET Specimen

As a numerical example we consider a single-edged-tension specimen (SET) as shown in Figure 2. We study the influence of the orientation of the main axes of anisotropy on the amplitude and orientation of the configurational forces at the crack tips. The main axis of anisotropy corresponds to the preferred direction **a** of the hyperelastic transversely isotropic energy (22). The material parameters are set to

$$\alpha_1 = 1\,\mathrm{N/mm}^2,\ \alpha_2 = 1\,\mathrm{N/mm}^2,\ \delta_1 = 1\,\mathrm{N/mm}^2,\\ \xi = 1\,\mathrm{N/mm}^2,\ \alpha = 10,\ \beta = 1,\ \gamma = 1. \tag{25}$$

In this 2D example the orientation of the preferred direction of the transversely isotropic material is described by the angle $\varphi = \measuredangle(\mathbf{a}, \mathbf{e}_1)$, whereas the direction of the configurational force at the crack tip A is characterized by the angle $\eta = \measuredangle\left(\mathbf{f}_A^{\text{config}}, \mathbf{e}_1\right)$. The height to width ratio of the SET specimen is chosen to be $H/W = 120\,\mathrm{mm}/40\,\mathrm{mm}$, the crack length to width ratio is set to $a/W = 0.5$ and the thickness of the crack is $d = 0.2$ mm, see Figure 2.

The specimen is discretized by 7886 six-noded triangular elements and the mesh is strongly refined in the vicinity of the crack tips as shown in Figure 3. In detail, we discretized a circular area around each crack tip with the radius $r = r_{10} = 0.52$ mm. The radii of the inner circles of finite elements of these domains are calculated by geometric series

$$r_i = \omega^{i-1} r_1, \quad i = 1, \ldots, 10, \quad \text{with} \quad r_1 = 0.1\,\mathrm{mm}, \quad \omega = 1.2. \tag{26}$$

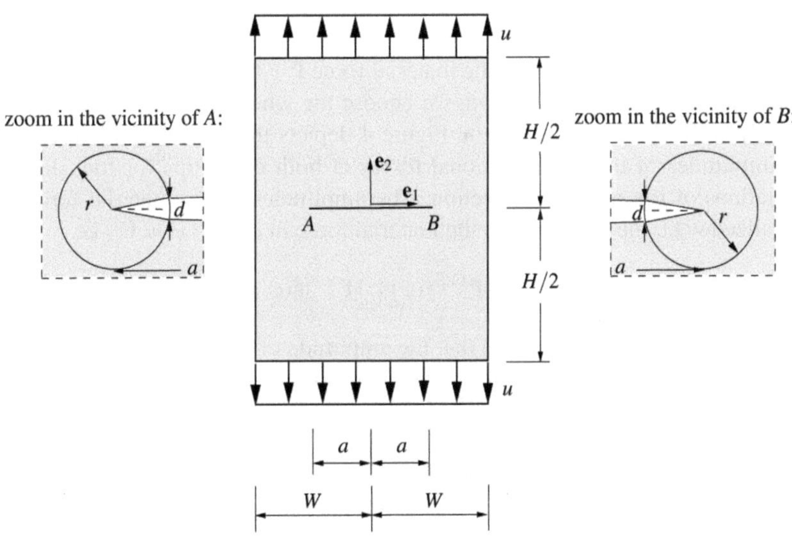

Fig. 2 Undeformed SET specimen, boundary conditions.

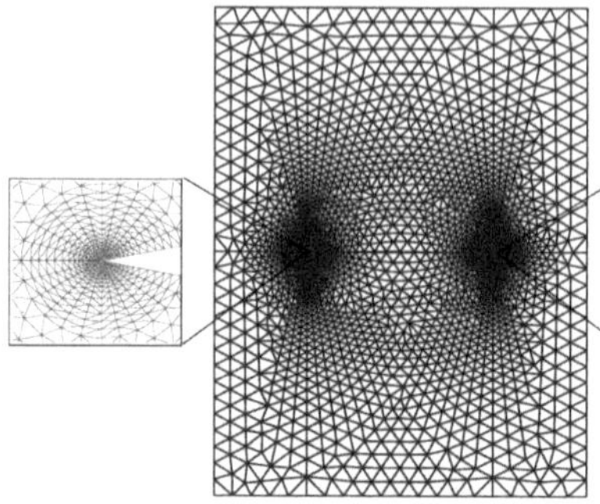

Fig. 3 Undeformed SET specimen, discretization with 7886 6-noded triangular elements.

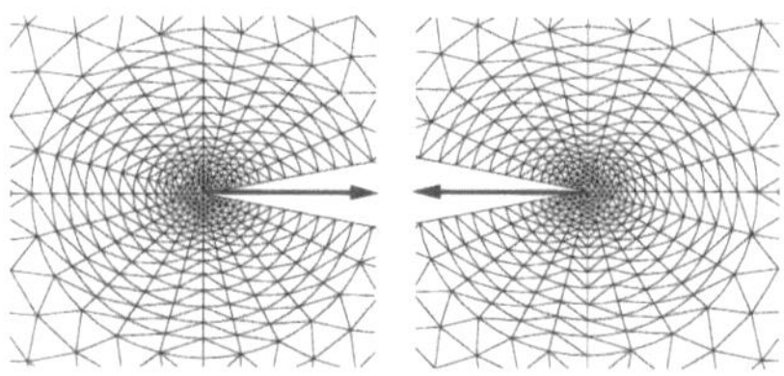

a) $\varphi = 0°,\ \eta = 0°,\ l^{\text{config}} = 1$

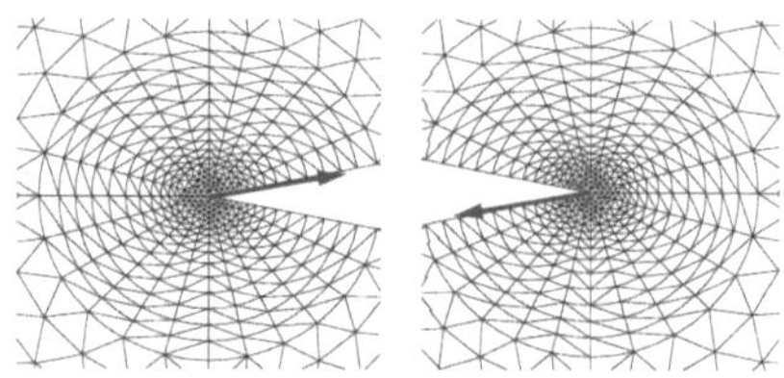

b) $\varphi = 15°,\ \eta = 9.66°,\ l^{\text{config}} = 0.859$

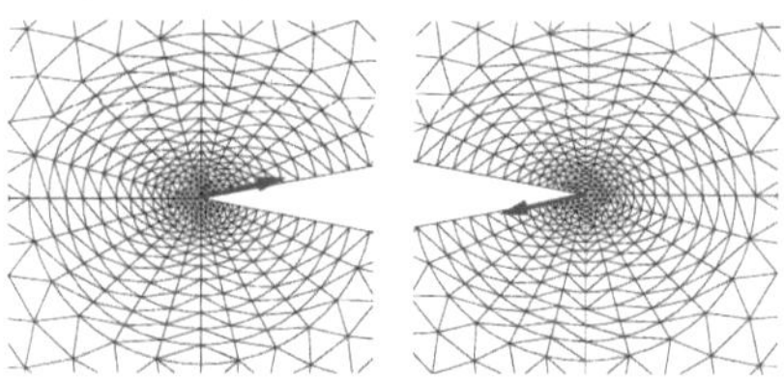

c) $\varphi = 30°,\ \eta = 12.52°,\ l^{\text{config}} = 0.504$

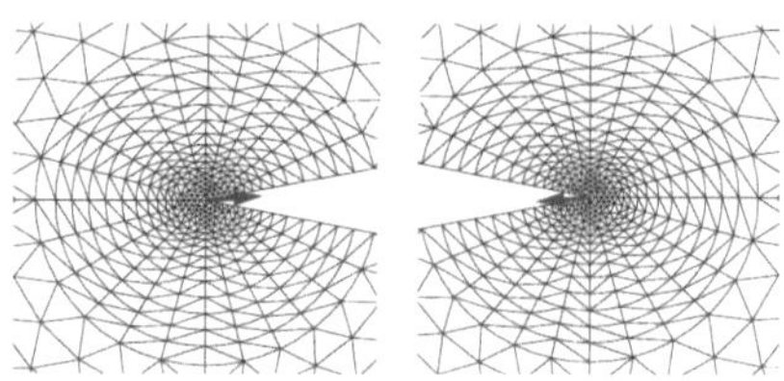

d) $\varphi = 45°,\ \eta = 4.25°,\ l^{\text{config}} = 0.292$

Fig. 4 Angle η versus angle φ. Ratios of norms of material forces.

For a good approximation of the material force $\mathbf{f}^{\text{config}}$ acting at crack tips A and B with the Material Force Method, we choose the whole circular area around the crack tip as the integration domain. Figure 4 depicts the results of the orientation and amplitudes of the configurational forces at both crack tips for four different orientations of the preferred direction. The amplitudes of the material forces are normalized w.r.t. the amplitude to the material force in case of $\varphi = 0°$, i.e.,

$$l^{\text{config}}(\varphi) = \|\mathbf{f}^{\text{config}}(\varphi)\| / \|\mathbf{f}^{\text{config}}(\varphi = 0°)\|. \tag{27}$$

Comparing these ratios we notice that the amplitude of the forces becomes significantly smaller by increasing the angle of anisotropy up to 45°. Regarding the changes of the angles φ and η Figure 4 shows the following: for $\varphi = 0°,\ 15°,\ 30°$ the angle of orientation of $\mathbf{f}^{\text{config}}$ rises from 0° to about 12°, whereas for $\varphi = 45°$ we obtain the angle $\eta \approx 4.25°$. Therefore, a high sensitivity of the configurational forces in amplitude as well as in orientation appears in the considered boundary value problem. A more detailed discussion on this topic and further results can be found in [22].

6 Conclusion

Configurational mechanics play an increasingly important role in both defect mechanics and computational mechanics. Since most contributions deal with the computation of configurational forces in isotropic solids, we have analyzed the influence of anisotropy on the material forces. In detail, we have investigated the change of amplitude and orientation of the configurational forces at crack tips due to varying main axes of anisotropy of a transversely isotropic hyperelastic material. For the description of the assumed hyperelastic material behavior we use an anisotropic polyconvex energy in order to guarantee the existence of minimizers. The anisotropy of the material is described in the framework of invariant theory. As a representative example we have considered a single-edged-tension specimen, where we first noticed a high sensitivity of the amplitude: Changing the angle of the preferred direction of the material about 45° yields a reduction of the origin amplitude of the material forces of approximately 70%. Furthermore, the orientation of the material forces is also sensitive to the change of orientation of the preferred direction.

Acknowledgement

Financial support from DFG (research grant SCHR 570/6, Ne 902/2) is gratefully acknowledged. We thank Dipl.-Ing. Dominik Brands and Dipl.-Ing. Alexander Schwarz for their support.

References

1. Ackermann, D., Barth, F.J. and Steinmann, P., Theoretical and computational aspects of geometrically nonlinear problems in fracture mechanics. In: *Proceedings of the European Conference on Computational Mechanics*, Vol. 385. Munich, 1999.
2. Ball, J.M., Some open problems in elasticity. In: *Geometry, Mechanics and Dynamics*. Springer, New York, 2002, pp. 3–59.
3. Ball, J.M., Convexity conditions and existence theorems in non-linear elasticity. *Archive for Rational Mechanics and Analysis* **63**, 1976, 337–403.
4. Balzani, D., Polyconvex anisotropic energies and modeling of damage applied to arterial walls. Ph.D. Thesis, Institut für Mechanik, Abteilung Bauwissenschaften, Fakultät für Ingenieurwissenschaften, 2006.
5. Balzani, D., Neff, P., Schröder, J. and Holzapfel, G., A polyconvex framework for soft biological tissues. Adjustment to experimental data. *International Journal of Solids and Structures* **43**(20), 2006, 6052–6070.
6. Boehler, J.P., A simple derivation of respresentations for non-polynomial constitutive equations in some cases of anisotropy. *Zeitschrift für Angewandte Mathematik und Mechanik* **59**, 1979, 157–167.
7. Braun, M., Configurational forces induced by finite-element discretization. *Proceedings of the Estonian Academy of Sciences, Physics, Mathematics* **46**, 1997, 24–31.
8. Ciarlet, P.G., *Mathematical Elasticity. Vol. I: Three-Dimensional Elasticity*. North Holland, 1988.

9. Ehret, A. and Itskov, M., A polyconvex hyperelastic model for fiber-reinforced materials in application to soft tissues. *Journal of Materials Science* **42**, 2007, 8853–8863.
10. Gross, D. and Seelig, T., *Bruchmechanik: Mit einer Einführung in die Mikromechanik*. Springer, 2001.
11. Itskov, M. and Aksel, N., A class of orthotropic and transversely isotropic hyperelastic constitutive models based on a polyconvex strain energy function. *International Journal of Solids and Structures* **41**, 2004, 3833–3848.
12. Itskov, M., Ehret, A. and Mavrilas, D., A polyconvex anisotropic strain-energy function for soft collagenous tissues. *Biomechanics and Modeling in Mechanobiology* **5**, 2006, 17–26.
13. Kienzler, R. and Herrmann, G., *Mechanics in Material Space with Application to Defect and Fracture Mechanics*. Springer, 2000.
14. Knowles, J. and Sternberg, E., On the failure of ellipticity and the emergence of discontinuous deformation gradients in plane finite elastostatics. *Journal of Elasticity* **8**, 1978, 329–379.
15. Liu, I.S., On representations of anisotropic invariants. *International Journal of Engineering Science* **20**, 1982, 1099–1109.
16. Markert, B., Ehlers, W. and Karajan, N., A general polyconvex strain-energy function for fiber-reinforced materials. *Proceedings in Applied Mathematics and Mechanics* **5**, 2005, 245–246.
17. Morrey, C., Quasi-convexity and the lower semicontinuity of multiple integrals. *Pacific Journal of Mathematics* **2**, 1952, 25–53.
18. Müller, R., Gross, D. and Maugin, G.A., Use of material forces in adaptive finite element methods. *Computational Mechanics* **33**, 2004, 421–434.
19. Müller, R., Kolling, S. and Gross, D., On configurational forces in the context of the finite element method. *International Journal for Numerical Methods in Engineering* **53**, 2002, 1557–1574.
20. Müller, R. and Maugin, G.A., On material forces and finite element discretizations. *Computational Mechanics* **29**, 2002, 52–60.
21. Scherer, M., Denzer, R. and Steinmann, P., On a solution strategy for energy-based mesh optimization in finite hyperelastostatics. *Computer Methods in Applied Mechanics and Engineering* **197**, 2008, 609–622.
22. Schröder, J., Ebbing, V. and Steinmann, P., Configurational forces in anisotropic solids. *Archive of Applied Mechanics*, 2009, in preparation.
23. Schröder, J. and Neff, P., On the construction of polyconvex anisotropic free energy functions. In: C. Miehe (Ed.), *Proceedings of the IUTAM Symposium on Computational Mechanics of Solid Materials at Large Strains*. Kluwer Academic Publishers, 2001, pp. 171–180.
24. Schröder, J. and Neff, P., Invariant formulation of hyperelastic transverse isotropy based on polyconvex free energy functions. *International Journal of Solids and Structures* **40**, 2003, 401–445.
25. Schröder, J., Neff, P. and Balzani, D., A variational approach for materially stable anisotropic hyperelasticity. *International Journal of Solids and Structures* **42**(15), 2005, 4352–4371.
26. Schröder, J., Neff, P. and Ebbing, V., Anisotropic polyconvex energies on the basis of crystallographic motivated structural tensors. *Journal of the Mechanics and Physics of Solids* **56**(12), 2008, 3486–3506.
27. Steinmann, P., Application of material forces to hyperelastostatic fracture mechanics. Part I: Continuum mechanical setting. *International Journal of Solids and Structures* **37**, 2000, 7371–7391.
28. Steinmann, P., Ackermann, D. and Barth, F.J., Application of material forces to hyperelastostatic fracture mechanics. Part II: Computational setting. *International Journal of Solids and Structures* **38**, 2001, 5509–5526.
29. Steinmann, P. and Maugin, G.A. (Eds.), *Mechanics of Material Forces*. Springer, 2005.
30. Steinmann, P., Scherer, M. and Denzer, R., Secret and joy of configurational mechanics: From foundations in continuum mechanics to applications in computational mechanics. *Zeitschrift für Angewandte Mathematik und Mechanik*, 2008, accepted for publication
31. Thoutireddy, P. and Ortiz, M., A variational r-adaptation and shape-optimization method for finite-deformation elasticity. *International Journal for Numerical Methods in Engineering* **61**, 2004, 1–21.

Authors Index

MIX
Papier aus verantwortungsvollen Quellen
Paper from responsible sources
FSC® C105338

If you have any concerns about our products,
you can contact us on
ProductSafety@springernature.com

In case Publisher is established outside the EU,
the EU authorized representative is:
Springer Nature Customer Service Center GmbH
Europaplatz 3, 69115 Heidelberg, Germany

Printed by Libri Plureos GmbH
in Hamburg, Germany